Springer
Berlin
Heidelberg
New York
Barcelona
Budapest
Hong Kong
London
Milan
Paris
Santa Clara
Singapure
Tokyo

The German Advisory Council on Global Change

(Wissenschaftlicher Beirat der Bundesregierung Globale Umweltveränderungen)

(Members as at July 1, 1995)

Prof. Friedrich O. Beese
Agronomist: Director of the Institute of Soil Science and Forest Nutrition at the University of Göttingen (Institut für Bodenkunde und Waldernährung an der Universität Göttingen)

Prof. Gotthilf Hempel
Fishery biologist: Director of the Centre for Marine Tropical Ecology at the University of Bremen (Zentrum für Marine Tropenökologie an der Universität Bremen)

Prof. Paul Klemmer
Economist: President of the Rhenish-Westphalian Institute for Economic Research, Essen (Rheinisch-Westfälisches Institut für Wirtschaftsforschung in Essen)

Prof. Lenelis Kruse-Graumann
Psychologist: Specialist in "Ecological Psychology" at the Open University, Hagen (Schwerpunkt "Ökologische Psychologie" an der Fernuniversität Hagen)

Prof. Karin Labitzke
Meteorologist: Institute for Meteorology at the Free University Berlin (Institut für Meteorologie der Freien Universität Berlin)

Prof. Heidrun Mühle
Agronomist: Head of Department of Agricultural Lands at the Environment Research Centre Leipzig-Halle (Projektbereich Agrarlandschaften am Umweltforschungszentrum Leipzig-Halle)

Prof. Hans-Joachim Schellnhuber (Vice Chairperson)
Physicist: Director of the Potsdam Institute for Climate Impact Research (Potsdam-Institut für Klimafolgenforschung)

Prof. Udo Ernst Simonis
Economist:Department of Technology–Work–Environment at the Science Centre Berlin (Forschungsschwerpunkt Technik–Arbeit–Umwelt am Wissenschaftszentrum Berlin)

Prof. Hans-Willi Thoenes
Technologist: Rhenish-Westphalian Technical Control Board, Essen (Rheinisch-Westfälischer TÜV in Essen)

Prof. Paul Velsinger
Economist: Head of the Department of Regional Economics at the University of Dortmund (Fachgebiet Raumwirtschaftspolitik an der Universität Dortmund)

Prof. Horst Zimmermann (Chairperson)
Economist: Department of Public Finance at the University of Marburg (Abteilung für Finanzwissenschaft an der Universität Marburg)

German Advisory Council on Global Change

World in Transition:

Ways Towards Global Environmental Solutions

Annual Report 1995

with 36 Illustrations

Springer

GERMAN ADVISORY COUNCIL ON GLOBAL CHANGE (WBGU)
Secretariat at the Alfred-Wegener-Institute
for Polar and Marine Research
Columbusstraße
D-27568 Bremerhaven
Germany

Acknowledgements:

External contributions and corrections to this report are gratefully acknowledged from

Dipl.-Pol. Frank Biermann, LL.M., Science Centre Berlin (Wissenschaftszentrum Berlin)
Prof. Dietmar Bolscho, Division of Educational Sciences 1, University of Hanover (Fachbereich Erziehungs-wissenschaften 1, Universität Hannover)
Ass.jur. Gudrun Henne, Free University Berlin, Faculty of Law (Freie Universität Berlin, Juristische Fakultät)
Prof. Gerd Michelsen, Institute for Environmental Sciences, University of Lüneburg (Institut für Umweltwissen-schaften, Universität Lüneburg)
Dr. Sebastian Oberthür, Society for Political Analysis, Berlin (Gesellschaft für Politikanalyse Berlin)
Prof. Volker von Prittwitz, Society for Political Analysis, Berlin (Gesellschaft für Politikanalyse Berlin) and Max-Planck Institute for Societal Research, Cologne (Max-Planck-Institut für Gesellschaftsforschung, Köln)

CIP Data aplied for
Deutschland / Wissenschaftlicher Beirat Globale Umweltveränderungen: Annual report / German Advisory Council on Global Change - Berlin ; Heidelberg ; New York ; Barcelona ; Budapest ; Hong Kong ; London ; Milan ; Paris ; Santa Clara ; Singapore ; Tokyo : Springer.
Früher im Economica-verl., Bonn. - Dt. Ausg. u. d.T.: Deutschland / Wissenschaftlicher Beirat Globale Umwelt-veränderungen: Jahresgutachten...
1995. World in transition: ways towards global environmental solutions. - 1996
World in Transition: ways towards global environmental solutions / German Advisory Council on Global Chan-ge. - Berlin ; Heidelberg ; New York ; Barcelona ; Budapest ; Hong Kong ; London ; Milan ; Paris ; Santa Clara ; Singapore ; Tokyo : Springer, 1996
(Annual report / GermanAdvisory Council on Global Change ; 1995) Dt. Ausg. u.d.T.: Welt im Wandel: Wege zur Lösung globaler Umweltprobleme
ISBN-13: 978-3-642-80182-2   e-ISBN-13: 978-3-642-80180-8
DOI: 10.1007/978-3-642-80180-8

Translation: Spence & Meadows, Bremen
Cover design: E. Kirchner, Heidelberg using the following illustrations
Clouds, M. Schulz-Baldes
Rain Forest, M. Schulz-Baldes
Flags of different nations, Agency Tony Stone
International Congress Centre Berlin, Superbild Berlin, H. Wiedel
Opening Address, Bundesbildstelle Bonn
School, Gesellschaft für Technische Zusammenarbeit (GTZ) GmbH. Ali Paczensky

SPIN 10503212   32/3137-5 4 3 2 1 0 - Printed on workprint 100% recycled paper

Outline of Contents

Contents

Boxes

Tables

Figures

Summary

Introduction

The Berlin Climate Conference in the spring of 1995 was another demonstration that individuals and societies will have to change their ways of thinking if human-induced climate change is to be counteracted. This process must apply for all aspects of global change; the key trends have not diminished over the last few years, on the contrary, they have become more threatening than ever before.

The German Advisory Council on Global Change (Wissenschaftlicher Beirat der Bundesregierung Globale Umweltveränderungen [WBGU]) describes in its 1995 annual report "Ways Towards Global Environmental Solutions". Although ultimate solutions have not yet crystallized in many areas, the Council proceeds on the assumption that, if those involved are willing and take appropriate action, problems can be solved, i.e. that irreversible and disastrous development is not inevitable. Whether these solutions are actually striven for is still an open question, since major reorientations are required at the local, national and global level.

Two paths must be taken in parallel. Firstly, *societal conditions* for the solution of global environmental problems must be changed; achieving these conditions at individual and institutional level represents a major challenge for governments and societies. Secondly, *international arrangements* relating to various global environmental problems have to be adopted and/or strengthened by democratic process, and implemented with appropriate measures.

The Societal Conditions for Solving Global Environmental Problems

Environmental Awareness and Environmental Education

Most international declarations and conventions for combating global environmental problems and their consequences demand a strengthening of *environmental awareness* among the population and measures relating to environmental education. Global environmental politics will only fulfill its tasks if the decision makers in the individual nations are supported by a population whose environmental awareness and willingness to behave in an environmentally appropriate way permits them to demand and assert the solutions to global environmental problems. Not until the idea of sustainable development is firmly anchored in the consciousness of people can strategies for behavioral change be effective. What therefore are required are worldwide and far-reaching measures of environmental education.

People's *perception* of environmental problems is one important requirement for changes of environmentally harmful forms of production and consumption. "Environmental awareness" has long since escaped the confines of the industrialized countries, although there still are substantial disparities between individual countries. However, there is as yet no worldwide survey system for the continuous recording of environment-related perceptions and attitudes. Since such information is of decisive importance for measures aimed at changing behavior, efforts to develop such an instrument as part of the *Human Dimensions of Global Environmental Change Programme* (HDP) should be given support by Germany.

Environmental *education* is an important tool for abandoning environmentally harmful forms of behavior, and for learning environmentally appropriate behavior. Criteria for sound environmental education involve learning from personal and conveyed experience in everyday situations (*situation orientation*), learning in connection with one's own direct actions (*action orientation*), and incorporation of the subject matter into the sociopolitical context (*problem orientation*). In spite of numerous political declarations of intent, initiatives and programs, environmental education worldwide must still be declared as underdeveloped, particularly with respect to global environmental problems. This should however not blind us to the substantial differences existing between individual countries. In the industrial countries, where environmental education has attained a relatively secure status, both in the formal educational system and outside of it, a local, regional or national perspective in environmental education still prevails. In the developing countries, on the other hand, considerable structural shortcomings exist in the educational systems, resulting in a very weak and insecure status of environmental education. For this reason, great importance is attached to the educational commitment of *nongovernmental organizations* (NGOs).

Recommendations:

- Strengthen environmental education within the overall conception of national environmental policy.
- Provide support for networking environmental education carried out by governmental and nongovernmental institutions.
- Consistently incorporate environmental education into development policy programs and projects.
- Strengthen international organizations (e.g. UNESCO) in order to implement international agreements in specific educational contexts.
- Tackle educational issues within the framework of international conferences, such as the second Conference of the Parties to the Convention on Climate Change.
- Strengthen environmental education as a research field of environmental sciences (analysis and comparative assessment).

Exchange of Know-how and Technology Transfer

Reinforcing of technology transfer from industrial countries to developing countries ranks among the classic demands of development policy, and meanwhile has become an established component of international environmental agreements. The Council emphasizes that such technology transfer must be regarded as an exchange of know-how in a broader sense, in which industrialized countries can also learn from developing countries (two-way traffic). This applies not only to the values and social structures of other cultures, but also to adapted technologies, such as forms of soil management, irrigation techniques or types of forest use. For this reason, the formation of knowledge and the reactivation of traditional knowledge should each be supported in the developing countries.

The transfer of know-how is predominantly effected via market and competitive processes, through granting property rights and access to existing and newly acquired know-how. Until now, such exchange of knowledge has mainly occurred between industrialized countries. As experience in Asia's fast growing economies shows, consistent educational reform and development of own research capacities are major prerequisites for exchange, and should therefore form the basis for measures and programs in the industrial countries.

Deficits exist regarding the framework for competition of enterprises with global operations, the restructuring of patent law, the right to intellectual property and the application of liability law in the exchange of information. Satisfactory coordination between environmental and development policy, on the one hand, and industrial and trade policy, on the other, has been lacking to date, and will have to be given greater attention in the future.

Institutions and Organizations

Global environmental policy institutions primarily exist as horizontal self-coordination between nation states because of the lack of higher hierarchical control levels. They use both *direct* and *indirect control instruments*. International institutional arrangements and practices change within the scope of a process, which may result in the formulation and implementation of more effective targets and measures.

In accordance with the basic principle of national sovereignty, environmental policy depends on the approval of nations in each individual case. Accordingly, the *decision-making process* traditionally takes place in the form of negotiations. Decision making is thus characterized by differing interest structures in the individual nations, and is usually a complicated and protracted process. The implementation of international action programs that have been agreed upon is also a complex process, and in most cases can only be monitored on the basis of corresponding reports by the nation states. Even if violations against agreed arrangements are detected, compliance with the respective arrangements can only be enforced under very specific conditions.

However, a variety of institutional innovations have been initiated in the course of the internationalization of environmental policy since the mid-80s. They include the setting up of institutions for a transfer of finance and technology from North to South, as a form of *direct control*, as well as certain changes that have been made in process, resource and organizational control by means of *indirect control*.

Important institutional arrangements in the form of funds have been set up for *finance and technology transfers* such that – as in the case of the Montreal

Protocol and the Convention on Climate Change – the environmental protection obligation on the part of the developing countries is tied to a transfer obligation of the industrial countries on a legally binding basis. If the North does not pay, the South is relieved of its obligations.

In addition, a partial change in traditional direct control is taking place in environmental policy, resulting in a growing preference for forms of indirect control in accordance with the concept of *sustainable development*. These innovative approaches include the development of human and institutional capacities (*capacity building*) in developing countries, resource transfer to poorer nations, new rights to participation for nongovernmental agencies, and agreement on procedures that promote a reconciliation of interests without the need for a hierarchical regulatory framework, and which facilitate and accelerate both decision making and implementation.

Germany has played an important role in the formulation of global environmental policy in only a few sectors to date; indeed, many an opportunity for exerting influence has been wasted. Nevertheless, it has a significant potential for influencing the further development of global environmental policy by virtue of its economic and technological strength, its political importance, especially in the European Union, as well as its high degree of environmental awareness and broad benevolent support from the general public. In some cases of global agreements Germany has shown a relatively strong financial commitment, e.g. with regard to the GEF. In view of this background, a more active role on the part of Germany in providing for an institutional innovation of global environmental policy has considerable prospects of success.

Recommendations:

FURTHER DEVELOPMENT AND MODIFICATION
- Improve the decision-making mechanisms of environmental conventions and protocols.
- Utilize economic instruments for the protection of global environmental assets (taxes, charges and tradeable emission permits).
- Grant extended rights to information and participation on the part of NGOs.

EXTENSION
- Institute international environmental audits for states and industries.
- Establish an International Court for the Environment or activate the International Court of Justice in The Hague for environmental matters.

FUNDAMENTAL RESTRUCTURING
- UNEP could obtain the status of a strong UN or-

ganization.
- Set up a Global Environmental Organization akin to the World Trade Organization.

ROLE OF GERMANY
- Create sufficient (which in most cases means greater) capacities for the diagnosis and therapy of global environmental problems.
- Encourage the commitment and strengthen the competence of German representatives in international institutions directly and indirectly related to environmental issues.

Growth and Distribution of the World's Population

The growth and location of the world's population are key determinants of global environmental change. The annual increase of approx. 95 million people, spreading urbanization, particularly in developing countries, and growing international migration pressure in the direction of Europe and North America set the context for global environmental policy.

The long-term forecasts regarding *population increase* have been slightly corrected downward in recent years. However, this must not be taken as a reason for a letup in efforts to further reduce growth rates. Rather, given the slowdown in fertility decline and the delay in "demographic transition", the opposite conclusion should be drawn: for the very reason that there is cause for hope of success through the initiated efforts, the latter must be reinforced.

The quantitative increase in *international migrations*, and refugee flows in particular, is alarming. While roughly 50 million people (i.e. 1% of the world's population) lived outside of their native country in 1989, the total number of transboundary migrants only a few years later had already reached a figure of more than 100 million.

A total of approx. 83% of worldwide population growth is accounted for by *urban regions*, i.e. the urban population will increase by some 75 million people annually over the next decade. Cities will be subjected to tremendous pressure as a result of the population growth and immigration. The rapid expansion of cities will give rise to immense social and environmental costs. If it is not possible to put a halt to this process, many cities will "collapse".

Recommendations:

In accordance with the "Rio Declaration" and AGENDA 21, the Council views the following as the most important objectives:
- long-term stabilization of world population by

- combating poverty (provision of care to the aged), and providing equality for women,
- recognizing the right to family planning as an individual human right, and improving family planning opportunities,
- reducing child mortality and improving education and training.
- prevention and reduction of forced migration through
  - international cooperation in coping with international migration flows,
  - efforts to intensify awareness of the consequences of uncontrolled migration and urbanization processes.
- creation of functional urban structures through
  - specification of regional planning that allows for a harmonization of "environment and development",
  - creation of polycentric instead of monocentric structures of regional development.

International Conventions Aimed at Solving Global Environmental Problems

The Framework Convention on Climate Change – Berlin and its Aftermath

Despite urgent warnings by scientists about disturbing trends, there has been a further increase in the use of fossil fuels and hence in the level of CO_2 emissions on a worldwide scale (IEA, 1994). There is no empirical evidence for a change in this trend, nor can any such change be anticipated, one reason being the rapid growth in world population and the quantitative expansion of the world economy. With this background, the first Conference of the Parties to the Climate Convention in 1995 to fulfill the hopes of many observers has been declared a failure. Firstly, the Conference did not adopt a protocol, and secondly, the wording of the Berlin Mandate gives rise to worries that the substance of the protocol to be adopted in two years might not match up to original expectations. On the other hand, there is no denying that the Berlin Conference, by acknowledging the inadequacy of existing commitments and adopting the Mandate to draw up a protocol, has taken the next steps for an effective climate protection policy. What is important now is that commitments and targets be upheld and developed further in a determined manner, in order that the Climate Convention becomes a powerful instrument of global climate policy.

In its 1995 annual report, the Council presents various scenarios for the global reduction of CO_2. They were calculated with the help of mathematical-physical models, featuring a new modeling approach, a "backwards mode" (*"inverse scenario"*). By analyzing the environmentally and economically tolerable stresses induced by climate change, a so-called *"tolerance window"* is calculated for an admissible degree of climatic change, from which the *maximum CO_2 emissions* are then derived. The main conclusions of the scenarios are:

1 Continuation of current emissions (business-as-usual) would reach the limits of the tolerable climate window in less than 30 years, which than would require such drastic reductions within a short period of time that the structures and technologies capable of enabling such reductions are barely conceivable.

2 The Council therefore considers an emission profile in which global CO_2 emissions are reduced by around 1% annually over the next 150 years, following a transitional period of about 5 years, to make sense and be feasible for implementation.

3 For physical and chemical reasons, even a permanent regulation of global anthropogenic CO_2 emissions at a constantly low level is inevitably bound up with serious impacts on the climate system. In the very long term, i.e. over several centuries, anthropogenic CO_2 emissions from the use of fossil fuels must be reduced to zero, even if those resources were inexhaustible. However, the climate system provides a considerable degree of liberty regarding how the emission profile is to be shaped.

One can assume that the reduction commitments resulting from these demands for the time being will be restricted to the Annex I countries (industrialized countries). In order to make a system of rigid national quotas more flexible, the Council recommends deploying *Joint Implementation* as an instrument, which could possibly be extended into a system of internationally tradeable emission entitlements. By applying these instruments, the necessary emission reductions could be achieved more cost-effectively, while at the same time facilitating access to energy-efficient technologies for the developing countries.

Recommendations:

- In line with its scenario, the Council recommends that measures be introduced within a very short implementation period that will reduce global CO_2 emissions by 1% per annum.
- The other greenhouse gases are to be included in the reduction strategies as soon as possible. Research into global warming potentials must be strengthened and crediting mechanisms worked out in order to reduce avoidance costs while retaining the same level of environmental effectiveness.
- Germany's self-imposed CO_2 reduction commitment: the self-commitment announced by the Federal Chancellor at the Berlin Conference represents a toughening of the national reduction target and hence an even greater challenge. The Council therefore recommends that the Interministerial Working Group on "CO_2 Reduction" adapt its catalogue of measures to this new target. Such an analysis would have to examine, in particular, the opportunities for Joint Implementation projects and the progress that can be made by reducing other greenhouse gases besides CO_2.

The Montreal Protocol – An Example for Successful Environmental Policy

The emissions of the main anthropogenic source gases which cause the formation of chlorine and bro-

mine in the stratosphere (such as CFCs, carbon tetrachloride, halons and methyl chloroform) have slowed down considerably. This is attributable to the *Montreal Protocol* and the amendments thereto. The increase in Freon-11 in 1993, for example, was 25 to 30% less than in the 1970s and 1980s. The maximum contamination by chlorine and bromine in the troposphere was probably in the year 1994, but will not occur for another 3 to 5 years in the stratosphere (IPCC, 1994). Due to the longevity of ozone-depleting substances, the stratospheric ozone layer will not be able to regain its original state until the middle of the next century.

Stratospheric ozone depletion of approx. 3% per decade is the cumulative impact of regionally and temporally different trends (WBGU, 1993). Over the tropics and subtropics (30° N to 30° S), i.e. in about half the Earth's atmosphere, no significant ozone depletion has as yet been measured. Depletion is therefore all the more severe in the other regions, with ozone depletion particularly drastic during the spring months over the Antarctic continent (the so-called "*ozone hole*"). However, there is also a marked tendency towards depletion over mid and high latitudes in Europe in the order of 5% per decade.

Recommendations:

- Germany should provide an adequate level of financing to the multilateral fund for the protection of the ozone layer, within which framework selected partner countries may be granted support. Measures for China and India, the two largest consumers of CFCs in this group, could prove especially efficient.
- Additional efforts must be made to achieve a rapid end to the production and consumption of CFCs and HCFCs in all countries, including the developing and *newly-industrialized nations*. The recommendations of UNEP should seriously be taken up and the relevant measures put into effect.

The Convention on the Law of the Sea – Towards the Global Protection of the Seas

The *Convention on the Law of the Sea*, which went into effect on November 16, 1994, offers a global framework based on international law for combining existing individual regimes. It might thus provide the foundation for a functional global regime of marine protection. Representing an important step forward, this "*constitution of the oceans*" designates environmental protection as the *basic standard* for all forms of marine use and requires the Parties to implement the relevant regulatory frameworks as minimum international standards, or at least take them into consideration with respect to terrestrial sources of emissions. However, an *integrated environmental management of the seas* still appears to be a long way off. Even in the endangered regional waters of the industrial countries only partial improvements have been attainable up to now, and in the developing countries there continues to be a lack of the required financial and technical resources, which the industrialized countries, in turn, still do not seem willing to provide to the required extent. If international measures for the protection of the seas are not carried out, however, far-reaching and, in some cases, irreversible damage can be expected in view of the continued rise in population in the coastal regions, the growth of industrial production and increasing pollution in the large river catchments.

Recommendations:

- *Transport function:* Measures for determining substandard ships and illegal oil discharges should be intensified within the framework of the *Paris Memorandum.*
- *Disposal function:* Germany should commit itself to negotiating an integrated *"International Convention on Protection of the Seas".* This convention could combine the various regional sea programs, particularly in developing countries, and contribute towards financing appropriate environmental programs of the developing countries with the help of a separate financing mechanism (*"Blue Fund"*).

  The Council recommends that the German Federal Government again take the initiative and declare the North Sea to be a protected special region so as to put a complete stop to discharges of oil and chemicals.
- *Resource function:* The Council recommends that efforts be made within the framework of the *International Sea-Bed Authority* (to be set up in Jamaica) in order to prevent commencement of commercial deep-sea mining operations prior to joint determination of their environmental impact. An arrangement similar to the *Madrid Environment Protection Protocol* to the *Antarctic Treaty* is envisaged here.

  Due to the tremendous threat to global fish stocks, Germany should commit itself to restrictive regulation of *fishing quotas* and contribute to the implementation of the principle of maximum sustainable yield, both Europe-wide and globally.

The Desertification Convention – A First Step Towards the Protection of Soils

In its 1994 annual report, the Council focused in detail and at length on the problem of soil degradation. The analysis showed that soils are the vulnerable thin skin of the Earth for which serious *"illnesses"* can be diagnosed worldwide. These *"illnesses"* represent a serious threat to the Earth's population and biosphere that in some parts of the world is already dramatic.

The *Desertification Convention* adopted in 1994 has created an important framework by defining certain basic requirements for *combating desertification*, for example, increasing efficiency of bilateral and multilateral cooperation, intensive exchange of data and mutual information between donors, involvement of the population in support measures, strengthening support through transfer of research and technology, taking local circumstances into consideration and providing for active participation of recipient countries.

However, the Convention is somewhat programmatic in character, while binding operational and specific financial consequences were not fixed. The significance of the Convention lies more in the political and psychological sphere than in specific development programs. The Council also regrets that the wording of the Desertification Convention does not go much further than mere declarations of intent. The only new and additional source of finance mentioned therein is the *GEF*, and that only with considerable restrictions. Nor was the 0.7% of GNP target for development aid included in the Convention. In the opinion of the Council, which has repeatedly demanded that development aid be substantially replenished, there is no solid financial basis for genuinely combating desertification. Nevertheless, the coming into force of the Desertification Convention means that important issues relating to bilateral and multilateral cooperation for the regions specified in the Convention will be affected.

Recommendations:

- The formal ratification of the Desertification Convention should be accomplished as quickly as possible. In certain circumstances the exertion of diplomatic influence on other nations to ratify the convention as soon as possible could be considered .
- The objectives of the first Conference of the Parties to the Convention should be defined very soon, whereby extension to a global soil protection convention should be striven for.

- In accordance with preventative crisis management, those countries that are particularly threatened by the combination of poverty, desertification and political conflicts should be given preferential support.
- Research into desertification must place greater focus on the networking of individual disciplines, both in the support of national research facilities in the respective countries and support for international agricultural research, as well as within the German research community. In view of the increasing importance of international conventions for national policymaking, more thought should be given to developing a *convention-oriented research*.

The Biodiversity Convention – The Implementation Is Yet to Come

The *Convention on Biological Diversity* is the first internationally binding Convention that applies a *trans-sectoral* approach to the protection of global biodiversity. The objective is not simply that of nature conservation, but also "the sustainable use of biological resources, and the equitable sharing of the benefits arising out of the utilization of genetic resources". Access to genetic resources is also established as a principle of international law. The *First Conference of the Parties* in Nassau in 1994 succeeded in establishing the basis for further work. The next step is to implement the convention in the contracting parties, for which the production of national reports on the status of biodiversity and the development of strategies for integrating the convention's objectives into national policymaking are of particular importance.

It is too early as yet to assess the success of the Biodiversity Convention, since no detailed results can be expected at this stage of the convention's process. One positive aspect is that the financing mechanism is already being applied to projects aimed at achieving the objectives of the Biodiversity Convention, and that the Conference of the Parties decided on eligibility criteria. The Council considers it important for the future negotiation process that a *protocol on biosafety* be formulated and adopted without delay, that an *instrument* be developed for protecting forests and that FAO's *"International Undertaking on Plant Genetic Resources"* be adapted to the Biodiversity Convention. The public discussions relating to the Convention have sharpened awareness in society for the seriousness of species and biotope losses. This is all the more important, in that attaining the Convention's objectives cannot be left entirely to national authorities, but also requires the active support of environmental organizations and the public at large.

Recommendations:

- Rapid development of a German strategy for implementing the Convention.
- Support for the clearing-house mechanism of the Biodiversity Convention, for example by setting up an institution in Germany for facilitating the exchange of information and technology transfer.
- Support for research in the field of bioprospecting.
- Support for the expansion of research infrastructure in developing countries: compilation of biodiversity inventories, nature conservation management, capacity-building for independent exploitation of national genetic resources and the creation of focal points in the countries of origin, aimed at improving the effectiveness and control of access to genetic resources.

Protection of Forests: Protocol or Convention?

A reversal of the global trends towards *loss* and *degradation of forests* is not foreseeable at the present time. This makes the lack of binding instruments of global environmental policy for the protection of the forests based on international law all the more aggravating. After the failure to draw up such a document at the UNCED in Rio de Janeiro in 1992, where only a nonbinding *"Forest Declaration"* was adopted, this issue continues to be of utmost importance to the current situation. On the one hand, the issue of forests could be treated in a separate convention (*Forest Convention*); on the other hand, it would be possible to regulate the use of forests in a protocol on the basis of the Biodiversity Convention (*Forest Protocol*).

Given that forests are an integral element of "biological diversity", immediate action is required in view of the dramatic pace of their destruction. Since the Biodiversity Convention has already gone into effect, a *"Forest Protocol"* would presumably take less time for negotiation than drawing up a completely new *"Forest Convention"*, whose basic objectives would first have to be agreed upon. Moreover, a regulation of forest use separate from the Biodiversity Convention may lead to decisive weakening and marginalization of that convention.

Recommendations:

- The Council recommends that the German Federal Government commit itself to a *Forest Proto-

col within the framework of the Convention on Biological Diversity.

The Gatt/WTO-Regime – The Greening of World Trade

The economies of the world are becoming increasingly integrated, as evidenced in particular by increasing international trade, the globalization of production and markets, and the growing number and importance of multinational corporations. The consequences of these developments are among others intensified international division of labor and increasing international exchange of goods. The institutional framework for regulating international trade is the *General Agreement on Tariffs and Trade* (GATT), which after the completion of the latest round on the reduction of tariffs was turned into the *World Trade Organization* (WTO). Trade-environment interactions have become more prominent in the internationale debate. Particularly negative environmental effects may result from a higher level of transport, increasing resource consumption or the shifting of polluting industries to countries with lower environmental standards. Positive environmental effects, on the other hand, can be anticipated if growth effects create financial scope for more environmental protection, if the exchange of goods leads to the diffusion of low-emission technologies, or if a higher level of environmental awareness is generated via the transfer of knowledge associated with the exchange of goods and production factors.

The agreements reached in 1994 at the end of the *"Uruguay Round"* have brought about important changes, above all the inclusion of protection and preservation of the environment and the principle of sustainable development as key objectives in the Preamble to the Agreement establishing the WTO, the dismantling of product-related subsidies in agriculture and textiles, and the reform of dispute settlement procedures. In the view of the Council, however, the integration of environmental issues and considerations into the GATT/WTO-regime has yet to be accomplished.

Recommendations:

- There is an urgent need for defining the term eco-dumping and for the efficient organization of the dispute settlement procedure.
- In the past, there has been far too little coordination between GATT rules and international environmental agreements. Where conflicts arise between trade arguments and environmental agreements, the latter should have priority.

- States must show they have valid grounds if they wish to impose trade sanctions on other states for environmental reasons. These include all international environmental problems of a transboundary nature, such as the ozone, climate, water and soil problems. Trade measures must relate to specific environmental objectives, be applied only to the extent necessary, and may not arbitrarily or unjustifiably discriminate between nations.
- Should these environmental reforms within GATT/WTO fail to be realized, the Council recommends the establishment of an independent environmental organization to which the surveillance of existing environmental agreements should be assigned, as well as, in the long term, additional competencies for enforcing and further developing international environmental agreements.

General Conclusions and Recommendations

Solving global environmental problems demands, first of all, the improvement of certain *societal conditions*. Three basic concepts appear to be especially important:

1   Global environmental policy can only achieve its goals if there is an increase in *environmental awareness* and the willingness to act in an environmentally sound manner. Efforts in environmental education must be improved worldwide through the involvement of both public and private education systems. This includes strengthening institutions that support environmental education worldwide, e.g. UNESCO.

2   The scope available for combating global environmental problems is critically dependent on *population trends*. The slight leveling off of world population growth should be seen as justifying even greater efforts, since there is indeed hope of success in this domain. Of central importance in this connection are the eradication of poverty, the improvement of the social and societal position of women, and the guaranteed provision of care to the aged.

3   Population growth, poverty and environmental degradation are causing increased migration pressure in many regions of the world. *Migration flows* continue to be directed at neighboring regions, but Europe also will be directly affected to an increasing degree in the future. The Council demands that the causes of migration be addressed more vigorously than before in the countries of origin. German development aid must not be allowed to shrink any further, but instead must be increased significantly over the long term.

The second approach to solving global environmental problems involves the formulation and implementation of *international agreements*. The following general conclusions relate to this second approach:

1   Faster progress in global environmental policy can often be achieved when states or groups of states willing to take action assume a *vanguard role* with respect to certain solutions. The Council recommends that the implementation of a system of internationally binding tradeable CO_2 emission entitlements be started as soon as possible within the European Union. The pilot phase for Joint Implementation of the Climate Convention should be started without delay and actively supported by Germany.

2   The system of international environmental agreements must be adequately expanded and further improved. There are conventions in place with respect to climate, biodiversity, desertification and the law of the sea, which now have to be fully implemented. In addition, agreements on forests and soils must be formulated and implemented. Regarding the protection of forests, the Council recommends a regulation in the form of a protocol under the Biodiversity Convention. The Desertification Convention should be made part of a wider convention on the protection of soils.

The Council appeals for environmental reform of the GATT/WTO regime. Should the WTO fail to give adequate consideration to environmental concerns, the Council recommends the creation of a new international environmental organization.

Presumably, conventions are unsuitable for dealing with population growth and environmental education. However, the stated objectives and measures in the various environmental conventions and other international agreements should be linked to each other more strongly than is currently the case and checked for incompatibilities.

3   The instrument of international agreements should be developed further, it is the precondition for progress in global environmental policy. This does not mean that further development needs to be based solely on formalized conventions with specialized institutions and multilateral funding. The Desertification Convention, for example, would not have been achieved without the bilateral funding option.

The Council wishes to emphasize that the main trends of global change – population growth, climate change, loss of biological diversity, degradation of soils, and scarcity of freshwater – show no signs of amelioration, and in some cases are worsening still further. The need for solutions to these global problems is therefore more urgent than ever before.

Introduction

A

At the 1995 Climate Conference in Berlin it became clear again that a change is needed in attitudes and in actual behavior in order to avoid and mitigate dramatic climate changes and their impacts. This new thinking has to lead to changes in individual and social action on the one hand, and to improved national and international laws and agreements on the other hand. In this year's report the Council uses these two approaches to address global environmental problems; it thus represents a follow-up to its first report of 1993.

"Ways Towards Global Environmental Solutions" is an ambitious and, at the same time, cautious title. There is an implicit assertion that solutions for global environmental problems are in sight. Indeed, the Council assumes that the current problems can be solved, given the necessary will to do so, i.e. it believes that irreversible catastrophic developments are not inevitable. Whether these roads will in fact be taken is another question, however, because this requires a significant reorientation on the local, national and global scale. In this respect the title of the report is cautious.

The fruits that are beginning to emerge from *international agreements* are encouraging. The agreements on the protection of the ozone layer (*Montreal Protocol*) and on the protection of the seas (*Convention on the Law of the Sea*) are successful examples of global policy. Also, the long-neglected loss of biodiversity has received increased attention through the *Convention on Biological Diversity*. The main part of this report is devoted to these international agreements. An analysis of the status and necessity for development of individual agreements in *Chapter C* shows that there are still great problems to overcome. The report takes a close look at these obstacles and offers respective recommendations. In taking this approach, one should not be discouraged by the fact that, for example, the original steps regarding climate policy provided for in Rio de Janeiro in 1992 have not been implemented for the *Berlin Conference of the Parties*. What is important is that political negotiation processes have been initiated at the international level and further work is now being carried out to find adequate solutions. Because of the urgency of the necessary steps, the Council gave a special statement on the occasion of the Berlin Conference, urging far-reaching measures with respect to climate protection (WBGU, 1995).

Numerous necessary steps regarding global environmental policy have yet not been taken. This contradiction is in part due to the fact that the *social prerequisites* for solving global environmental problems have not been met. These prerequisites and the corresponding measures for meeting them are summarized in Chapter B of the report:

- Environmental awareness and environmental education must be reinforced worldwide in order to support global environmental policy.
- Exchange of know-how and technology transfer have to be implemented in order to enable countries to develop in an environmentally sound manner.
- Institutions and regulations have to be modified, both nationally and internationally, in order to provide incentives for environmentally conscious behavior.
- The increase and unfavorable distribution of the world population must be reemphasized as a central cause of numerous global environmental problems.

The more specific issues regarding the ways of solving global environmental problems will have to be left for future reports. One important issue is that of the conditions of successful international agreements, including global environmetal diplomacy. It would be important, for example, to analyze the conflict between economic development and environmental pollution with regard to growing global water problems. In its next report, however, the Council will first comment on the status and perspectives of research on global change in Germany.

Societal Conditions for Solving Global Environmental Problems

B

1.1
Introduction

Most international environmental declarations and conventions contain a demand for reinforcing the environmental awareness of the population and taking measures concerning environmental education. These processes are perceived as important societal prerequisites for changing patterns of production and consumption, and thus promoting environmentally sound lifestyles, on a long-term basis (*Box 1*). Several chapters of AGENDA 21, for example, demand promotion of individual and collective modes of behavior that are suitable for ensuring a process of sustainable development. This ambitious task of extensively altering environmentally harmful behavior and of promoting environmentally sound modes of behavior can only be tackled successfully if the citizens of a country and of all nations attach high priority to questions of environmental protection. Only when the necessity of sustainable development is implanted firmly in the consciousness of people with their various roles and positions in society can strategies for changing behavior become effective.

Since a solution to global environmental problems cannot be brought about by local or national measures alone, but has to involve international trade and political negotiation as well, it is imperative for trade and negotiation partners to know something about the significance of environmental problems in the society's scale of urgency and about the approaches to "learning" environmentally sound modes of behavior that are available in the educational landscape of a country.

This report examines the perception and assessment of environmental problems and the "learning" of environmentally sound or the "unlearning" of environmentally harmful modes of behavior in the context of global environmental changes and from a worldwide perspective. Two basic complexes of questions have to be clarified in this connection:

1 Environmentally relevant behavior is influenced by a wide range of different factors that addition-

ally interact with one another. Since the respective importance of these factors also depends on context variables, simple cause-and-effect attributions are not possible (*see Section B 1.3*). However, an adequate perception of the problems and their emotional weighting are undoubtedly important prerequisites for changing environmentally relevant behavior. Therefore, it is necessary to know what information is available on the perception of environmental problems in individual countries. Such information can currently be gained from sociological survey research in different countries as well as from comparative national studies concerning the change in values in different societies (*see Section B 1.3*).

2 Environmentally relevant modes of behavior are learned or forgotten again from one's childhood. Different institutions and agencies in the field of environmental education are involved in this life-long learning process. Thus the question here concerns the significance of environmental education in different countries, whether through state institutions or through NGO initiatives, and the extent to which global environmental problems have become part of such educational activities. In view of insufficient documentation, however, this report will not present a comprehensive assessment on the theory and practice of environmental education. Instead, it will provide initial assessments based on the available specific analyses and derive from them proposals for further development of environmental education as a subject of research and as an environmental policy strategy (*see Section B 1.4*).

1.2
Environmentally Related Behavior and Its Determinants

The Council dealt with various theoretical principles and empirical approaches to explaining and altering environmentally relevant modes of behavior in detail in its 1993 annual report (WBGU, 1993). The

BOX 1

Recommendations in International Declarations and Conventions on "Environmental Awareness and Environmental Education"

AGENDA 21

Chapter 4: Changing consumption patterns
- Focusing on unsustainable patterns of production and consumption
- Developing national policies and strategies to encourage changes in unsustainable consumption patterns

Chapter 23: Strengthening the role of major groups

Chapter 24: Global action for women towards sustainable and equitable development

Chapter 25: Children and youth in sustainable development

Chapter 26: Recognizing and strengthening the role of indigenous people and their communities

Chapter 27: Strengthening the role of nongovernmental organizations: partners for sustainable development

Chapter 28: Local authorities' initiatives in support of AGENDA 21

Chapter 29: Strengthening the role of workers and their trade unions

Chapter 30: Strengthening the role of business and industry

Chapter 31: Scientific and technological community

Chapter 32: Strengthening the role of farmers

Chapter 36: Promoting education, public awareness and training
- Reorienting education towards sustainable development
- Increasing public awareness
- Promoting training

Framework Convention on Climate Change

Art. 6: Education, training and public awareness
at national/regional level:
- Promote and facilitate the development and implementation of educational and public awareness programmes on climate change and its effects.
- Promote and facilitate public access to information on climate change and its effects.
- Promote and facilitate public participation in addressing climate change and its effects and developing adequate responses.
- Promote and facilitate training of scientific, technical and managerial personnel.

at international level:
- Cooperate in and promote the development and exchange of educational and public awareness material on climate change and its effects.
- Cooperate in and promote the development and implementation of education and training programmes, including the strengthening of national institutions and the exchange or secondment of personnel to train experts in this field, in particular for developing countries.

Convention on Biological Diversity

Art. 12: Research and Training
- Establish and maintain programmes for scientific and technical education and training in measures for the identification, conservation and sustainable use of biological diversity and its components and provide support for such education and training for the specific needs of developing countries.

Art. 13: Public Education and Awareness
- Promote and encourage understanding of the importance of, and the measures required for, the conservation of biological diversity, as well as its propagation through media, and the inclusion of these topics in educational programmes.
- Cooperate, as appropriate, with other States and international organizations in developing educational and public awareness programmes, with respect to conservation and sustainable use of biological diversity.

Rio Declaration

Principle 8:
- States should reduce and eliminate unsustainable patterns of production and consumption.

Principle 10:
- Environmental issues are best handled with the participation of all concerned citizens, at the relevant level.
- At the national level, each individual shall have appropriate access to information concerning the environment ....., and the opportunity to participate in decision-making processes.

- States shall facilitate and encourage public awareness and participation by making information widely available.

Convention to Combat Desertification

Art. 3: Principles

- Taking decisions on the design and implementation of programmes to combat desertification and/or mitigate the effects of drought with the participation of populations and local communities.
- Development of cooperation among all levels of government, communities, nongovernmental organizations and landholders to establish a better understanding of the nature and value of land and scarce water resources in

affected areas and to work towards their sustainable use.

Art. 19: Capacity building, education and public awareness

- Promotion of capacity building – that is to say, institution building, training and development of relevant local and national capacities – in efforts to combat desertification and mitigate the effects of drought.
- Undertaking and supporting public awareness and educational programmes ... to promote understanding of the causes and effects of desertification and drought.
- Establishing and strengthening networks of regional education and training centres to combat desertification and mitigate the effects of drought.

most important statements in that report are summarized and expanded on below.

There is no single overall theory of environmentally related behavior but there is a whole range of theoretical and empirically supported models for conceptualizing such behavior (including both environmentally harmful and environmentally sound modes of behavior) and its determinants. What is common to these models is the large number of factors that influence environmentally relevant behavior. It is difficult to weight these individual factors, particularly because the determinants in question interact with each other. For this reason it appears meaningful to view potential conditions of environmentally relevant behavior, including

- perception and assessment of environmental circumstances,
- environmentally relevant knowledge and information processing,
- attitudes and value orientations,
- incentives to act (motivational and reinforcement factors),
- opportunities and offers to act,
- perceived consequences of actions (feedback) and
- perceived actions of reference groups and model persons,

as relatively equal in importance and to take them into account in each case when designing specific intervention measures (WBGU, 1993). At any rate, one must avoid the all too frequent presumption that "environmental awareness" (as one conditioning factor) can be equated to the actual environmental behavior.

Depending on the disciplinary perspective (psychology, sociology, other empirical social sciences), the term "environmental awareness" is defined on

the basis of different aspects, several of which are presented here for illustrative purposes (*Box 2*).

In addition to this list of viewpoints and proposed definitions, which is by no means an exhaustive one, there are numerous ad hoc definitions from demoscopic surveys where responses to single questions are frequently interpreted as an expression of "environmental awareness". Such surveys in particular, pawned off as instantaneous, diagnostic snapshots, tend to dominate the social discourse and thus contribute to "watering down" the term even more.

The lack of clarity regarding "environmental awareness" might suggest doing away with this term in a scientific context completely and, instead, for example, directly referring to the three components of a psychological concept of environmental awareness (cognition, affect, intention) as determinants of environmentally related behavior. As a rule, this is precisely what is attempted within the framework of the various conceptualizations of environmentally related behavior. Such an approach becomes necessary in any case when an operational definition is required for the collection of data.

The model developed by Fietkau and Kessel (1981) can now be regarded as an almost "classic" theoretical approach to conceptualizing "environmental behavior", to which many researchers have made reference explicitly or implicitly since then. On the basis of plausibility considerations, the authors postulate a "framework of order", within which the factors of environmentally relevant knowledge, environmentally relevant attitudes/values, offered modes of behavior, incentives to act, perceived behavior/consequences and environmentally relevant behavior are intertwined in an intricate web of interrelationships (*Fig. 1*), though without any weighting of these factors.

Another theoretical conceptualization originates from Stern and Oskamp (1987) and is based on a heuristic multilevel causal model that implies a kind of "logical" sequence of factors (from background factors to observable effects of behavior), again assuming dynamic feedback effects between the individual levels (particularly through learning and self-justification effects in the sense of reducing contradictions in the "conceptual structure" of individuals, so-called cognitive dissonances) (*Tab. 1*).

In contrast to these theoretical models, there are empirically supported conceptualizations. Urban's "structural model of environmental awareness" (1986), for example, is based on a theoretical three-component concept of environmental awareness (value orientations, attitudes, willingness to act) and determines statistically significant interrelationships between conditional factors on the basis of questionnaire data (*Fig. 2*). This provides indications of the diversity of the factors to be taken into account (e.g. education, age, occupational sector, offered modes of behavior) as well as of the sometimes only moderate interrelationships between these factors. Urban himself developed his model further (Urban, 1991), resulting in, among other things, considerable changes in the composition of the determinants.

In summary, one can say that environmentally related behavior has been conceptualized in a number of ways, both with regard to content and methodologically. A single model of behavior has not yet materialized. Consequently, social and behavioral science research as well as policy-related applications are still compelled to assume a wide range of influencing factors that can only be defined in concrete terms with regard to specific intervention plans.

1.3
Social Perception of Problems and Environmental Protection: Empirical Findings

By virtue of the international declarations and conventions adopted in Rio de Janeiro in 1992 and in the following period, the governments of the Parties have documented the great importance attached to the areas of "environmental protection" and "development" in their respective national priorities. But what are the priorities of the people with regard to those problems perceived as most pressing? What influences are they subjected to? Do they change over time? Can general trends be discerned? What differences exist between individual nations and cultures and what can they be attributed to?

A first approach to answering these questions can be provided by a comparative analysis of sociological survey data on the perceived (relative) "importance" of the problems of a society (agenda-setting surveys);

BOX 2

"Environmental Awareness": Viewpoints and Proposed Definitions

– Environmental awareness in an everyday sense: sensitivity to environmental crises; fears, dissatisfaction and concern with regard to negative changes in the environment.

– Environmental awareness as basic ecological understanding: "ecological awareness of problems" (Billig, 1994).

– Environmental awareness as an individual value orientation: attaching great value to nature and to environmental protection (Lantermann and Döring-Seipel, 1990).

– Environmental awareness as an individual attitude composed of three components: cognitive (knowledge, rational assessments), affective (concern, expressions of feelings) and conative component (behavioral intentions) (Spada, 1990, with reference to Rosenberg and Hovland, 1960).

– Environmental awareness as an attitude, with the addition of the component of actual behavior (Schahn and Holzer, 1990).

– Environmental awareness as an awareness of ecological responsibility with components of ecological thinking, ecological control and moral notions (Lecher and Hoff, 1993).

– Environmental awareness as the sum of the notions, opinions, feelings and intentions shared by a society with regard to people-environment topics (sociological notion of environmental awareness).

Table 1
An approximate causal model of resource use with examples from residential energy consumption.
Source: Stern and Oskamp, 1987

Level of causality	Type of variable	Examples
8	Background factors	Income, education, number of household members, local temperature conditions
7	Structural factors	Size of dwelling unit, appliance-ownership
	Institutional factors	Owner/renter status, direct or indirect payment for energy
6	Recent events	Difficulty paying energy bills, experiences with shortages, fuel price increases
5	General attitudes	Concern about national energy situation
	General beliefs	Belief households can help with national energy problems
4	Specific attitudes	Sense of personal obligation to use energy efficiently
	Specific beliefs	Belief that using less heat threatens family health
	Specific knowledge	Knowledge that water heater is a major energy user
3	Behavioral commitment	Commitment to cut household energy use 15%
	Behavioral intention	Intention to install a solar heating system
2	Resource-using behavior	Length of time air conditioner is kept on
	Resource-saving behavior	Insulating attic, lowering winter thermostat setting
1	Resource use	Kilowatt-hours per month
0	Observable effects	Lower energy costs, elimination of drafts, family quarrels over thermostat

suitable surveys have been conducted, at least in the industrialized countries, for some time now.

Such survey data is often difficult to access because the surveying institutes are usually privately run. In these cases one is forced to fall back on secondary sources that often contain gaps in their description of the methodological approach. This applies particularly to data from countries outside of Germany that were not collected within the framework of international survey programs. In addition, it can be presumed that systematic, continuous surveying has, in fact, only been carried out in the western industrialized countries up to now, and even there this practice has not gone beyond the preliminary stage.

1.3.1
Methodological Problems of Survey Research

The methodological limits of survey studies often emerge very quickly upon closer analysis so that, despite their omnipresence in social discourse, statements made on the basis of survey data can only be meaningfully interpreted through critical consideration of the marginal methodological conditions. This caveat applies more or less to all statements based on survey data, and even, for example, to attempts to monetarize environmental assets by means of the expressed willingness to pay (*see below*) and is, for this reason, expressly considered before the presentation of the respective study results. A good illustration can be provided by an example from recent German agenda-setting surveys (*Box 3*).

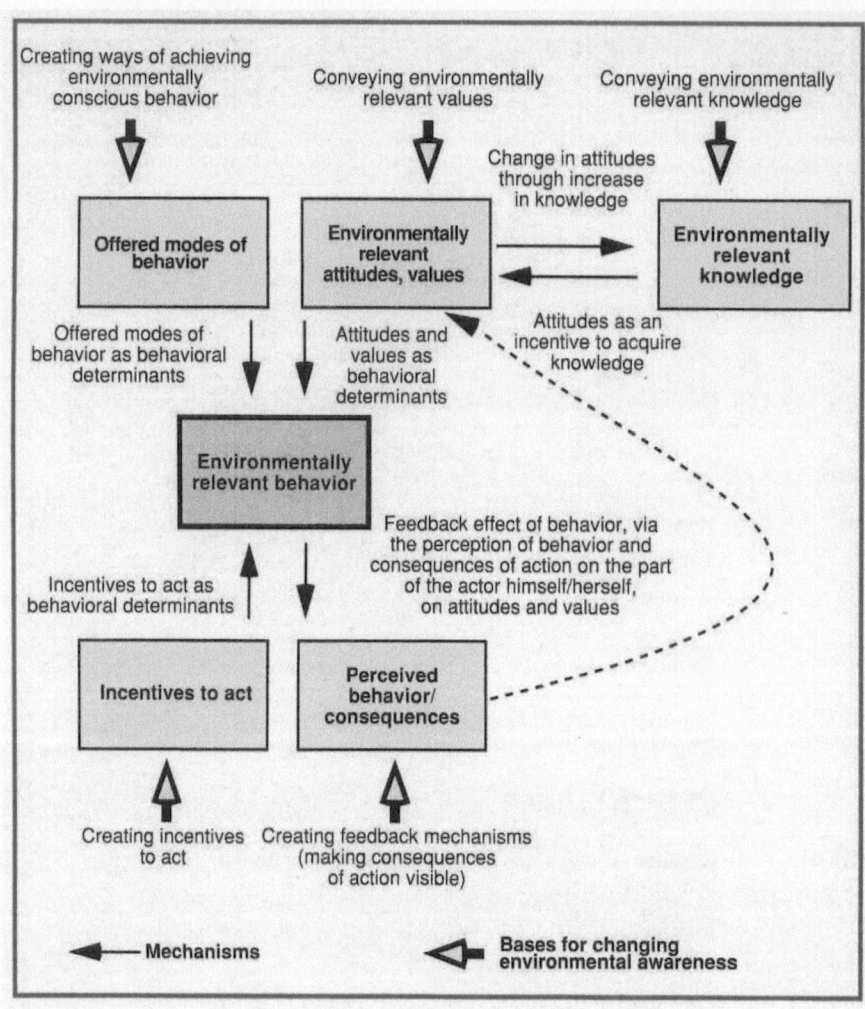

Figure 1
Model for
conceptualization of
"environmentally relevant
behavior".
Source: Fietkau and
Kessel, 1981

Under the premise that the research institutes mentioned are reputable institutions conducting serious research work, the striking differences in the results can be predominantly attributed to the respective methodological approach. In particular, the way in which the question is formulated (open versus closed) is obviously a significant factor.

If no alternative responses to choose from are given and the surveyed persons are asked to name issues, problem areas or the like spontaneously (open formulation of the question), aspects of environmental protection normally rank in the middle of their "list of priorities" at best. Current individual preferences are revealed more clearly with this approach, though in a distorted manner due to the involuntary application of availability heuristics ("media topicality": WBGU, 1993). The result may vary, depending on whether the designation of a number of equally weighted issues or a ranking is requested, and the number of responses requested also plays a role. Finally the study results can also be influenced by whether the categories for allocating the responses

are formed a priori or after the survey, how many there are, the degree of differentiation, etc.

If, on the other hand, the surveyed persons are given a list of issues to be evaluated within the framework of a closed formulation of the question, socially desirable developments such as environmental protection receive high values (effect of "social desirability" effect). However, closed formulations of questions must also be differentiated as to whether the surveyed persons have to choose a certain number of given alternative responses or rank them. Frequently each of the alternatives have to be evaluated on the basis of scales (e.g. from "very important" to "very unimportant"). The concrete formulation and degree of differentiation of the scale (graduation) as well as the presence of remaining categories ("don't know", "no opinion") may have an influence on the results. Just the selection of the given information (number, degree of differentiation, choice of words, etc.) represents a central problem when using a closed formulation of the question.

Figure 2
Statistical estimate of a
theoretically specified
causal model of
environmental awareness
and environmental
behavior (data base:
questionnaire study with
216 grown-up Germans
surveyed).
Source: Urban, 1986

Urban conceives
environmental awareness
as a combination of
environmentally related
value orientations,
attitudes and
corresponding willingness
to act, which is supposed to
predict self-reported
environmentally relevant

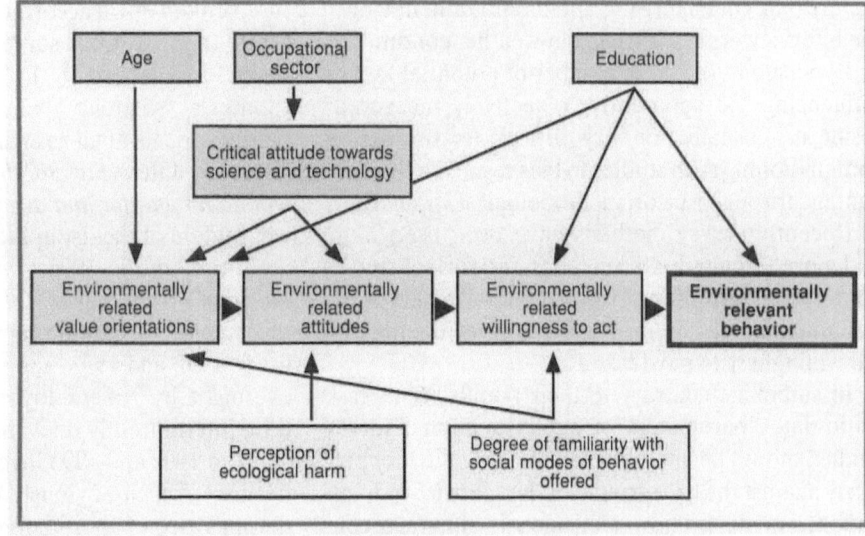

behavior. The variables of age, occupational sector, education, critical attitude towards science and technology, perception of ecological pressure as well as familiarity with socially organized modes of behavior proved to be statistically significant factors (level of significance: 5%; influence exerted in the direction indicated by the arrows) in a causal, path analysis model. The path coefficients and the explained variance in the targeted variables regarding environmental awareness and environmental behavior are not presented. All influences are positive, with the exception of the relationship between the variables of age and environmentally related value orientations.

BOX 3

**Methodological Problems in Survey Research –
an Example**

According to a study by the Sample-Institut in May 1994, 83% of the Germans surveyed (in eastern and western Germany) considered the solution of problems related to "pollution" to be "important" or "very important" (Sample-Institut, 1994). These issues were ranked 2nd among important social problems, following "unemployment". Those surveyed had to classify various given issues on a scale of five from "not important at all" to "very important" (closed formulation of the question).

At the same time (May 1994), but with a different methodology, the *Forschungsgruppe Wahlen* (Election Research Group "ZDF-Politbarometer") determined that "unemployment" (55%), followed by "asylum issues/foreign citizens"

(31%) and "right-wing radicalism" (17%), were the "most important issues" for the old German States. "Environmental protection" follows here in 5th place with only 13% (Forschungsgruppe Wahlen, 1994). The following ranking resulted for the new German States: "unemployment" (79%), followed by "law and order" (17%), "problems connected with unification" (14%) and "asylum issues/foreign citizens" (10%). All other issues mentioned were below 10% and the issue of "environmental protection" was not in the top ten at all. Thus the (western states) value of 13% can be considered as representing the maximum for the entire Federal Republic of Germany. The procedure followed by the Forschungsgruppe Wahlen (Election Research Group) differs from that of the Sample-Institut in that the surveyed persons themselves were able to state up to two "most important issues" without being given any to choose from (open formulation of the question).

Besides the aspects mentioned, the arrangement of the questionnaire within a study and related sensitization and anchoring effects, the exact formulation and the spatial and time reference framework offered to the surveyed persons play a role with regard to methodology. Further influencing factors include

the representativeness of the respective sample (as a rule 1,000-2,000 persons are surveyed and conclusions regarding the entire population are drawn from the result) as well as many other marginal conditions of the studies that can scarcely be checked completely (e.g. characteristics and modes of behavior of the interviewers or even current weather conditions).

By virtue of the large number of potential factors influencing the apparently "objectively measured" result, it is usually not very difficult for the parties commissioning such studies to bias results to their advantage through a priori methodological stipulations.

Recently survey methods have been used more and more frequently within the framework of studies on the monetarization of environmental assets or environmental damage, primarily to survey the individual willingness to pay (*Box 4*).

In summary, contrary to their popularity as fast, up-to-date "barometers of public opinion", survey studies should be analyzed and interpreted very precisely against the background of the specific study situation, no matter how meticulously they are conducted. A comparative analysis of contemporary studies which use different methods may be helpful, though it takes away much of the (at first sight) meaningfulness of each individual study. In any case an ideal way of determining "actual" attitudes of the population is not yet in sight. Both of the basic approaches briefly described (open versus closed formulation of the question) lead to their own problems. In the face of all criticism, however, it must be emphasized that surveys cannot be dispensed with as a study method of the social sciences. Rather, they often represent the only way of recording perceptions, attitudes, mental states and modes of behavior at low cost and with reasonably meaningful results.

1.3.2
Perception of Problems over Time
(Longitudinal Analyses)

A comparative analysis of "priority lists" of the population in Germany on a long-term basis shows that the issue of "environmental protection" has generally gained significance in the course of time. According to some studies, it even held a top position for a certain time among those issues considered important. On the other hand, substantial fluctuations can be observed, some at relatively short time intervals. The issue of "environmental protection" must, therefore, be seen in connection with other issues as well as with political and other daily events. In addition, method-related distortions have to be taken into account (*see B 1.3.1*).

For several years the Forschungsgruppe Wahlen (Election Research Group) has been surveying the "most important issues" for the German population in monthly representative surveys ("ZDF-Politbarometer", 1992-1995). The results for the period from January 1992 to April 1995 produce the following pic-

ture, broken down according to old and new German States (approx. 1,000 surveyed persons in each case; open formulation of the question; up to two responses): whereas "environmental protection" hardly appears at all as an important issue in the new German States (*Fig. 3b; the "unemployment" problem dominates far and away here*), it ranks in the lower middle of the list in the old German States (*Fig. 3a; rise in spring 1995 – an effect of media reports touching on the Berlin Climate Conference?*). Here the issue of "asylum/foreigners" was at the top of the list in 1992 and 1993, and was then replaced almost overnight by "unemployment" in autumn 1993. A third intermittently relevant issue in western (!) Germany in 1992 and 1993 involved "problems with unification", which obviously lost their significance for the majority of those surveyed towards the end of 1993, however. In spring 1995 the concern over jobs continued to dominate the problems perceived by those surveyed, in eastern Germany even more than in the old German States. In western Germany "environmental protection" has attained a new peak since July 1992, in third place after "asylum/foreigners" (with an upward trend), while this issue currently virtually plays no role at all among "important problems" in eastern Germany.

If, by contrast, one looks at a series of annual surveys of the *Institut für Praxisorientierte Sozialforschung* (Institute for Practice-Oriented Social Research, ipos) for the *Federal Ministry of the Interior* (ipos, 1993) (1990-1993; West versus East; at least 1,000 surveyed persons in each case; closed formulation of the question; on a scale of four for each question), a completely different picture results: in this case "effective environmental protection" remains at the top of the list among the "very important tasks and goals" in the period under study in the old German States, with a somewhat downward trend, followed closely by "creating jobs" and "preventing abuse of rights to asylum" (both increasing in importance, at 65% roughly equal to "environmental protection" in 1993) as well as "combating drugs" (somewhat downward trend). In the new German States, on the other hand, the significance of "environmental protection" was clearly dropping in the period 1990-1993 (from just under 80% to just under 70%) while "creating jobs" has dominated since 1991, followed by "combating crime". Both issues were still lagging behind "effective environmental protection" in 1990.

If a kind of preliminary conclusion were drawn on the basis of the jumble of available longitudinal data (different institutes and clients, survey intervals and methods, etc.), it would reveal that "environmental protection" clearly is not the issue of the Germans at present, despite a substantial upswing, particularly during the 80s. The principle of "charity begins at

BOX 4

Survey Methods for Analyzing the Willingness to Pay

The demand for consideration of the benefits of future generations, as has been raised in the discussion over sustainable development, leads to the growing importance of option, legacy and existence values. Attempts to make these benefit components, which are not directly reflected in market data, and other psychosocial aspects of environmental changes (so-called intangibles) accessible for a monetary assessment frequently make use of the individual willingness-to-pay for analytical purposes (Schulz, 1985; Schluchter et al., 1991; Hampicke et al., 1991). The willingness of the surveyed persons to pay a certain maximum amount of money for an improvement in the quality of the environment is taken as an approximate value of the current damage to the environment. The aim of such analyses is usually to obtain a monetary estimation of changes in benefits due to environmental changes in order to enable a comparison with the costs of appropriate environmental policy measures and, at least in principle, an internalization of the social and ecological costs. Since analyses of the willingness to pay in this sense are based on hypothetical markets, they fundamentally cannot be checked against reality.

Survey methods to determine the willingness to pay (contingent valuation methods) can be divided into willingness-to-pay and willingness-to-sell approaches. Different cognitive mechanisms can be assumed for the responses, depending on whether people are asked about their willingness to pay money for obtaining or preserving an asset or about the necessary compensation payments for the retention or worsening of a certain state of the environment. Willingness-to-sell surveys should fundamentally result in higher assessments than willingness-to-pay surveys because individual income restrictions do not come into play here.

After noncritical application of contingent valuation methods for a long time, a whole range of methodologically critical studies has now been submitted in view of the great political relevance of the study results (summary: Hausman, 1993).

These studies primarily attempt to reveal potential sources of systematic distortion.

They see key problems in the specific design of the respective questionnaire or interview. The problems discussed above, for example, result according to the way the question is formulated. Moreover, those surveyed often have no previous experiences regarding the assessment of environmental assets or damage in terms of money. This makes the method susceptible to distorting "anchor and sequence effects" as well as to the influences of inadequate or ambiguous descriptions of the facts to be assessed. In addition, contingent valuation methods place considerable demands on the abstraction capacity and imagination of those surveyed, particularly because of their hypothetical nature. The hypothetical nature of the situation also poses the question of the possible strategic behavior of the persons surveyed, which may lead to an overestimation of the "actual" willingness to pay. Available studies on this aspect assess this danger as relatively slight, however. Empirically determined factors that additionally influence the reaction of those surveyed include age, the level of information and personal concern.

In view of the great number of methodological problems involved in interpreting willingness-to-pay data from surveys, the frequently cited advantages of flexibility and feasibility lose their relative importance. Willingness-to-pay surveys are even classified as a method for merely surveying attitudes by some researchers because of the high correlations found to other scales (Kahneman et al., 1993). If one ignores this more fundamental criticism of the monetarization of environmental assets or damage, the method of contingent valuation still remains an important instrument in economically oriented research since many benefit aspects are either hardly or not accessible at all by means of other methods. Another positive aspect in line with the demand for more participation, such as in the field of environmental education, is the fact that those concerned themselves are asked about their assessment of environmental assets or damage in direct surveys, a task that is otherwise frequently passed on to "experts" who do not necessarily have better assessment standards.

home" was a more determining factor due to the concern about jobs. It remains to be seen whether this relative ranking will change in the near future in view of the emerging economic recovery, which according to many experts will not necessarily bring about a decline in unemployment. The recently observed rise in the importance attached to "environmental protection" would allow such an interpretation, at least in western Germany. For the time being, however, one can assume that although a general sensitivity for environmental problems exists in the Federal Republic of Germany, it clearly competes with other values, a fact that certainly has an influence on environmentally relevant behavior and on the acceptance of environmental policy measures.

Recent time lapse data from the USA are provided by a *Gallup Institute* study series in which the "currently most important problems for this country" were surveyed at four points in time in the years 1991 to 1993 (Newport, 1993) (approx. 1,000 surveyed persons in each case; open formulation of the question; up to three responses). Similar to the methodologically comparable "Politbarometer" studies of the Forschungsgruppe Wahlen in Germany (*Figs. 3a and 3b*), "environment" ranks far behind at a relatively constant level here (2-3%) whereas economic problem areas, such as "economy in general" and "unemployment", dominate over the entire period (particularly at the beginning of 1992 with 42% and 25%, respectively). At the end of the survey period examined, however, social issues, such as "crime" and "health care" (at the end of 1993: 16% and 28%, respectively), clearly gained ground, presumably due to the domestic policy debate in the USA at that time.

In spite of the similarity in trends, the absolute values of the Gallup studies are only conditionally comparable to those of the Forschungsgruppe Wahlen for methodological reasons. The U.S. data, for example, are categorized in a more differentiated manner: in addition to "economy in general" and "unemployment/jobs", there are five other categories into which the (spontaneous) responses of those surveyed to the issue of "economic problems" can be classified as well as a total of 15 categories with other "noneconomic problems".

The significance of global environmental problems for the survey results concerning the awareness of problems cannot be determined as yet on the basis of the available survey data.

1.3.3
Perception of Problems in a National Comparison (Cross-sectional Analyses)

Based on available data, a national comparison of the political priorities currently set by the population is easiest to make for the European Union. Identical surveys for the latter were last conducted in April 1994 by national social research institutes in the 12 Member States at that time under the direction of *Market and Opinion Research International* (MORI) (MORI, 1994). *Fig. 4* shows the responses to the question as to what the interviewed persons think will be "the main problems for the European Union next year" (500-1,000 surveyed persons per country; closed formulation of the question, multiple responses).

Overall, the study determined that the issue of "unemployment" dominates the perception of problems by the population in Europe, too, with 30% on average, though the figures vary widely in the individual Member States: the variation ranges from 11% (Denmark) to 56% (Ireland). Ignoring the two remaining categories, "don't know" and "others", the issue of "economic recession" follows in second place for the Europe of the Twelve overall, followed by "European unification", "Bosnia/Yugoslavia" and "Monetary Union/standard currency". The issue of "environment/pollution" ranks only sixth with an average of 5%. Here again there are differences between the individual countries, with a variation ranging from 0 or 1% for Greece, Spain and Great Britain to 15% for Luxembourg.

If one solely looks at the respective national rankings, the issue of "environment/pollution" has the highest position in Luxembourg (2nd place), in Germany it occupies 4th place, followed by Denmark and the Netherlands (5), Belgium and Ireland (6), Portugal (7) and Italy (8). Those surveyed in France, Greece, Spain and Great Britain regard "environment/pollution" as a "main problem of the EU next year" to the least extent: this issue is ranked 10th in their national "list of priorities". In addition to differences in the economic strength of the EU Member States, sociocultural differences may contribute significantly to the perception and assessment of environmental problems. Admittedly, the available data can, at best, contribute to such an analysis of causes through the formulation of hypotheses. At present there are only few data concerning the perception of environmental problems on the part of the population that are comparable worldwide or at least for several countries on different continents and/or in different culture groups. Individual studies are available for various countries but can only be compared

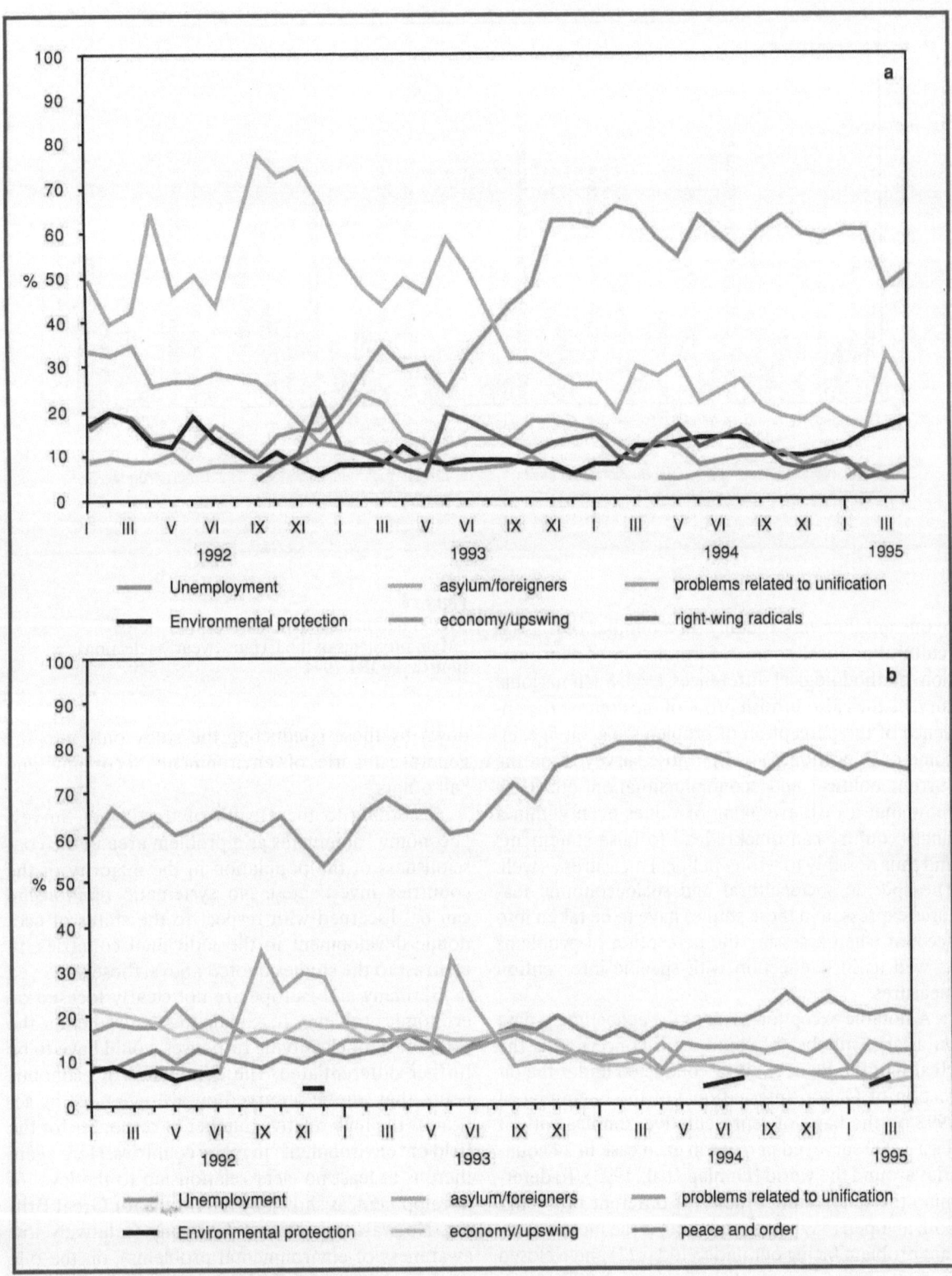

Figure 3
Forschungsgruppe Wahlen time series "Most
important problems" (selection):
a) Federal Republic of Germany-West
b) Federal Republic of Germany-East.
Source: Forschungsgruppe Wahlen, 1992–1995

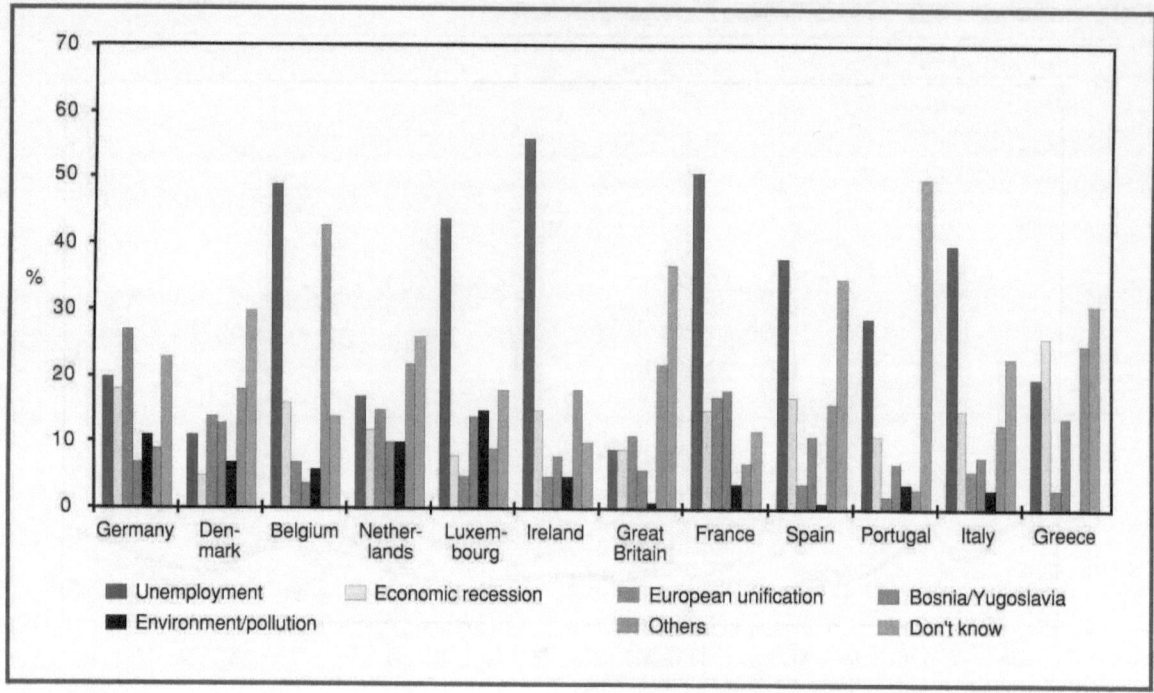

Figure 4
MORI national comparison, EU:
"Main problems of the EU next year" (selection).
Source: MORI, 1994

to one another very roughly for a number of reasons (cultural/political context, formulation of the question, methodological differences, etc.). Such national surveys thus also furnish proof of the context dependence of the perception of problems, e.g. on age, income or educational level of those surveyed, on the current political and economic situation, etc. They show that a rash averaging of values even within a single country can quickly lead to false statements and thus possibly to wrong political measures as well. The specific sociocultural and socioeconomic features expressed in these studies have to be taken into account when assessing the perception of problems as well as in connection with specific intervention measures.

A notable exception as far as the collection of data on a large number of countries is concerned is the Health of the Planet Survey conducted under the direction of Gallup International at the beginning of 1992 on the basis of representative samples with at least 1,000 surveyed persons in each case in 24 countries around the world (Dunlap et al., 1993). To determine the relative importance of different problems, both an open ("What do you think is the most important problem facing our nation today?") and a closed formulation of the question were used ("For each one [out of a list of issues and problems], please tell me how serious a problem you consider it to be in our nation"). *Fig 5a* shows the results of the open formulation of the question for 21 countries, though the spontaneous responses of those surveyed with respect to the "most important problem" was broken

down by those conducting the study only into the general categories of "environment", "economy" and "all others".

According to the Health of the Planet Survey, "economy" dominates as a problem area in the consciousness of the population in the majority of the countries investigated. No systematic relationship can be discerned with respect to the status of economic development in the individual countries. In contrast to the studies quoted above, those surveyed in Germany and Europe are not clearly focused on economic problems. To explain this contradiction, the categories for classifying responses would have to be further differentiated. The dominance of economic issues that can be observed on average must be set against the high relative number of responses for the field of "environment" in many countries. Here again there is at least no clear relationship to the level of development, as shown by the results for Great Britain, Norway and western Germany (relatively low awareness of environmental problems), on the one hand, and those for Turkey, Chile and India (relatively great awareness of environmental problems), on the other hand. Rather, in nearly all countries environmental problems are spontaneously designated as the "most important problem" by those surveyed.

If one also examines the proportion of surveyed persons who classify the given issue of "environmental problems" as a "very serious problem" on a scale of four (*Fig. 5b*; closed formulation of the question), Germany together with South Korea heads the list, followed closely by Mexico, Russia and Turkey, whereas those surveyed in the Netherlands, Denmark and Finland attach the least importance to environmental problems.

While certain parallels to an open formulation of the question can be observed for many countries (though expectedly at a higher level), Finland, Ireland and the Netherlands, in particular, do not fit the pattern. Although those surveyed there spontaneously think of environmental problems as the "most important problem", they do not classify them as a "very serious problem" to the degree expected when other issues are given for assessment at the same time. Indications of the reasons for these "discrepancies" can be found in some of the responses to other issues. In the Netherlands, for example, only a few problems are classified as "very serious" at all.

A study conducted on behalf of the Japanese newspaper *Yomiuri Shimbun* in Japan, the USSR, the USA, Germany, France and Great Britain (Yomiuri Shimbun, 1991) shows for 1991 what problems "threaten the international community in general" in the opinion of those surveyed in these six countries (*Fig. 6*). Since reference is usually only made to the respective country in which a survey is conducted, such a question cannot be found at all, or only rarely, in any comparable studies carried out up to now. From a global point of view, however, it is of great interest.

A closed formulation of the question with specification of 10 problem areas produced quite inhomogeneous results for the issue of "destruction of the environment" in the six countries studied (on the basis of 1,000 to 2,000 surveyed persons in each case): whereas this issue is regarded as a threat to the international community by 63% of those surveyed in Japan, it ranks only 5th in France at 27%. This cross-sectional analysis may serve as proof of the fact that clear differences can definitely be made out as far as perceiving individual problem areas as "threatening" is concerned, even against the economically (with the exception of the former Soviet Union) and culturally (with the exception of Japan) relatively uniform background in the selected countries. Accordingly, the perception of problems as a behavior-influencing factor that cannot be neglected must always be interpreted in the respective space-time context.

Despite the meager or nonuniform worldwide data and regardless of methodological shortcomings, there is, in summary, an observable trend that "environmental awareness" as a concern of the population in connection with the changing environmental situation cannot only be found in industrialized coun-

tries. Environmental problems are also regarded as very serious problems in numerous developing countries (Dunlap et al., 1993). This fact must be valued all the more considering the environmental issue being frequently overshadowed by the numerous pressing everyday problems there.

1.3.4
Values and Value Orientations: the World Values Survey

In the years 1981-1984 and 1990-1992 the World Values Survey (WVS) was conducted almost simultaneously in 43 countries worldwide (1990-92; in the 1981-84 period, however, only in 24 countries) under the direction of the World Values Study Group of the *Inter-University Consortium for Political and Social Research* (ICPSR). Although, taken together, these 43 countries represent nearly 70% of the world's population, the survey focuses on the countries of the "North". The aim of the study, in which an extensively standardized questionnaire was used, was to obtain data on value orientations and attitudes of the populations and to establish a basic data record for further parallel surveys at later times.

In the context of the WVS the term "values" is taken to mean everything that is important to people with regard to various aspects of their everyday lives (Zulehner and Denz, 1992). Ronald Inglehart, whose thesis of a change from "materialistic" to "postmaterialistic" values had a significant influence on the drawing up of the questionnaire as well as on initial publications regarding the WVS (Ashford and Timms, 1992; Ester et al., 1993; Inglehart and Abramson, 1994), played a major role in setting up the WVS. However, since Inglehart's hypothesis of a value change has, in the meantime, been the subject of criticism by many social scientists for different reasons (see, for example, Klages, 1992), the description here deliberately does not refer to this paradigm.

Figs. 7a and 7b present those results from the second stage of the WVS (1990-92) for 24 countries that are directly related to the environmental issue (World Values Study Group, 1994). Each shows the percentage rates of agreement of those surveyed in response to statements given independently of each other. The two response categories "strongly agree" and "agree" in a scale of four were combined into one category called "agreement". The relatively large number of "don't know" responses in some national samples are not taken into account for reasons of clarity. A detailed interpretation will be abstained from at this point because cultural, socioeconomic and other background data would also have to be considered for this purpose. Moreover, the above de-

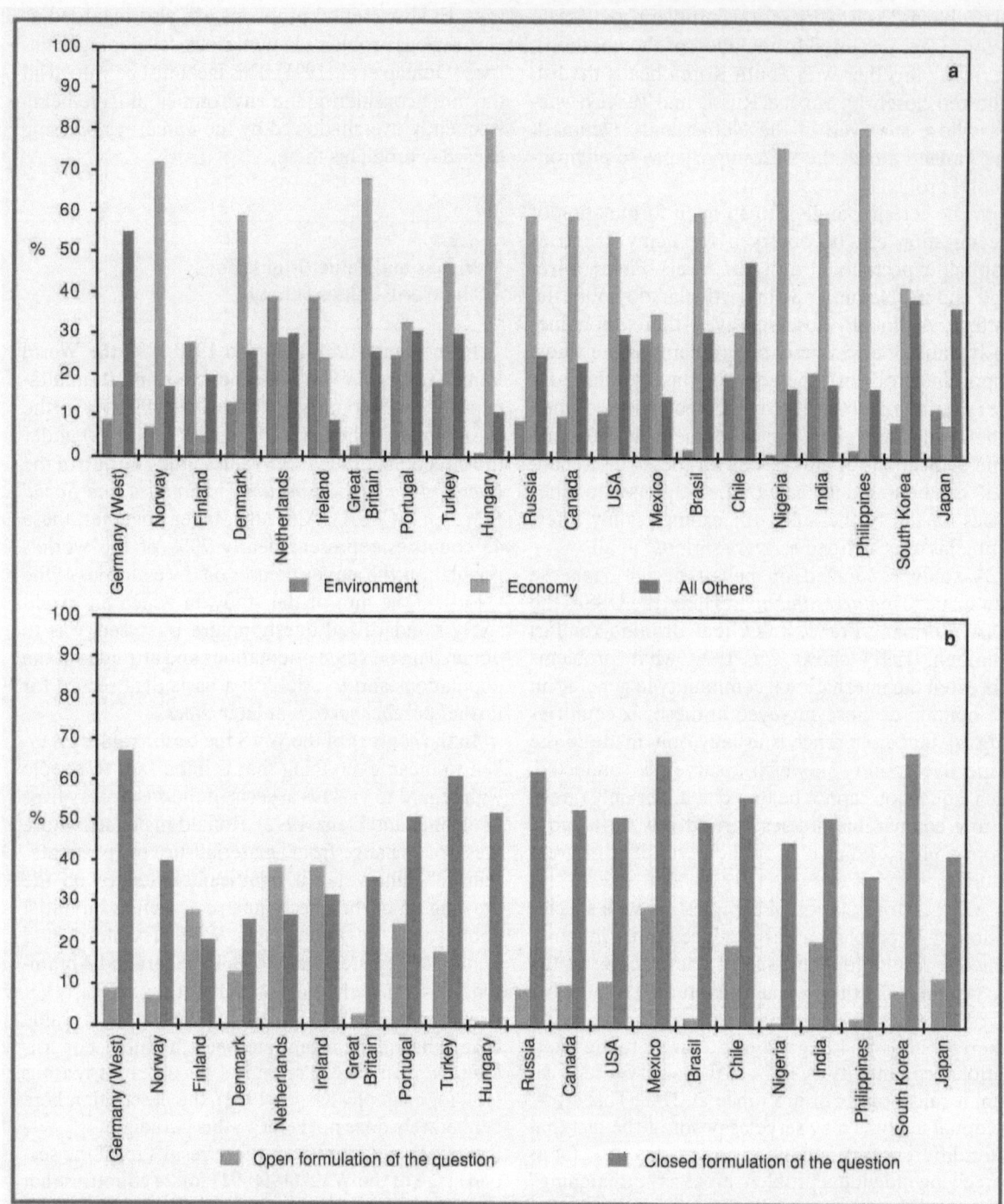

Figure 5
Health of the Planet Survey – national comparison
a) "What do you think is the most important problem facing our nation today?" *(open* formulation of the question)

b) "Environmental issues" as a specified item on a scale of four: classification as "very serious problem" *(closed* formulation of the question).
Source: Dunlap et al., 1993

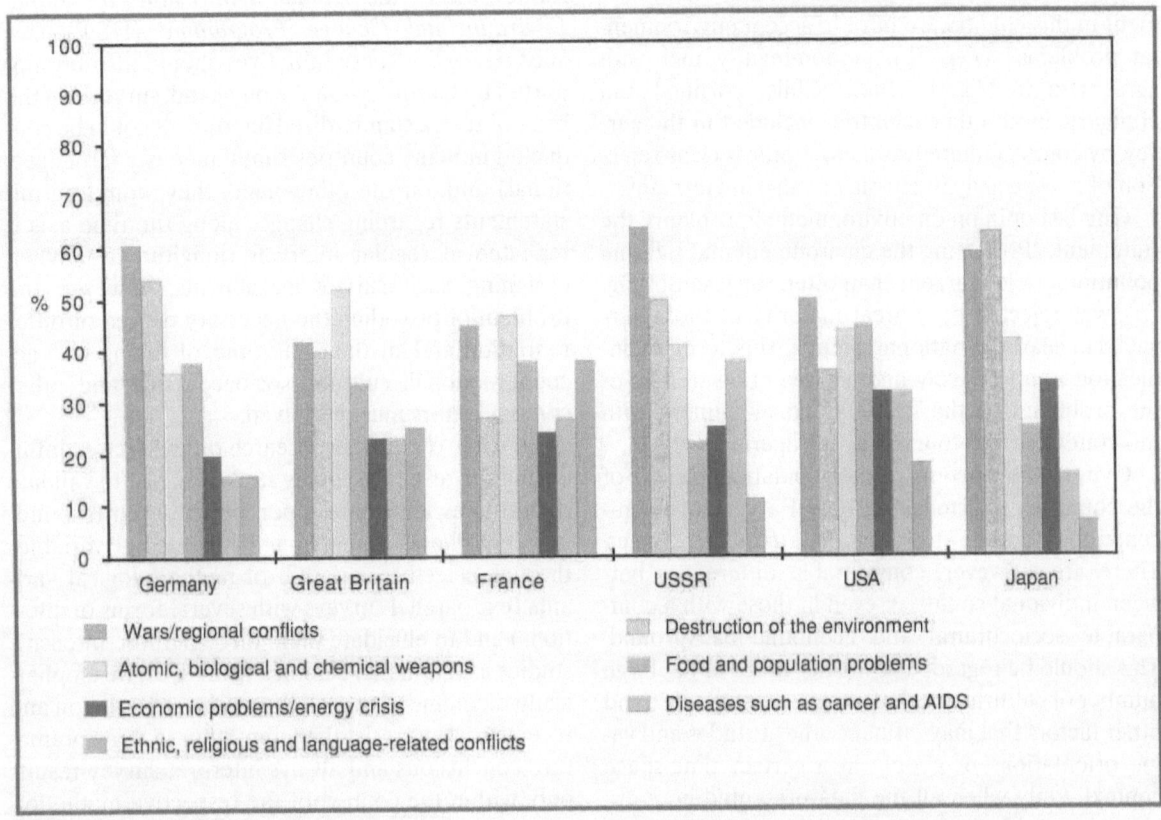

100
90
80
70
60
% 50
40
30
20
10
0

          Germany   Great Britain   France   USSR   USA   Japan

■ Wars/regional conflicts   ▨ Destruction of the environment

▨ Atomic, biological and chemical weapons   ■ Food and population problems

■ Economic problems/energy crisis   ▨ Diseases such as cancer and AIDS

■ Ethnic, religious and language-related conflicts

Figure 6
Yomiuri Shimbun – cross section of industrial countries:
"Problems threatening the international community".
Source: Yomiuri Shimbun, 1991

scribed methodological criticism also applies to this survey (*see Section B 1.3.1*).

The statement, "I would give part of my income if I were certain that the money would be used to prevent environmental pollution" (*Fig. 7a*), met with more agreement than rejection in each of the countries included in the WVS. The least number of surveyed persons agreed with this statement in western Germany (less than 50% as opposed to 60% in eastern Germany), followed by Japan and Hungary. The citizens in other European countries, but also in developing or newly industrializing countries showed a considerably higher degree of agreement in some cases (over 80% in Turkey, South Korea, Chile, Denmark and the Netherlands, for example).

For the statement, "I would agree to an increase in taxes if the extra money is used to prevent environmental pollution" (*Fig. 7a*), the rates of agreement were similar regarding the response pattern, but lower in almost every case. Agreement and rejection here were more or less equal among surveyed persons in western Germany (comparable to Ireland and Japan, for example). A negative response dominated strikingly in Hungary while China displayed a very high rate of agreement. In most other countries under study those in agreement with tax increases for a specific (environmental) purpose outweighed those who rejected the latter.

The statement, "The Government has to reduce environmental pollution but it should not cost me any money" (*Fig. 7a*), produced an inconsistent picture for the countries included here, with clear majorities in Portugal and Italy (for agreement) and in the Netherlands and Denmark (for rejection). A slight East-West divide can be observed again for Germany, where on average roughly 50% of those surveyed agreed with the statement: in the old German States most people agreed while fewer surveyed persons in the new German States were of the opinion that environmental protection should not cost them anything.

"All the talk about pollution makes people too anxious" (*Fig. 7b*). This opinion was held to the greatest extent by those surveyed in Hungary and Chile. In Russia, the Netherlands, western Germany and China, on the other hand, the ratio between agreement and rejection was strikingly almost the opposite, indicating a high degree of sensitivity for the seriousness of environmental problems in these countries.

The statement, "If we want to combat unemployment in this country, we have to accept environmental problems" (*Fig. 7b*), predominantly met with agreement in Nigeria, India, Chile, Portugal and Hungary. In all other countries included in the survey, by contrast, there was a more or less clear rejection of it, especially in Russia but also in Germany.

One last opinion on environmental problems: the statement, "Protecting the environment and fighting pollution are less urgent than often suggested" (*Fig. 7b*), was rejected by a great majority of those surveyed in nearly all national samples. This, too, is an indication of a relatively high degree of awareness of the problems, on the whole. Clear agreement with this statement was found only in Nigeria.

Overall, the reactions of the population in many of the countries selected here reveal a very problem-conscious attitude towards environmental issues. There are, however, considerable differences between individual countries, even in those with a comparable sociocultural and economic background. This should be regarded as further proof of the large number of cultural, socioeconomic, demographic and other factors that may influence the attitudes and value orientations of people in a certain time-space context. Only when all the data presented here are viewed together with other variables (also those of the WVS itself) would further clarification be possible. The willingness of those surveyed to contribute to overcoming environmental problems, financially as well (through less income or indirectly via a tax increase), is significantly high in many countries. This potential should be taken notice of in national and international policy to a greater extent than to date and taken advantage of when implementing appropriate programs and measures.

1.3.5
Summary

The Health of the Planet Survey and the World Values Survey provide valuable contributions to the subject matter of this report. However, they cannot and should not blind us to the fact that a standardized worldwide survey system for recording environmentally related attitudes and perceptions of problems is still lacking. Although corresponding surveys are conducted in many countries, the problems related to the methodology and formulation of the question in such surveys (*see Section B 1.3.1*) make it impossible to compare the respective results between countries or over time.

Based on this, mention must be made of the current efforts for developing a standardized *Global Omnibus Environmental Survey* (GOES) within the

framework of the *Human Dimensions of Global Environmental Change Programme* (HDP) (*see* WBGU, 1993). These initiatives should also be supported by Germany. On the one hand, surveys on the basis of such a standardized instrument could be conducted in many countries simultaneously (cross-sectional) and, on the other hand, they would permit statements regarding changes along the time axis if repeated at regular intervals (longitudinal). When designing such survey instruments, however, the problem of providing the necessary degree of differentiation and, at the same time, of taking into account national, cultural, socioeconomic and other context factors must be solved.

To carry out further research on the factors influencing the results of survey studies and/or to validate such studies, it appears important to test corresponding hypotheses concerning influencing variables through selective variation of methodological variants (e.g. parallel surveys with several forms of questions) and to elucidate them through more intensive studies using smaller samples. Moreover, we emphatically recommend that all those who commission and receive surveys resist the temptation to draw premature conclusions and always interpret survey results only within the context of the respective methodology.

1.4
Environmental Education

Environmentally relevant modes of behavior are learned from childhood (*see WBGU, 1993*). Whether intentionally or unintentionally, a number of educational institutions or socialization agencies, ranging from family and friends to school, vocational training and university as well as the place of work and the neighborhood, are involved in this constant learning and socialization process that takes place throughout one's life span. In view of this, environmental education represents an important way of unlearning environmentally harmful behavior and learning environmentally sound modes of behavior within the framework of global people-environment relations. AGENDA 21 expressly demands a "reorientation of education to sustainable development" (*Chapter 36*) and additionally attaches importance to environmental education in the discussion surrounding the concrete definition of the concept of sustainable development.

In the Council's view "environmental education" encompasses all conceivable areas of education, i.e. going beyond the field of school education that has usually been the focus of educational policy debates up to now. The definition of the UNESCO/UNEP

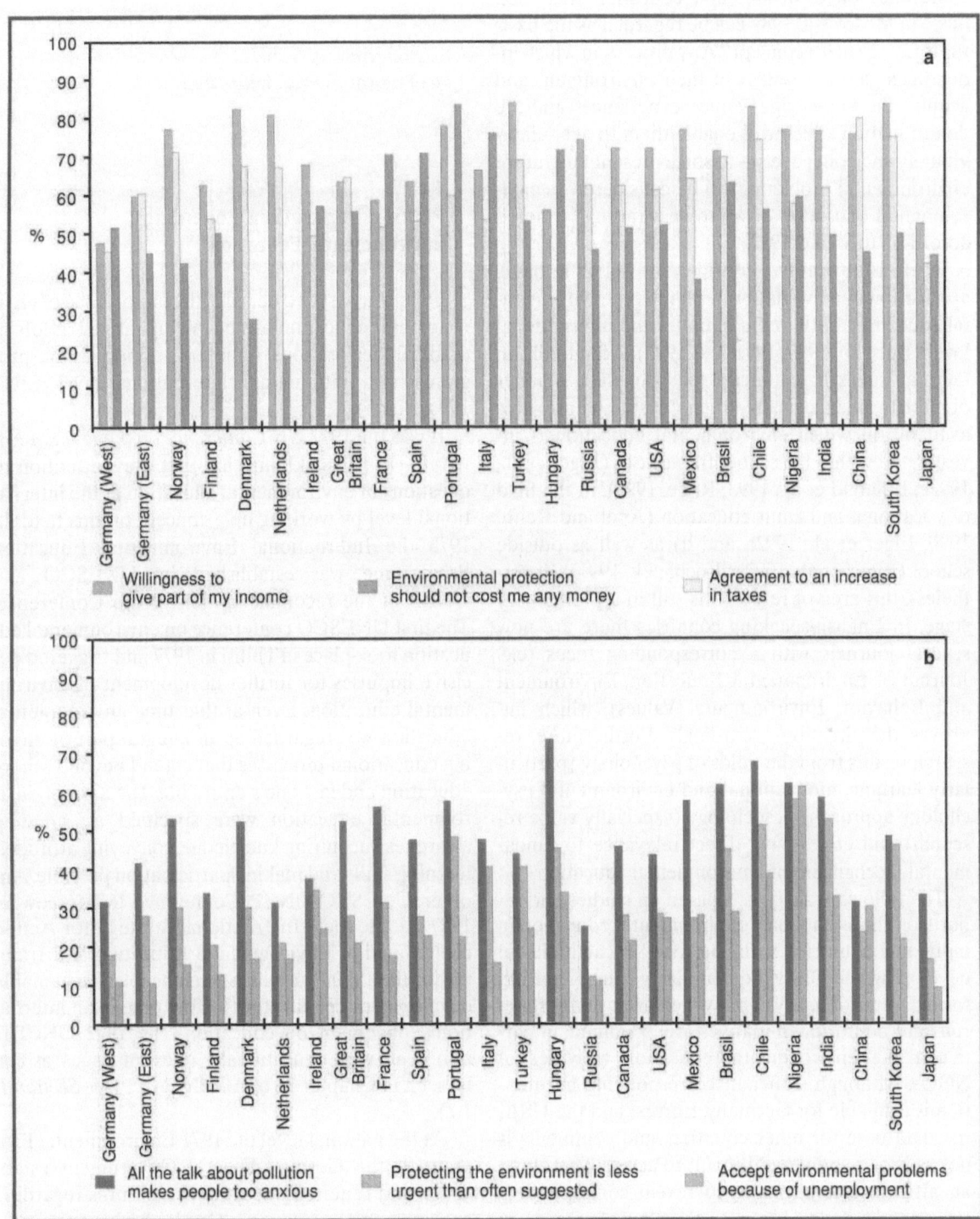

Figure 7
World Values Survey – approval of statements.
Source: World Values Study Group, 1994

conference on environmental education that took place in Moscow in 1987 can be regarded as the basis for forming such a concept: "Any process, in which individuals gain awareness of their environment and acquire the knowledge, values, experiences and the determination which will enable them to act – individually and collectively – to solve present and future environmental problems, will be considered as environmental education or awareness raising" (quoted according to Vinke, 1993).

There is obviously a considerable deficit in basic, application and evaluation research on environmental education, both at the national and international level (Bolscho, 1986, 1991 and 1993). Admittedly, in German-speaking countries, for example, a number of mostly pedagogically oriented studies are now available in which environmental education is investigated within the school framework (Elger et al., 1992c; Eulefeld et al., 1993; Rode, 1995), in the field of vocational and adult education (Apel and Reith, 1990; Elger et al., 1992a and d) as well as outside school (Elger et al., 1992b; Kochanek, 1991). Nevertheless, this area of research is still in a preliminary stage. In English-speaking countries there are now several journals with a corresponding focus (e.g. Journal of Environmental Education, Environment and Behavior, Environmental Values), which improves the situation accordingly. Furthermore, research results from the fields of psychology (particularly learning, motivational and environmental psychology approaches), sociology (especially value research) and ethics have direct relevance to fundamental mechanisms of environmental education.

The following analysis is based on studies and reports on the situation of environmental education in individual countries, with the focus on activities in developing and newly industrializing countries. Both formal educational systems, which are usually state-run, and nonformal initiatives are examined, in particular the environmental education activities of NGOs. Although sufficient corresponding information is available for Germany, Europe and the USA, the data base for other countries and continents is not as broad and more difficult to access. Therefore, an attempt will be made to reveal corresponding trends and basic problems on the basis of exemplary findings. In contrast to the primary distinction between different groups of countries according primarily to economic and technological aspects, as was introduced in the Council's 1993 annual report (WBGU, 1993), a simplified categorization into "developing countries" (including newly industrializing countries) and "industrialized countries" is justifiable in connection with education.

1.4.1
Development, Standards and Problem Areas of Environmental Education

1.4.1.1
Political Declarations Regarding Environmental Education

Over the past two decades the importance of environmental education has been emphasized in lots of political declarations of intent, initiatives or programs on a national and international level (SRU, 1994; *Box 5*).

It was the 1972 UN *Conference on the Human Environment* in Stockholm that first drew attention to questions of environmental education at the international level by working up a concept on this field. In 1975 the International Environmental Education Programme was established by UNESCO and UNEP at the recommendation of the Conference. The first UNESCO conference on environmental education took place in Tbilisi in 1977 and triggered decisive impulses for further development of environmental education. Even at that time environmental education was regarded as an integral part of ongoing educational processes that extend beyond school education and last one's entire life. The aims of environmental education were specified as: creating awareness, acquiring knowledge, conveying attitudes, learning skills and making participation possible. Another UNESCO/UNEP Conference in Moscow in 1987 adopted an "International Strategy for Action in the Field of Environmental Education and Training for the 1990s", in which the concept of sustainable development appears for the first time in an international document on education. The 1992 UNCED also dealt with educationally relevant issues, as can be seen in Chapter 36 of AGENDA 21 (*see Section B 1.1*).

At the national level the 1971 Environmental Program of the German Federal Government established and required educational measures regarding environmental protection. On the basis of this program, environmental topics were increasingly incorporated into curricula and school books, though the main emphasis was on describing pollution rather than examining causes, background and political consequences. In recent years the reports of the *Sachverständigenrat für Umweltfragen* (Council of Experts for Environmental Issues, SRU) (SRU, 1994) and of this Council (WBGU, 1993 and 1994) have expressly established a link between sustainable development and environmental education.

These activities, declarations, decisions and recommendations show that the general significance of environmental education has clearly been recognized within the framework of national as well as international policy. A common feature of all initiatives is the concern for the environment as well as the realization and the resulting demand that environmental education must be viewed as a part of environmental policy and corresponding educational measures have to be taken. The impression conveyed up to now, however, is that the various political recommendations have more of a symbolic function rather than seriously pursuing the goal of concretely defining and implementing environmental education at the various levels and in the various fields (*see also* Schneider, 1993).

1.4.1.2
Environmental Education in Practice: Criteria, Standards and Deficits

In the course of the over-twenty-year history of environmental education it has become accepted that environmental problems cannot be solved by administrative, technical or economic means alone. On the other hand, there is agreement that the conveyance of knowledge is not sufficient with regard to environmental education and that environmentally sound action depends on a wide range of influencing factors instead (WBGU, 1993). Today the principle of sustainable development seems to provide a suitable reference framework for taking educational concepts into consideration. Accordingly, educational initiatives are beginning to face up to this concept. The standards or criteria for successful environmental education that have emerged in the course of time (*Box 6*) also seem suitable for giving life to the concept in the field of environmental education. Due to the lack of evaluation studies that might furnish reliable results on the actual effects of environmental education measures, however, it is not easy to reconstruct generally valid conditions for successful environmental education.

Against a background of different traditions and cultures, people encounter a wide variety of situations in their daily lives in which many examples of environmental problems manifest themselves. Environmental education should start by looking at such experiences and asking about the combinations of causes and the problems involved as well as offering possible concrete forms of action to take (situation orientation). Within the context of extensively "invisible" global environmental changes or those looming in the future, however, such fields of experience frequently have to be created in the first place with the help of suitable media (simulation).

As a rule, environmental problems are characterized by conflicts of interest, varying assessments and diverging ideas about the consequences of actions. Therefore, environmental education must be integrated in the social and political context. The political background has to be examined so that decision-making criteria can be offered to the learners for their own actions. It is this very problem orientation, however, that repeatedly leads to conflict between environmental education and politics and society in the respective country because central areas of society, such as the prevailing lifestyles as well as the organization of work, housing and transport, are called into question. Social expectations and influence primarily leave their mark on formal educational institutions whereas informal NGOs, at least in western democracies, are given greater freedom for action in most cases.

To recognize these conflicts of interest, to understand them and, if possible, become involved must be one of the aims of environmental education. Accordingly, learners should be enabled to take self-determined, responsible action. The existence of participatory structures, therefore, promotes successful environmental education. However, people must also have participatory abilities in the form of communicative and social competence for understanding, weighing up possibilities and negotiating compromises that again have to be promoted within the framework of educational initiatives.

Action orientation can take on many different forms in environmental education. Within the school context, for example, it ranges from the investigation of a body of water or the ecological shaping of the school and the drawing up of an information brochure on a current environmental problem to participation in an environmental initiative.

Environmental education in the sense of holistic learning is not only viewed in the context of "nature-oriented education", as has emerged recently. It is not enough, for example, to address solely aspects of environmental problems related to the natural sciences or to technology. Ethical, social and societal aspects of people's everyday lives also have to be discussed. Recognition of the complexity of global people-environment relations includes at the same time the ability to reflect and critically question the actions of both individuals and society. In order to motivate learners to adopt environmentally sound behavior, emotional aspects (e.g. pride or fear) should be taken into account, which are often neglected (Lantermann et al., 1992; Vinke, 1993).

From the point of view of sustainable development in particular, environmental education is aimed

BOX 5

**National and International Political
Environmental Education Initiatives**

Year	National	International
1971	Environmental program of the German Federal Government	
1972		UN Conference on the Human Environment in Stockholm
1977		UNESCO Intergovernmental Conference on Environmental Education in Tbilisi
1978	UNESCO follow-up conference in Munich	
1980	Conference of German Ministers of Education resolution on environmental education in schools	European UNESCO Regional Workshop on Environmental Education in Essen
1986	BMBW symposium "Environmental education: A task for the future"	
1987	1987 report of SRU	UNESCO/UNEP International Group on Environmental Education and Training in Moscow
	BMBW work program "Environmental Education" BLK resolution on incorporating environmental issues into the educational system	
1988	BLK program "Vocational Environmental Education"	Decision of EC Council of Ministers on environmental education
1990	Enquete Commissions "Education 2000" and "Protecting the Earth's Atmosphere"	
1991	Recommendations on vocational environmental education made by the Federal Institute for Vocational Training (BIBB)	
1992		UN Conference on Environment and Development in Rio de Janeiro (UNCED)
1993	1993 report of WBGU	
1994	1994 report of SRU 1994 report of WBGU	
1995		Special WBGU report on 1st Conference of Parties to the Framework Convention on Climate Change

at anticipatory learning. What is required is a way of thinking that takes into consideration expected developments in and mutual influences of the natural sphere and the anthroposphere in the shaping of our present lifestyle despite all uncertainty (Dörner, 1987). The result is an expansion of the individual "preference horizon" through the addition of the dimension of future needs as well as of the (hypothetical) preferences of future generations.

This shows that in environmental education the conveying of knowledge, up to now frequently the only step taken, especially within the framework of formal educational systems, is merely regarded as one requirement, though an important one, for learning more environmentally sound behavior (WBGU, 1993). The conveying of knowledge in this context must never be understood as purely the transmission of information, but always as enabling people to

BOX 6

Criteria for Successful Environmental Education

- learning from experience (*situation orienta-tion*),
- integration into social and political context

(*problem orientation*),
- participation,
- learning through action (*action orientation*),
- holistic learning,
- anticipatory learning,
- relative significance of conveying knowledge.

think in terms of solving problems, which includes thinking beyond the local and regional situation (Dörner, 1989).

Essentially environmental education is always aimed at changing modes of behavior. By implicitly altering attitudes and dispositions to taking action, it can also indirectly contribute to responsible environmental behavior through a change in social values. Such efforts are considerably hindered by the fact that there is (still) no political consensus on the objective of this altered behavior within a national or international framework. As a result, the controversies that are well-known in connection with practical environmental policy are, to a certain extent, reflected in the efforts being made in the educational field.

Attempts to reach a consensus on environmental policy goals is a matter for national and international political discourse. Environmental education can and must be involved in this process, thus inevitably making it political education as well. For this reason it will always act in the space between diverging interests. An additional difficulty is that, even if such a consensus on general guidelines for action existed, sufficiently reliable knowledge provided by the social and behavioral sciences on the various conditional factors of human action is either still lacking or has not been sufficiently taken into account to date.

1.4.1.3
Environmental Education in Industrialized Countries

A critical assessment of (state-run) environmental education activities in Germany can be found in the most recent report of the Sachverständigenrat für Umweltfragen (Council of Experts for Environmental Issues, SRU, 1994) so that a detailed discussion is not necessary at this point. According to this report, the didactic criteria that have long been demanded for environmental education are gradually being adopted in German schools. Teachers are increasingly prepared to accept and implement basic standards of appropriate environmental education, in spite of the

many institutional and subject-related barriers (shortage of time, inflexible forms of organization, inadequate further teacher training, dominance of thinking in terms of specific subjects). There is, for example, a greater tendency to use environmental situations that are important in the local and regional environment of the learners as the basis (*situation orientation*). In addition, greater attempts are being made to enable learners to deal with environmental problems, for example, by orienting learning processes to restructuring the school or immediate environment (*action orientation*). Finally, the social context of environmental problems is also increasingly taken into account (*problem orientation*). A comparison of the curricular structure of environmental education between 1985 and 1990/91 shows an extension that includes, in particular, subjects outside of the natural science field, such as religion, art and history as well as languages (Eulefeld et al., 1988 and 1993; Bolscho, 1989 and 1993).

The results of the project Environment and School Initiatives (ENSI), in which, among other things, didactic concepts were to be developed for improving environmental education (OECD, 1993), may provide information on the status and standards of environmental education in the OECD countries. According to this study, certain standards of environmental education are accepted and striven for in all 19 countries concerned. In the national projects, for example, reference to local and regional environmental developments is emphasized as the primary objective. This situation orientation is given a specific national or regional accent while the institutional structures of the educational system in addition to subject-related aspects influence practical implementation: in projects in southern European countries, for instance, the issue of "environment and tourism" plays an important role. Frequently there is a focus on the field of "waste" and usually a specific action orientation is striven for, such as in the form of pupil participation in keep the landscape clean campaigns. Northern European countries, on the other hand, often orient environmental education projects to aspects of protecting natural resources. In Norway, for example, there is a project in which pupils are involved in surveying and monitoring systems regarding water quality. As far as problem orientation is concerned (not only in the OECD countries participating in ENSI), there is a tendency of excluding this criterion. In most cases, the immediate region remains the sole reference point for environmental education and the originally meaningful criterion of situation-oriented environmental education is not extended beyond the local situation. The frequent demand, "Act locally, think globally", is thus more of a theoretical principle than practical reality at the moment.

Whether these developments, which apply to the majority of the OECD countries, already indicate a "reorientation of environmental education to sustainable development" must remain open for the time being. The fact that the need and difficulties of practical implementation of such a reorientation have been recognized is shown by a critical note in the Hungarian ENSI national report (House, 1994): "The basic conflict in environmental education is that it wishes to make children accept the attitudes that are in direct contrast to the models presented by the modern consumer society."

1.4.1.4
Environmental Education in Developing Countries

When the term "education" is used, certain social developments and structures are usually assumed, at least with regard to school education, e.g. an institutionalized educational system in an open, democratic society as well as sociocultural structures on which the educational objectives are based. Educational and socialization processes in other regions of the world, however, are difficult to analyze with such a Eurocentric view of education. It seems more meaningful in this context to describe the objectives of environmental education pragmatically as "man's/woman's approach to dealing with his/her social and natural environment".

The basic problems of environmental education in developing countries can be described on the basis of available reports and studies from three points of view:
- Is there a social policy interest in environmental education?
- Can formal educational systems open themselves to environmental education?
- Has the field of environmental education, defined as a critical corrective to social developments, to be viewed more outside of formal educational systems?

ENVIRONMENTAL EDUCATION AND SOCIAL POLICY INTEREST

Top priority, particularly in developing countries, is frequently attached to economic, i.e. industrial, development in most cases. It is argued that impairment of environmental resources must initially be accepted, and that poverty is "the greatest enemy of the environment". In this nation, only when a certain economic and technological level has been reached, can efforts be devoted to environmental protection: "First development, then environmental protection". There are many indications that this way of thinking also has an effect on plans and projects for intensifying environmental education in developing countries, at least to the extent that environmental education takes place within the framework of state educational systems (for the Caribbean region *see* Hickling-Hudson, 1994).

An environmental advisor of the Jamaican government points out another aspect that is significant for environmental education projects in other developing countries as well: "Since the environment movement is being carried by those who are white, wealthy and young, the perception among needy and disfranchised, in both poor and rich countries, is that environmentalists care more about plants and animals than about people" (according to Hickling-Hudson, 1994). The conclusion can be drawn that environmental education in developing countries only has a chance if people perceive it as important for their everyday life. This importance, however, cannot be forced on them "from the outside" but must develop on its own against the background of traditions and interests of those concerned and within the framework of participatory structures.

An example of how global environmental problems and programs aimed at solving them are perceived by developing countries not as self-determined but as determined by others is the case of species protection. After pointing out that 70% of the Indian population works in agriculture, Vandana Shiva concludes that the availability of seeds for these people symbolizes "democracy, freedom and ecology" as "the foundation for sustainable agriculture". If foreign corporations patent property rights and rights of use in order to market species diversity and utilize it as raw material, then this represents "bio-imperialism" (Koch, 1995). Therefore, environmental education projects based on global interrelationships should not dispose of developing countries pretending global environmental protection, and should not ignore the interests of the people concerned and their perception of the environmental situation: "The imperialistic category of global is disempowering at the local level" (Shiva, 1993).

ENVIRONMENTAL EDUCATION AS PART OF SCHOOL EDUCATION

The question of whether formal educational systems can incorporate environmental education must be viewed against the background of harsh criticism sometimes heard of the development of educational systems in developing countries: they are alleged to be largely "dysfunctional" because they primarily provide training for the "modern sector" and thus do not take into account the traditions and cultures of the respective countries (Goldschmidt, 1993). "The subject matter and curriculum has little or nothing to do with the everyday lives of those concerned. Learning does not help to satisfy the basic needs, to cope with everyday life"; therefore, quantitative improvement is not meaningful "without a qualitatively completely altered basic educational system" (Datta, 1992). Treml (1992), too, refers to a "failure in the development" of the educational system: "Education, from which so much was expected, did not prove to be the lever for development that was hoped for, but was often an obstacle to development because it produced or reinforced social and cultural disparities instead of reducing them".

In view of this and similar analyses, environmental education in developing countries seems to have the best chance if it is integrated into an ecologically oriented basic education system. Basic education generally encompasses formal school education for children and adolescents under 15 as well as educational programs for adolescents and grown-ups who do not attend school. On the basis of basic education, "the cultural identity of the individual and of society" (BMZ, 1992) can develop, thus meeting an important requirement for participation, which is in turn a key criterion for successful environmental education. Moreover, environmental education as a part of basic education could also lead to qualitative improvements in basic education, an assumption implicitly made in much of the discussion over environmental education in the industrialized nations. Some of the initiatives in developing countries are already showing that this strategy is certainly meaningful (*see Section B 1.4.3.2*), especially since at least a portion of the children and adolescents in developing countries are acquiring elementary skills, despite all the shortcomings and problems that exist, particularly in primary school.

However, the training of the teaching staff and their view of the school's tasks may be a barrier to the integration of environmental issues into the curricula of state-run schools, not only in developing countries. A survey of teachers in Jamaica showed that although they verbally attached great importance to environmental education in school, they placed the main emphasis in daily teaching practice on conveying traditional knowledge of the subject matter (Taylor, 1988). Within the framework of environmental education, such knowledge usually concerns conservation, without taking adequate consideration of cultural, demographic and socioeconomic aspects.

ENVIRONMENTAL EDUCATION THROUGH NGOs

Large portions of the world's population, particularly in developing countries (e.g. poor, unemployed or illiterate persons and women), can either not be reached at all, or only with difficulty, via school or vocational training measures of the formal educational system. Since the formal educational systems in developing countries, moreover, have proved to be largely incapable of innovation, the genuine field of environmental education should perhaps be regarded as lying outside of these systems. There is extensive agreement that action- and problem-oriented learning quickly reaches its limits within the framework of formal educational systems. This is why the educational work carried out by NGOs frequently has the function of being a necessary addition as well as a critical corrective, due especially to its clearly higher acceptance among the population. Few reliable findings are available regarding its actual influence, however.

In the 60s and 70s NGOs often had the aim of acting completely independently of the central government in developing countries in matters involving education. They frequently found themselves in the situation of offering services that could not be provided by the state (e.g. basic educational projects in crisis areas). Due to their capacity and their view of themselves, most organizations were unable to perform the required tasks in the long run, however. Archer concludes from this: "Making this organic link between the experiences in grass-roots microprojects and macrolevel lobbying work is perhaps the greatest challenge faced by the NGOs today" (Archer, 1994).

An empirical survey on NGOs in Venezuela shows how difficult this is. According to a local NGO staff member, the biggest problem is "obtaining new members" who "have to deal with a society in which environmental awareness hardly exists" (Gonzalez, 1992). Small NGOs, however, lack funds and the necessary organizational framework so that their political effectiveness is limited.

De facto, NGOs work together with state educational systems in many areas today, e.g. in further training courses for teachers. In spite of reservations on principle – NGOs are often afraid that the state will take over in such cases – the prevailing view is that cooperation is necessary. It is an illusion that "educational experience outside of the school can replace learning at school"; what is necessary is a "combination of formal and non-formal places of learning" (Karcher, 1994).

Because of their double function of providing both local and global perspectives, NGOs are involved in both the local and the global (environmental) policy context. This also involves the basic political conditions that in some cases place considerable restrictions on the function of NGOs as a critical corrective to state environmental policy objectives and strategies in many developing countries. Although the governments do admit to positive aspects of the NGOs, they see a number of serious problems in connection with the latter. In this notion, NGOs tend to take a simplified view of environmental problems, their staff is not sufficiently qualified and has too little experience, etc. (Ganapin, 1991). Fundamentally different positions emerge here between governmental and nongovernmental environmental policy (and thus environmental education as well). It can be assumed that NGOs are frequently criticized by the government due to their role in opposition to a one-sided, economically oriented development at the expense of the environment and the population. Because they are involved in political and economic fields of conflict, their possible role ranges between two poles, that of "mediator between the environment and politics, between the population and the government" and that of "one of the players in the corridors of power" (Hoering, 1994). The manner in which specific tasks are assigned to this role depends on the political situation in the respective country. Environmental protection and democracy frequently prove to be "two sides of the same coin" (Hoering, 1992).

Furthermore, environmental problems and the respective country's openness to the latter differ very greatly, even from a global point of view. An important aspect for successful environmental education, for example, concerns the main environmental problems faced by a country, both locally and nationally, e.g. overexploitation of natural resources, tourism, intensive agriculture or industrialization. A study of the environmental movement and environmental education in Venezuela shows, for example, that oil production in this country overshadows all other environmental problems to a certain degree (Gonzalez, 1992).

1.4.2
Activities Concerning Environmental Education in Germany: Global Aspects

In past years a number of educational initiatives have been established by governmental and nongovernmental institutions in Germany that deal with global environmental problems, in particular those connected with the climate, and with approaches to sustainable development. Several examples will be examined in detail below.

1.4.2.1
Activities in Connection with Schools

The issue of "environment and development" still plays a minor role, even in comparison to other environmental issues, in the formal educational system, which is characterized by a governmental framework and direct or indirect influence. Global environmental questions have only marginal importance in schools, and even more in the field of vocational training (Bolscho et al., 1994; Eulefeld et al., 1993). Although an increase in courses related to the environment has been observed at universities and colleges in recent years, this trend cannot be designated as an integration of the issue of "environment and development" into the individual courses of study (Wissenschaftsrat, 1994).

As a follow-up to the work of the Enquete Commission of the German Bundestag on "Preventive Measures to Protect the Earth's Atmosphere" (1990a, b and c), a commission of experts worked up proposals for implementation of their recommendations in the educational system (BMBW, 1990). It proved to be very difficult to include global environmental changes directly in educational practice; especially the fact that they cannot be experienced and assessed to an adequate extent poses a crucial problem in the context of environmental education. Therefore, global environmental problems should be depicted such that the learners can perceive a connection to their everyday experiences. The Commission recommends that schools include subject matter that is important for protection of the Earth's atmosphere in existing curricula and then establish a link between all such subject matter. According to these concepts, the focus of environmental education in schools should not be placed on conveying traditional, subject-centered knowledge. However, the recommendations of the Commission in this respect, such as in favor of interdisciplinary environmental education, are not very extensive and more of a general nature (BMBW, 1990).

Pilot projects for conveying basic knowledge on the protection of the Earth's atmosphere and pointing out practical ways of taking action (particularly with regard to energy) have been launched at different levels and take these recommendations into consideration. In Hamburg, Lower Saxony and Schleswig-Holstein, for example, a pilot project of the Commission of the Federal and Länder Governments

(BLK) on "Energy Use and Climate" was initiated in August 1991 (BLK Modellversuch, 1991). Numerous teaching materials were developed and further teacher training courses were conducted in connection with the project in these three German States. Many of the approaches pursued in connection with the pilot project show that topics oriented to global interrelationships can even be integrated into lessons in an action- and situation-oriented form within the framework of existing school and subject structures .

Further examples of treating the topic of "Protection of the Earth's Atmosphere" in a school context include other pilot projects (Illing et al., 1994; Landsberg-Becher, 1991; Lehrer- und Schülergruppe, 1990), recommendations of the Boards of Education on environmental education (e.g. Ministry of Education and the Arts in Lower Saxony, 1993) as well as papers in teaching journals (e.g. Horlacher and Urban, 1992; Künzel and Künzel, 1992; Zachow, 1993). It clearly appears less likely that teachers will be motivated to try out new subject matter and forms of instruction through the external conditions at schools; rather, immaterial incentives play a greater role, such as the possibility of cooperation with other schools, publication of the results of school work or the recognition frequently connected with such initiatives (Eulefeld et al., 1993). Reliable statements on the actual implementation of recommendations and proposals, and even less so on their effectiveness with respect to environmentally sound action, cannot be made as yet, however.

The present situation regarding further training for teachers can be regarded as inadequate in connection with environmental education (Eulefeld et al., 1988; Eulefeld, 1993). In addition to the lack of suitable interdisciplinary further training concepts and courses, especially concerning issues like "environment and development", there is frequently a shortage of financial and human resources, such as those necessary to cover lessons cancelled due to further training courses.

In the school context development education (also called: "development policy education" or "learning in connection with the Third World/One World") in contrast to environmental education already has a certain tradition as an interdisciplinary field of learning . Even though both fields have common concerns, especially from the point of view of "environment and development", the close interlinkage between environmental and development policy issues has yet to be sufficiently realized to date and opportunities for mutual enhancement are not taken advantage of (Scheunpflug-Peetz et al., 1992). This is largely due to the fact that interdisciplinary scope and perspectives first have to be created in curricula defined by subject-based structures. The didactic dis-

cussions in both areas of education to a large extent take place independently of each other although integrative concepts that would take greater account of the complexity of the problems are increasingly emerging, particularly in view of a certain "competition" between the two areas over interdisciplinary resources.

1.4.2.2
NGO Activities

Outside of the school NGOs have established a whole range of heterogeneous educational initiatives centered around the issue of "environment and development" that are usually not well coordinated with one another. Examples of such activities will be presented in the following.

The support and coordination of activities related to climate protection at the local or municipal level is the aim of the Alliance of European Cities with the Indigenous Peoples of the Rainforest for the Conservation of the Earth's Atmosphere (*Box 7*). The work carried out by this climate alliance is governed by the principle of "think globally, act locally".

The "Nordlicht" Climate Protection Campaign, initiated by psychologists at the University of Kiel, is attempting to put social science findings regarding environmental awareness and environmental behavior (WBGU, 1993) systematically into practice in an action-oriented educational initiative with a broad impact (*Box 8*).

Some NGOs carry out special environmental education work for children and adolescents with the aim of promoting environmental awareness and environmentally sound behavior on the part of this target group (*Box 9*).

As a result of the Rio Conference in 1992, most NGOs in Germany joined forces as the Forum Umwelt und Entwicklung (Forum for Environment and Development), among other things, in order to coordinate their work better. The Projektstelle Umwelt und Entwicklung (Project Center for Environment and Development) serves as a coordination instrument and place for exchanging commonly elaborated positions. It maintains contacts to organizations in developing countries and carries out coordination with international associations. One of its tasks is to point out to the German public the interrelationships between the environment and development and the consequences of the way of life and production methods of the industrialized countries for Third World countries.

Development policy issues have also long been established in a variety of ways in educational work outside of the school (e.g. on the part of churches, organizations and independent institutions) as well as in adult education. In these fields they are increasingly linked to ecological approaches. Within the framework of many activities there is an attempt, at the same time, to teach and live according to alternatives, on the basis of a critical appraisal of the lifestyle of the "North": people should practice in their own environment that which is deemed to be "sustainable" according to a global perspective. The youth organizations of churches and conservation groups are also increasingly devoting themselves to development policy issues with global ecological focal points and the same applies to initiatives concerning fair trade with countries of the "Third World" (Glöge, 1993; Pinzler, 1994; Durning and Ayres, 1994).

"Development centers", as they are called in the Netherlands, would be possible forums for greater exchange and mutual benefit between governmental and nongovernmental initiatives for ecologically oriented development education. The *National Commission for Development Policy Information and Public Awareness* (NCO), an independent foundation, is the umbrella organization of the 22 centers for Development Cooperation (COS) in the Netherlands that are supported by the *Ministry for Development Cooperation* (MEZ). Churches, trade unions, employers' organizations, agricultural organizations as well as women's and youth organizations are represented there, too.

These and similar approaches to development education and its extension to include ecological aspects have obviously only become known within a small community to date, however. A broader resonance to what are sometimes very remarkable conceptual drafts (Zahn, 1993) would be desirable.

The NGO activities in the field of environmental education only briefly described here make it clear that many initiatives with a global perspective have been launched in recent years. They stand out in stark contrast to comparable initiatives of institutionalized educational bodies that usually "remain stationary" at the national level. In addition, the educational initiatives of the NGOs are more action-oriented, cooperative and participatory. Overall, the activities of NGOs have clearly been underestimated or not well-known enough up to now. In the meantime, however, they are able to play an important role in supplementing school-based environmental education activities. Moreover, it may be possible for NGOs, through their environmental education work, to address new target groups that have not (yet) been reached by formal educational institutions.

BOX 7

Climate Alliance

The Alliance of European Cities with the Indigenous Peoples of the Rainforest for Preserving the Earth's Atmosphere, which now comprises roughly 350 cities, was established at the beginning of the 90s (*Klimabündnis/Alianza del Clima*, 1993). On the basis of a manifesto, the Climate Alliance pursues the goal of altering the wasteful lifestyle in the urban centers of Europe and, in particular, reducing the emission of gases threatening the climate. In addition, the cities support the peoples of the rainforest (especially in the Amazon region) in asserting and defending their rights so that they can contribute to sustainable use of the rainforest and protect the climate through their way of life.

The Climate Alliance performs an important educational function in that it establishes relationships between people living under very different conditions in cultures far removed from one another through efforts towards international cooperation and motivates these people to take action to achieve the common goal of climate protection. A project is currently in the process of assessing the various educational activities of the German cities in the Alliance. It shows already that most of these activities and measures concern the environmental issues (energy, transport, waste) while development policy ones have only been touched on in passing. A positive aspect is that many educational activities are being carried out in cooperation with other partners (local energy suppliers, adult education institutions, kindergartens, etc.) and thus address very different target groups.

Extensive information is not yet available on the activities of cities belonging to the Alliance in other European countries. Climate Alliance activities are primarily taking place in western and northern European countries, with the Netherlands standing out by virtue of a large number of interesting projects there.

1.4.3
Activities Concerning Environmental Education in Developing Countries: Global Aspects

1.4.3.1
Activities in Schools

As a rule, great importance is attached to the promotion of basic education in developing countries. In the "Sectoral Concept" of the *Federal Ministry of Economic Cooperation and Development* (BMZ) the "promotion of basic education in developing countries" is described as a guideline for projects in the field of education (though the highest priority is still given to vocational training in the respective budget items) (BMZ, 1992 and 1993). The World Bank, too, has declared its commitment to greater promotion of basic education, primarily through the support of deprived schools in rural regions (World Bank, 1991). Various projects demonstrate that the integration of environmental issues into the curricula of schools providing general education and, in particular, primary schools is increasingly recognized as a necessary task and that an attempt is being made to implement it (*Boxes 10 and 11*).

The *Deutsche Gesellschaft für Technische Zusammenarbeit* (German Agency for Technical Coopera-

tion, GTZ) has carried out projects for the promotion of basic education in several countries of Africa, Asia and Latin America (Bergmann, 1992). Even though environmental protection aspects were not the focus of these projects, a number of learning units on issues involving environmental and resource protection were included in the respective teaching materials and teachers' recommendations (e.g. erosion protection, farming suitable to the location). Generally, however, environmental education has increasingly been taken into account in the planning and implementation of development projects only in the past few years due to the significance of environmental problems in developing countries (Vinke, 1993).

1.4.3.2
NGO Activities

NGOs, particularly in developing countries, are becoming more involved in environmental education activities that are aimed at promoting environmental awareness and environmentally sound modes of behavior, as well as exertion of influence on environmental changes. The pedagogical strategies of NGOs vary according to ecological, political, economic and cultural context, as demonstrated by examples from the Philippines and Thailand (*Boxes 12 and 13*).

BOX 8

The "Nordlicht" Climate Protection Campaign

The Nordlicht Climate Protection Campaign (Prose, 1995) links a whole range of factors having a potential influence on behavior within the scope of a social marketing approach and simultaneously implements several of the above-mentioned criteria for successful environmental education work: people taking part in the campaign are given leaflets with practical tips for saving energy and reducing private automobile traffic. As a kind of self-commitment, they can choose what sort of voluntary action they wish to perform (for example, buying energy-saving bulks) and/or set themselves specific goals (e.g. reduction in the number of kilometers driven every month). In addition, they are supposed to provide for further distribution of the leaflets in their local area, initiating a "snowball effect", if possible with the support of

local sponsors. The activation of such local social networks is aimed at a credible, personal further recommendation of environmental protection measures, i.e. a multiplier effect. Finally, all activities are to be reported back to the initiators so they can be documented and processed for regular interim assessments as part of press work.

A subdivision of such assessments according to towns or town districts makes it possible to organize actual energy-saving competitions and additionally demonstrates that, taken together, "small campaigns" by many people for the promotion of climate protection can produce measurable effects. Within the described framework concept, which is open for creative modification by the participants, over 14,000 newly installed energy-saving devices and nearly 120,000 "kilometers saved" were reported for the entire Federal Republic of Germany within a short period (1994-95); the number of unreported cases is probably considerably higher.

The now concluded ozone campaign of the *Umweltstiftung WWF-Deutschland* was an attempt to illustrate local environmental problems and make use of the experiences of children and adolescents. Since "invisible" environmental conditions were made visible, the approach appears worth considering in connection with global environmental changes, too. Within the framework of the campaign tobacco plants were used as bioindicators in order to detect surface level ozone. The participants were given detailed informative and teaching material as well as suitable seeds.

Reports for the period from 1993 to 1995 are available on campaigns carried out by 145 schools, 14 youth groups and two universities. The studies conducted showed that 10% of the tobacco plants were damaged by surface level ozone. With regard to pedagogical aspects, the initiators stated: The results "do not convey the impression of an imminent environmental disaster. Nevertheless, they demonstrate the seriousness of the situation and leave no doubt about the necessity of combating the problem" (Umweltstiftung WWF-Deutschland, 1995).

The worldwide operating environmental organization *Greenpeace* also regards environmental education for children and adolescents as one of its tasks, frequently on the basis of contacts to schools. In its view environmental education

should contribute to concrete environmental protection. Accordingly, the didactic focus is placed on action orientation. In addition, environmental education is to be based on people's experiences and has to contain social, artistic, poetic and sensory elements as well. In addition, it is closely linked to cultural and social criticism and thus is at the same time to political education. It is supposed to promote holistic thinking and action as well as critical awareness and to prompt people to become active, both individually and together with others.

The "Greenteam" Project, initiated by Greenpeace Deutschland but also pursued in several other countries, is intended to encourage small groups of children and adolescents between the ages of 10 and 14 to tackle, work on and solve or at least defuse local environmental problems on their own (Greenpeace, 1994). The individual "Greenteams" composed of 5-7 young persons set their own goals. However, they frequently work together with an adult who supports them but is not supposed to have an influence on their specific actions and way of working. If they wish, the teams can receive support from Greenpeace at any time. The work of the individual "Greenteams" is coordinated centrally, contacts are established, further training is offered to support persons and working material suitable for children and adolescents is developed. The over 1,200 "Greenteams" that were established in Germany by mid-1994 bear witness to the great resonance to this concept.

1.4.4
Survey on the Status of Environmental Education in International Comparison

With the support of the Foreign Ministry, the Council requested information on the status and perspectives of environmental education in selected countries from German diplomatic missions at the end of 1994. The following (tentative) analysis of the material received from 41 countries includes written information from the embassies as well as various materials made available, in most cases, by official offices. The following aspects are taken into account here:
- legal framework conditions,
- governmental plans, programs and activities concerning environmental education,
- NGO activities,
- cooperation between governmental and nongovernmental institutions,
- availability of working materials, recommendations, etc. for environmental education.

With regard to the legal framework conditions for environmental education, there are references to and/or regulations concerning environmental education both in the environmental and educational legislation in most of the 41 countries. In many cases the legal framework is quite refined while implementation usually leaves a great deal to be desired. This applies particularly to younger states whose legislation is still in the process of being set up or has been established only recently. The most advanced are the legal framework conditions in the industrialized countries where environmental education is addressed not only within the school context, but also outside of

> **BOX 10**
>
> **Teacher's Manual of the African Social and Environmental Studies Programme**
>
> The teacher's manual entitled "Environmental Education for Sustainable Development for Primary School Teachers and Teacher Educators in Africa" was created at the initiative of the African Social and Environmental Studies Programme (ASEP) and with the participation of educational experts from 15 African countries (Muyanda-Mutebi and Yiga-Matovu, 1993). It contains specific suggestions for providing practice- and action-oriented environmental education as well as basic information for teachers. Current and politically controversial issues regarding the local and regional environment of the pupils are also dealt with, such as the question of clearing forests in favor of agricultural use.
>
> The ASEP initiative was implemented in a multinational project in Kenya, Tanzania and Uganda with the support of the GTZ. The project is to be continued with the participation of additional development services after a trial phase of two and a half years (GTZ, 1993).

schools as well as in connection with vocational training, whereas schools are the focus of attention in the developing countries.

Since the 1992 UNCED governmental plans and programs concerning environmental education have been expanded or first developed in several developing countries, in particular. In some of these environmental plans and programs there is mention of the connection between sustainable development and environmental education. This is especially the case in younger countries where, however, concrete implementation of the program seems to be difficult.

The commitment of NGOs in environmental education varies in the individual countries, tending to be stronger in more federal and/or industrialized countries than in more centrally governed states. Since some of the German embassies could not make any statements regarding NGO activities, however, reliable data cannot be provided in this regard. With the exception of Japan, the commitment of NGOs in environmental education is stronger in countries in which environmental education generally is at a high level. On the other hand, it must be recognized that NGO educational initiatives are frequently established with international support in countries that are less developed and do not carry out any activities of note in the field of environmental education. The focus there is on local or regional anchoring of the initiatives through efforts to integrate the population directly into the respective projects.

The cooperation between governmental and non-governmental institutions does not display a uniform picture either. Such cooperation does take place in the industrialized countries, though relatively unsystematically. Cooperation with NGOs is frequently sought by state institutions in the developing countries. Since the NGOs in these countries operate within an international context in many cases, one can presume that governmental organizations also wish to tap additional resources in this fashion.

As far as working materials and recommendations for environmental education are concerned, a substantial gap is perceptible between industrialized and developing countries. Corresponding materials are frequently provided in the developing countries with the support of relief organizations. This is usually done on a project basis since this is the best way to ensure that specific national and cultural conditions are taken into consideration and that there is participation on the part of those concerned.

Table 2 represents an attempt to conduct a comparative assessment of the environmental education activities for the countries included in the survey. The selected assessment yardsticks are of a tentative nature.

An analysis of the regional and/or continental distribution of environmental education activities produces the following picture: environmental education activities appear to be most advanced in the industrialized countries of Europe, North America and Oceania as well as in Japan. Compared to the northern, western and central European countries, the southern European countries have a somewhat less developed environmental education while the eastern European transformation countries are lagging significantly behind.

Environmental education in Asian countries displays a very heterogeneous level. Some countries are far advanced, others have just completed the beginning phase. A "boost" in environmental education can also be observed in newly industrializing countries, such as China, India, South Korea and Malaysia.

A "leap" in environmental education appears to be taking place in the countries of Central and South America, too. At least conceptual approaches are emerging, which indicate that environmental education is regarded as an important element of environ-

---

BOX 11

Primary School Course "Environmental and Agricultural Studies" in Zimbabwe

The primary school subject entitled Environmental and Agricultural Studies represents an attempt to integrate environmental education into the curriculum of general education schools in Zimbabwe. It was based on existing curricular structures in agricultural education. Since the subject matter and methods developed for the latter were well-suited for environmental education, references to the environment were added, a procedure that may be exemplary for many developing countries (Moyo, 1991; O'Connor and Turnham, 1992).

Since Zimbabwe's independence (1980) Environmental and Agricultural Studies have been a required subject at the country's seven-year primary school, and since 1982 projects have been carried out for implementation and evaluation of this subject by an institute of the national Ministry of Education in cooperation with the *Deutsche Stiftung für Internationale Entwicklung (German Foundation for International Development,* DSE). An evaluation study in 1990 (Lewin and Bajah, 1991) showed the limits of this approach. Criticism was directed at, among other things, focusing lessons on the teacher, lack of references to the local situation, infrequent use of overlaps with other subjects, structural problems as well as the little regard given to this subject by the parents of the pupils. In particular this disregard, which was also observed in agricultural education, might represent a significant obstacle for sustainable establishment of environmental education in Zimbabwe as well as in other developing countries. It additionally indicates the importance of the situation outside school for successful environmental education. Simplification of the curriculum, introduction of options regarding subject matter and intensification of further teacher training were some of the suggestions made for revising Environmental and Agricultural Studies.

BOX 12

Environmental Education Activities in the Philippines

The activities of the NGO Linkod Tao-Kalikasan (LTK; In the Service of the Human Earth Community) in the Philippines are aimed at informing the population about environmental problems as well as offering guidance on practical environmental actions. An LTK seminar program for multipliers deals with general perspectives of environmental development as well as with preparing plans for local educational initiatives. One striking aspect is the deliberately intensive integration of church groups and staff. In the view of the traditionally influential role of the Catholic Church in the Philippines as a moral authority (particularly in rural regions), it takes into account the historical and cultural development of the country and thus increases the program's chances of success (Bago and Velasquez, 1993).

The program carried out by the LTK for training so-called ecovolunteers within the framework of a pilot project is more oriented towards specific environmentally related activities. Ecovolunteers ought to be rooted in the community in which they work and to integrate the traditional knowledge comprehending environmentally compatible lifestyles and economic methods into their projects (Manalo, 1994). Here again an attempt is made to integrate environmental protection into existing sociocultural structures and retain indigenous knowledge or make innovative use of it.

---

mental policy. However, a stronger commitment to environmental education, particularly in the industrializing countries of South America as well as in Central American countries, will in all likelihood only be possible with outside support.

The most difficult situation for environmental education is in Africa, where it appears to play only a minor role, with the exception of South Africa and, to some extent, several north African countries. A tentative interpretation of this situation is the thesis: "The poorer the country, the less important is environmental protection". In any case, the overall status of environmental education in African countries does not seem to be very high. Nevertheless, a few countries have undertaken activities regarding environmental protection and environmental education,

BOX 13

Environmental Activities in Thailand

The Thai Project for Ecological Recovery (PER) pursues a pedagogical concept based on the (re)activation of indigenous knowledge of the population for solving problems (Permpongsacharoen, 1993). This attempt to integrate environmental education into the lives of the population in connection with their own interests is intended to overcome several barriers that hinder implementation of environmental projects, barriers that certainly play a role outside of Thailand, too. For example, people often feel powerless when they are confronted with large foreign companies with which the government has signed agreements for industrial use of the forest and see no sense in making a personal contribution (large-scale exploitation of the forest has even led to imitation effects). They also feel overwhelmed by the colossal magnitude of the problems, given the possibilities open to their small community.

Target groups frequently do not perceive the problems addressed in environmental protection projects as their own: the Community Firewood Project in northern Thailand, for example, was aimed at motivating the people living there to set up rapidly growing plantations and making use of wood obtained there as fuel instead of wood from tropical forests. Although plantations were subsequently set up, none of the trees there were felled. Only later did the environmentalists involved discover the reason for this behavior: Thai farmers traditionally only use thin, dead branches as fuel wood. To ensure that trees grow back, the trunks are left as intact as possible, but are not felled. The fact that an environmentally sound method of wood use was anchored in the indigenous knowledge of the local people was not taken into consideration in this project (Permpongsacharoen, 1993). Here, as in many other projects (*see, for example,* Jacobson, 1992), however, another sociocultural variable was ignored: the role of the woman. In regions in which subsistence farming is practiced, it is usually women who possess the knowledge on how to use and maintain local resources in the most reasonable way.

Nonetheless, the participative concept of education advocated by PER can also lay claim to successes. For instance, the construction of the Nam Choan Dam was prevented when a broad alliance of villagers, students, journalists, scientists, conservationists, women's groups and farmers put forward arguments against the project on the basis of their own respective knowledge, experience and interests and made use of their different ways of achieving an effect within the framework of a nationwide initiative.

at least programmatically, since the 1992 UNCED and have also formulated a demand for sustainable development. It seems doubtful, however, whether these programs and concepts can be implemented since these countries will hardly be in a position to achieve this through their own efforts alone.

In summary, a significant gap is perceptible between industrialized and developing countries with regard to environmental education on the basis of the survey conducted by the Council. Connections between economic level of development, environmental awareness and environmental education appear obvious, at least at first glance. However, specific national and cultural features must also be taken into account in a differentiated analysis.

1.4.5
Summary

A key problem in evaluating environmental education activities is the fact that there is currently a lack of systematic information on the extent of such activities in the different countries, especially with regard to global environmental changes, that could be evaluated for comparative purposes. Nonetheless, one can cautiously assert that, in spite of the many political declarations of intent, initiatives and programs in the course of the past 20 years, environmental education worldwide is still underdeveloped, based on its potential and requirements.

This general statement, however, must not blind us to the considerable quantitative and qualitative differences between industrialized and developing countries, between North and South, East and West, even within culturally and politically relatively homogeneous world regions, such as the EU. Environmental education in the industrialized countries of the North, for example, now has a relatively secure position both within and outside of the formal educational system. In most cases, however, it remains restricted to a local and national point of view that does not do justice to the new quality of complex global environmental changes.

Table 2
Level of environmental education activities for selected countries
Source: WBGU

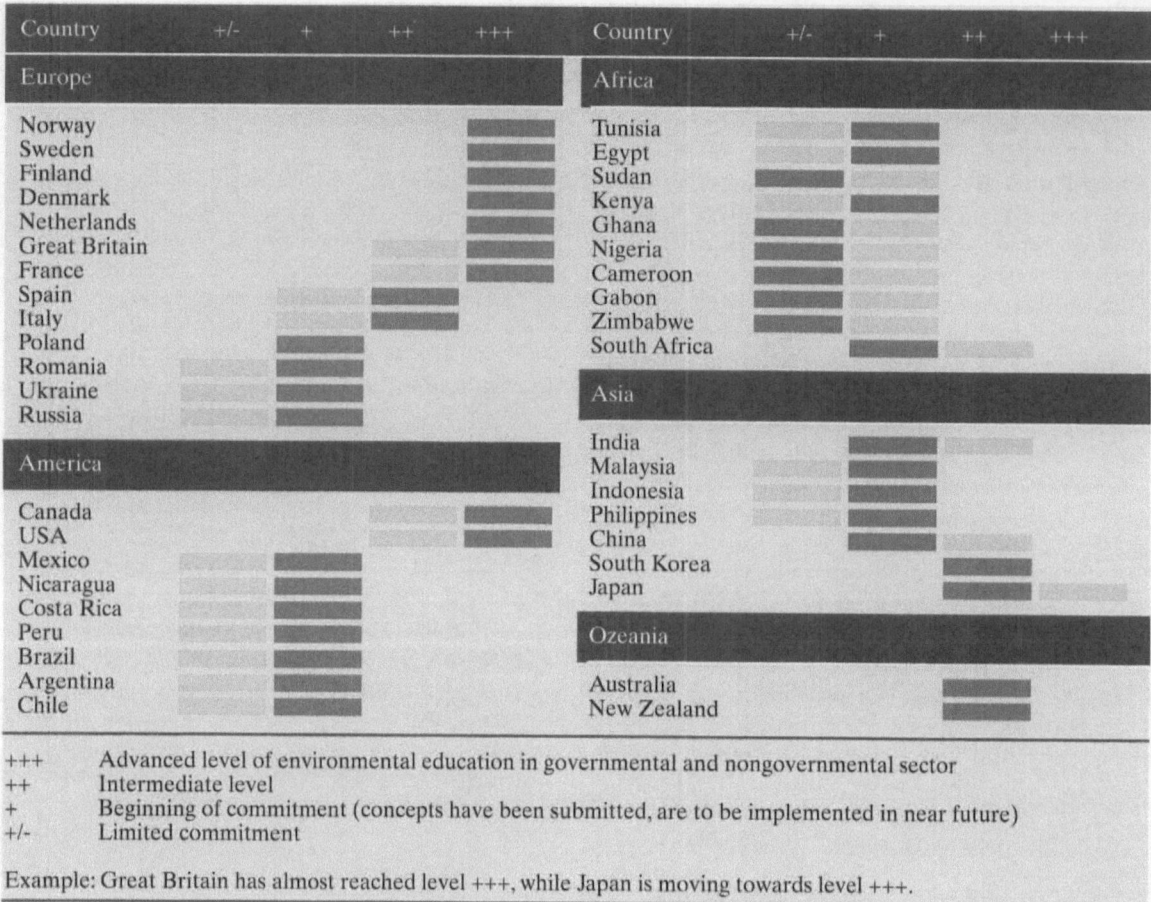

Country	+/-	+	++	+++	Country	+/-	+	++	+++
Europe					**Africa**				
Norway				▪	Tunisia		▪		
Sweden				▪	Egypt		▪		
Finland				▪	Sudan	▪			
Denmark				▪	Kenya			▪	
Netherlands				▪	Ghana		▪		
Great Britain				▪	Nigeria		▪		
France				▪	Cameroon		▪		
Spain			▪		Gabon		▪		
Italy			▪		Zimbabwe		▪		
Poland		▪			South Africa			▪	
Romania		▪							
Ukraine		▪			**Asia**				
Russia		▪							
					India		▪		
America					Malaysia		▪		
					Indonesia		▪		
Canada				▪	Philippines		▪		
USA				▪	China		▪		
Mexico		▪			South Korea			▪	
Nicaragua		▪			Japan				▪
Costa Rica		▪							
Peru		▪			**Ozeania**				
Brazil		▪							
Argentina		▪			Australia			▪	
Chile		▪			New Zealand			▪	

+++   Advanced level of environmental education in governmental and nongovernmental sector
++   Intermediate level
+   Beginning of commitment (concepts have been submitted, are to be implemented in near future)
+/-   Limited commitment

Example: Great Britain has almost reached level +++, while Japan is moving towards level +++.

In countries of the South, especially in developing countries, by contrast, structural problems in state educational systems frequently complicate the implementation of environmental education. Consequently, great importance is attached to projects carried out by NGOs in the field of education. Due to the large number and limited fields of work of these organizations, however, it is relatively difficult to assess the situation of environmental education work performed by NGOs in many countries of the South. Available reports and studies provide numerous indications that NGOs in many of these countries have difficulties in sensitizing the population to environmental problems in view of the urgency of other problems and due to the general political situation.

The analysis of national and international developments in the field of environmental education indicates that global environmental changes have been given little attention to date in relation to the medium- and long-term threats they pose. This may be due to their complex nature, the low degree of perceptibility in most cases, and the enormous space and time dimensions involved in these problems. In view of the findings from the social and behavioral sciences, there is immediate need for innovations in subject matter and methodology as well as in the setting of appropriate priorities for all types of environmental education.

The necessary scientifically supported innovations in the field of education are impeded, however, by the fact that empirically reliable statements cannot yet be made with regard to conditions that promote or hinder the success of environmental education. In initiatives and projects for environmental education, therefore, at least those criteria should be pursued that have proven to be crucial in practice for successful environmental education work in a large number of approaches over the course of time. They primarily include learning from concrete experiences in everyday-life contexts (situation orientation), learning in connection with one's own direct actions (action orientation) and integration of the subject matter to be taught into the social and political context (problem orientation).

Furthermore, environmental education should not limit itself to the conveying of knowledge, but should include emotional and motivational aspects as well as stimulate anticipatory thinking and acting, based on a holistic approach. Finally, elements of participation and communication have proven to be important for environmental education in order to enable the learners to act independently and on their own authority. They thus represent essential prerequisites for an environmentally and socially sound development.

1.5
Research Recommendations

The conception of suitable programs for overcoming environmentally harmful modes of behavior requires the availability of comprehensive and reliable information on the population's perception of environmental and other problems. For this reason the Council first calls for

- establishment of a worldwide, internationally comparable social monitoring system, with which environmentally relevant perceptions, attitudes, motivations and modes of behavior can be recorded, analyzed and evaluated at different levels of aggregation, at regular intervals and with an optimized, culturally adapted methodology. Greater participation in international research programs in the field of the social and behavioral sciences is recommended for this purpose, particularly in the *Human Dimensions of Global Environmental Change Programme* (HDP). Data from such social monitoring are necessary as a supplement to long-term ecological observation and as an input for models serving Earth system analysis and can be used for cross-national monitoring of the advances made with regard to sustainable development.

In the field of environmental education the Council sees a pressing task in the

- systematic compilation, analysis and comparative assessment of findings regarding environmental education measures at the national and international level, while taking into consideration approaches, projects and experience both in formal educational systems and in the nonformal field (NGOs).

In the opinion of the Council, such a continuous documentation and analysis of the status quo should go hand in hand with

- further elaboration and empirical substantiation of criteria for successful environmental education, while emphasizing global aspects, particularly through

- national and international evaluation studies on environmental education measures with respect to their actual effects on human behavior and on the environment,
- intensification of worldwide basic research on differential aspects of environmental education within the context of global environmental changes and against the background of the respective basic cultural, socioeconomic and technical framework conditions (cultural and group-specific features, context dependence of educational measures, etc.),
- promotion of interdisciplinary research and exchange processes between basic and applied research in the field of environmental education, departing from the previous practice of extensively concentrating on schools.

1.6
Recommended Action

The Council feels that it is fundamentally necessary, both at the national and international level,

- to place greater weight on environmental education as an integral element of environmental policy, in particular by
  - substantially strengthening the field of education within the framework of the overall concept of state environmental policy,
  - developing concrete strategies for the implementation of the various political recommendations and declarations of intent regarding environmental education,
  - making environmental education a regular subject of discussion within the framework of international environmental policy negotiation processes,
  - strengthening international organizations (particularly UNESCO) in order to implement international agreements in specific educational contexts.

In the view of the Council, the operational side requires

- placing priority on the promotion of environmental education measures that meet established and monitorable criteria of successful environmental education so as to bring about a constant qualitative improvement of environmental education and readaptation of subject matter to the globally altered environmental situation. This could be achieved, in particular, by
  - implementing scientifically elaborated criteria for environmental education in the curricula of the formal educational system and increasingly integrating the areas of "environment" and

"development",

– working up interdisciplinary concepts for the training of teachers and multipliers,

– supporting nongovernmental projects in the field of environmental education (NGO projects, municipal initiatives).

The Council sees significant opportunities for improvement at the level of organizational and communication structures. The aims should be

• greater networking both of governmental along with nongovernmental environmental education projects and of projects between the two sectors so as to bring about better coordination and joint use of resources and thereby increased efficiency in the field of environmental education overall. Appropriate measures might include

– setting up one or more information, coordination and service offices for environmental education projects carried out by governmental and nongovernmental organizations,

– further support of the work performed by the Projektstelle Umwelt und Entwicklung (Project Center for the Environment and Development) in order to network NGO activities within the framework of environmental education,

– setting up forums for international exchange of experience, such as in connection with EXPO 2000.

Particularly in view of the deficient situation regarding environmental education in developing countries, the Council recommends

• consistent integration of environmental education into environmental policy programs and measures by

– providing developing countries with material and human resources support for the implementation of governmental plans and programs on environmental education within the framework of development cooperation,

– generally linking the promotion of development projects to environmental education measures,

– promoting nongovernmental projects concerning environmental education in selected developing countries (especially in Africa and South America) and in central and eastern European countries,

– promoting projects for basic education in developing countries, while emphasizing the connection between "environment" and "development".

In all these recommendations for action regarding environmental education it must be kept in mind that in most cases success from the appropriate measures can only be expected in the medium to long term. That is why it is even more important to initiate necessary steps immediately, even though clear attribution of success to certain individual measures is rarely possible. All environmental education measures must always take into account the respective cultural, socioeconomic and demographic conditions and contexts, including the crucial problems that may differ completely in individual countries or regions. For this reason environmental education measures should be implemented, based on the specific country and target group.

2.1
Formation, Dissemination and Adaptation of Know-how

Research on developing countries and more recent analyses of economic growth in an international comparison of countries point to special features that are important with regard to overcoming global environmental problems:
- A high variance can be observed in the national growth and/or development rates over time as well as in a regional comparison (Wolf, 1994).
- Differences in economic development prove to be surprisingly stable over time and do not show any strong convergence tendencies (Wolf, 1994).
- In spite of this determination of relative stability in development differences, some countries (such as the "four small tigers", i.e. Hong Kong, South Korea, Singapore and Taiwan) were able to improve their relative position significantly in contrast to the overall trend and this position is still changing (Naya and Takayama 1990; Kulessa 1990, Draguhn 1991).
- Development or economic growth and pollution do not always have to correlate positively with each other. Below a certain level of development there is poverty-related pollution while above this level economic development and pollution may correlate positively, but do not necessarily do so (WBGU, 1993).
- Although there has been a positive correlation between economic development and the utilization of environmental resources, it is lower than is presumed in many cases, thus indicating that an explanation of global development and environmental problems requires more comprehensive approaches.

Today economic growth and development researchers extensively agree that an inevitable convergence does not exist between developed and less developed countries. It is becoming increasingly more evident that the observable relative constancy of development differences over time as well as the

catching-up process of some countries, contrasting to this constancy, have to be explained by principles governing the formation, dissemination and adaptation of know-how. This also applies to the variance of pollution and to the observed decoupling of economic growth and pollution. The global distribution of "technical know-how" plays a special role here.

Thus it is not surprising that inducement of technology transfer from industrialized nations to developing countries is one of the classic demands of development policy and of the developing countries. There is hardly a development policy agreement or conference in which the developing countries do not express their wish for unrestricted access to the modern technologies of the industrialized nations and for improvement of transfer conditions. An international code of behavior for technology transfer (UNCTAD Document TD/Code TOT 47) as well as a revision of the international patent system are intended to serve this purpose.

Technology transfer and/or technological cooperation is also of crucial importance for overcoming global environmental problems. For example, technology can contribute to the simultaneous enhancement of economic development and environmental relief. In general, technology can bring about a relative and perhaps even absolute decoupling of economic growth and pollution. This is why the *Convention on Climate Change* raises the demand for boosting technology transfer from the industrialized to the developing countries. In particular, as in Article 4(1)c, it requires that the former "promote and cooperate in the development, application and diffusion, including transfer, of technologies, practices and processes that control, reduce or prevent anthropogenic emissions of greenhouse gases not controlled by the *Montreal Protocol* in all relevant sectors, including the energy, transport, industry, agriculture, forestry and waste management sectors". AGENDA 21 emphasizes even more strongly in many places, particularly in Chapter 34, the necessity of increased technology transfer and/or technological cooperation as important starting points for overcoming the global development divide and for coping with national and global environmental problems. In the end the demand for increased transfer of technologies and know-how to ensure sustainable development was included in the final documents of the First Conference of the Parties in Berlin; the *Secretariat of the Convention on Climate Change*, for example, is requested to submit reports on the measures in this field for the follow-up conference (UN, 1995).

The Council, too, takes up these demands but will, for the following reasons,

– distinguish between the terms exchange of know-how and technology transfer,
– not only emphasize "unilateral" transfer, but also exchange between industrialized and developing countries, and
– break down "know-how" into the formation, dissemination and adaptation of know-how.

Experience shows that without a comprehensive exchange of know-how technology transfer is doomed to failure in many cases. In other words: overcoming global differences in development and environmental problems not only requires the transfer of knowledge on the specific technological system, including equipment, infrastructure, processes, know-how, complementary services, preliminary training, technological orientation of management, etc., i.e. imparting knowledge about technical options for taking action. Rather, development and environmental aspects regarding transfer of know-how also include the imparting of knowledge on (new) economic or social institutions, forms of conveying knowledge (such as new findings in didactics), rules, conventions, standards and values. A comprehensive concept of know-how plays a particularly important role in the theory of social evolution (Hayek, 1979). According to this explanatory approach, social development is the result of a sometimes desired but usually unplanned cultural process of selective application and duplication of "learned rules of behavior" among groups of people which facilitates survival un-

der the respective environmental conditions. These rules of behavior may also affect how people deal with their natural environment and be of a scientifically objectified and subjective nature (Popper, 1984a) since knowledge that was not formed purposefully can make significant contributions to coping with regional and global environmental changes. For this reason the Council advocates a broad definition of knowledge (*Box 14*).

From this point of view it is evident that one should refer to an exchange of know-how between industrialized and developing countries rather than to a transfer of know-how. This position is based on the insight that knowledge in industrialized countries is significantly influenced by the problems, experiences, values, etc. there and that it is, by all means, possible for industrialized countries to learn from developing countries. This not only applies to the values of other cultures and the forms of human coexistence that have come into being there, but also to the technologies found there, such as forms of soil management, irrigation techniques, types of forest use, etc. The necessity of such a reciprocal exchange of know-how will gain global importance for the implementation of sustainable development in the future.

Inclusion of an exchange of know-how defined in this manner in the context of global environmental changes may contribute to a reduction or even elimination of the anthropogenic global threats to the stability of ecological, economic and social systems with the help of greater information. Such know-how has to be subjected to a continuous process of monitoring and modification (Popper, 1984b) and must vary regionally due to the differences in people and in the circumstances of their lives; but it should be made available globally and as comprehensively as possible because it points out action-taking options. It is thus important not only to create new knowledge, but also to provide for its spatial dissemination, i.e. exchange of know-how.

BOX 14

Know-how

In this context know-how is the sum of the stock of
– information on scientific know-how and environmentally compatible technologies,
– know-how about institutions and the way in which organizations function in order to solve environmental problems,

– information on (environmental policy) problems and possible solutions and
– attitudes, values and religious explanations (Earth and living environment as Creation, legitimization of human interventions in nature, etc.)
that exists in a society and is gained on the basis of education and experience or is taught via the educational system or contact with researchers.

As mentioned above, the demand for a greater exchange of know-how and technology transfer is, above all, a result of the oppressive continuation of development differences. If this tendency had the characteristics of a principle, the disparities in affluence between the industrialized nations and the developing countries as well as the resulting global environmental problems (such as soil degradation) would continue to exist. To this extent, an increasing number of scientific disciplines, particularly economic and political science, are involved in an explanation of this phenomenon.

Today empirically oriented economic researchers largely agree that economic development is not only determined by the accumulation of production factors (real capital, for example) or the availability of resources, but in particular by the level of know-how (which also implies technical progress). However, this know-how is no longer viewed as autonomously and exogenously given, but - and this is crucial - as endogenously explained. In other words this means that new knowledge comes into being where economic growth is already taking place and self-reinforcing processes are observed. Furthermore, it is becoming evident that a distinction must be made between spreading, dissemination and adaptation of know-how (*Box 15*) and that the global dissemination of know-how by no means entails ability and application (WBGU, 1993).

The more recent theory of growth (Romer, 1986, 1987 and 1990; Lucas, 1993) and the evolutionary approach in economics (Witt, 1994) assume that the long-term development of human society or of individual nations depends very decisively on know-how

and that, as resources become scarcer, development bottlenecks can be overcome best through extended know-how. This know-how yields increasing marginal earnings as an input in the production of goods and comes from a research sector which is primarily based in the highly developed industrialized nations (islands of know-how formation according to Bernhard Felderer). The task of research is to develop more and more productivity-enhancing ideas through the deployment of human capital and existing know-how. A key assumption of the more recent theory of growth is that this constantly expanding know-how in industrialized countries can be privatized only to a limited extent, i.e. it can be used solely by the discoverer (limited internalization functions of patents, for example). Via positive external effects (so-called spillovers), therefore, research favors the respective regional environment or those regions that maintain a successful exchange of know-how with each other through intensive trade relations or an exchange of factors. To the extent that this spatial exchange of know-how mainly takes place between the highly developed industrialized nations, the global development divide (trap of underdevelopment) is reinforced and, at the same time, the ability to cope with environmental problems diverges. State-supported research, at least basic research, gains importance as soon as there is an opportunity for the formation of know-how independently of development or for endogenous technical progress, and research results have the nature of a collective asset. This demonstrates how important it is to promote know-how and technology transfer in order to overcome the

---

BOX 15

Key Issues in the Dissemination of Know-how

Formation of know-how

as a result of existing know-how, learning by doing, teaching values via upbringing and education (imparting curiosity, independent thinking, belief in progress, religious attitudes), cultural diversity, urbanity, ability to perceive problems, imparting know-how through training, access to information, ways of obtaining access (property rights), organization and support of research, size of enterprises, competitive climate, entrepreneurship (daring, willingness to take risks), government requirements, development of market demand, etc.

Spreading of know-how

via acquisition of property rights to know-how (rules for protecting intellectual property, patent law, financial opportunities), exchange of researchers, research cooperation and migration pattern of experts and specialists, location-related decisions made by research-intensive enterprises (such as multinational corporations), trade relations (exchange of goods, incl. consulting), development of telecommunications infrastructure, etc.

Adaptation of know-how

as a result of training level, application willingness (willingness to act), application opportunities (economic room for maneuver), economic necessity of taking action (reward through market demand, competitive pressure), etc.

global development divide and solve global environmental problems.

This alone is not sufficient to solve all problems, however. In addition, efforts have to be aimed at supporting the formation and adaptation of know-how in the developing countries. This involves more than the redistribution of existing know-how or improvement of access to existing stocks of knowledge. The formation of knowledge in the industrialized countries is, to a great extent, a reflection of the environmental problems prevailing there or expressed by the government, of the economic incentives applying there (market demand, promotion of research), of the values there and of the existing level of know-how there. This also means, however, that a transfer of know-how from the industrialized countries to the developing countries is, by no means, always tailored to the development and environmental problems prevailing in the latter countries. Rather, it is akin to cultural swamping and requires a wide variety of on-site adjustments. Therefore, it is additionally necessary to promote research in the developing countries themselves, reactivate existing know-how there, induce learning processes under the specific environmental conditions in the developing countries and provide the requirements for adaptation of know-how.

2.2
Opportunities and Limits of the Exchange of Know-how and Technology Transfer

2.2.1
Prerequisites

The extension and exchange of know-how and thus of technology transfer, too, are always based on individual cognitive processes within a society. Only transfer operations that take place between individuals as members of institutions (e.g., polytechnics, enterprises, consultants) as well as individual access to stored information (books, databases, etc.) can contribute to reducing deficits in know-how applicable to overcoming regional development barriers or coping with global environmental changes. If, for example, it is possible to coordinate the knowledge of individuals regarding certain causal sources of environmental problems with the knowledge of other individuals on the resulting effects, decisions on the necessity of remedial measures can be made more efficiently.

The decisive point here is that the transfer of know-how is carried out to a significant degree via

market and competitive processes, while at the same time maintaining entitlements and access to existing and newly acquired know-how. Exclusion of others from newly acquired know-how (such as via patent applications) is possible only to a limited extent and therefore triggers positive external effects in the region or in countries that exchange innovative goods and techniques as part of trade. Nevertheless, patent law is fundamentally recognized by most states and ownership of know-how can be acquired and rights to use of know-how bought. Such an arrangement creates incentives to further develop the existing level of know-how. These incentives may involve the prospects of monetary advantages in the form of a temporary monopoly edge with great profit-making opportunities, immaterial advantages through enhancement of one's scientific reputation or the satisfaction of altruistic personal aims (Kerber, 1991; Röpke, 1977). Experience gained in the highly developed countries shows that the patent system there can reinforce innovative activities and thus economic growth by granting the right to (exclusive) disposal of new know-how and permitting the profitable sale of the latest technical information (Häusser, 1995).

It is additionally becoming clear that there are several important prerequisites for the exchange of know-how and technology transfer. The minimum requirements include lowering the illiteracy rate and providing the population with a solid basic education for better exchange of information. It is not surprising that the "four small tigers" already mentioned met a key requirement for the exchange of know-how and technology transfer prior to their upswing through consistent educational reform and continual expansion of their polytechnics and universities. Polytechnics in particular can strengthen non-market-determined transfer and represent at the same time an important prerequisite for building up their own research with greater orientation to the regional environment and related problems. Moreover, rapid development of the telecommunication and transport infrastructure is necessary, as a rule, in order to enhance the opportunities for individual contact, support the exchange of researchers and, as already emphasized, guarantee property rights to new know-how.

It is usually not enough, however, merely to provide information or opportunities to access it. If information is to stimulate ideas and encourage recipients to apply and look for further information, additional requirements have to be met. This involves, above all, questions of motivation and willingness to adapt, especially regarding information having negative content, in order to induce corrections of traditional modes of behavior and production. In view of their diverging abilities and circumstances, individuals tend to react differently to material and immaterial incentives and thus alter their level of knowledge in various ways (Schumpeter, 1912; Nelson and Winter Gerybadze, 1982; Reichert, 1994; Athey and Schmutzler, 1994; Witt, 1994). To this extent, it must be ensured that the prevailing problems in the individual countries are accurately perceived in each case. The heterogeneity of the individual approaches to the extension of know-how should then form the basis for using a broad spectrum of knowledge within the society in order to solve specific problems and for constantly checking its usefulness (Erdmann, 1993). Adoption of successful problem-solving approaches of other individuals subsequently induces a possible contribution to overcoming global problems.

The provision of such requirements is particularly important for exchange of know-how and technology transfer with respect to global environmental changes. To ensure that these problems are perceived and people are open to the transfer of know-how in this field, there must first be an environmental awareness (*see Section B 1*), which in turn means that educational requirements must be met or a demand for environmental protection assets must exist for the purpose of creating economic incentives. This demand is especially important, both in the industrialized and in the developing countries. In contrast to other technological lines (such as microelectronics, new materials, etc.), the environmental technology industry is not so much a key technology focusing on a certain process or product as a "cross-sectional sector" that, at least at the beginning, is not faced with a rapidly expanding demand in the private sector. Experience shows (RWI, 1994) that it is primarily state-induced demand that ensures profitable production and arouses interest in technology transfer in this field. In other words: a market demand for environmental technology has to be triggered via regulatory laws or economic instruments. As a rule, the exchange of know-how and technology transfer suffer without the economic boost of market demand.

At this point the Council would like to recall that, because of these interrelationships, the exchange of know-how and technology transfer can be accelerated through agreements between the industrialized and developing countries regarding the permissibility of certain instruments. As shown elsewhere in this report, for example, the permissibility of so-called Joint Implementations based on bilateral or multilateral agreements could be designed such that a stream of capital and technology is triggered from the industrialized to the developing countries, thus improving the willingness to adapt in the latter. At the same time the resulting impulses would stimulate the search for regionally adapted technologies, such as in the field of solar technology.

The direct investments from enterprises in industrialized countries as well as the location-related and investment decisions of multinational corporations must also be seen in close connection with the above. It is fundamentally conceivable that such investments are made to avoid strict and expensive environmental protection requirements in the industrialized countries. However, experience shows that in the event that pollution is clearly caused by enterprises with international operations, they very quickly experience a loss in reputation affecting worldwide sales so that they are willing to meet high environmental standards, such as in the form of technology transfer to their new locations. To this extent, they are becoming increasingly important as supporters of know-how and technology transfer. However, this transfer is closely tied to location-related decisions that are oriented more to the environmental quality of the new location.

Transfer problems arise when the environmental problems or environmental policies of the developing countries differ from those of the highly developed industrialized countries and thus require another solution. Due to their problems and the resulting measures taken, the industrialized countries have obviously extended environmental technology know-how, particularly in the fields of air quality control, water quality policy and waste disposal. In the first two fields this led to the further development of filter technology and improvement in the efficiency of plants (such as power stations). Because of the analogous problems, the industrialized countries can help the developing countries with regard to power stations as well as the large cities in Asia, Africa and South America in tackling their water pollution and hygiene problems by means of state-supported technology transfer. The latter is thus becoming more and more important in connection with municipal supply and disposal systems. In order to improve technology transfer in accordance with the BOT principle (build, operate, transfer), operator models should be supported that locally set up, operate (learning by doing) and then hand over pilot projects to national bodies so as to motivate others to carry out similar projects in this manner.

In other relevant environmental fields, on the other hand, there is a need for so-called appropriate or even new technologies. This applies, in particular, to the increase of soil productivity with simultaneous prevention of soil degradation, the solution of the problem of water scarcity and pollution in thinly populated regions that do not allow modern water supply and disposal techniques and the island-like development of an energy supply system. This need for adaptation also results from the specific cultural and educational features with which implementation of the large-scale technology of developed industrialized nations is confronted. Such adaptations can, for the most part, only be carried out on site. A policy of exchange of know-how and technology transfer should be supplemented by long-term promotion of on-site basic research on environmental technology.

As has already been emphasized, the changes in the level of knowledge taking place in the respective economy through the inclusion of know-how will be used as a productive factor for explaining growth and development processes (Romer, 1990). This procedure takes into account that the knowledge gained by an individual is first applied to a problem by this person. This involves information that can be used by other individuals to solve other problems, without impairment of the "discoverer" in his or her possible uses (the level of knowledge as a collective asset or the fact that there are positive external effects). The spatial dispersion of these positive development effects (spillovers) is limited, as a rule, but it constantly creates new room to maneuver in other places for more productive use of existing resources. The spatial dispersion of knowledge is restricted here to states with similar economic, organizational and know-how structures since especially intensive trade activities or factor exchange movements (exchange of scientists, direct investments, etc.) take place between these states involving the transfer of know-how impulses that can be applied by both parties. Between countries with differing structures, however, a specialization tendency is intensified, promoting increasingly diverging tasks - development and application of know-how in countries with a higher per capita income and large foreign trade volume, application of standardized ways of using know-how in states with a lower per capita income (Grossman and Helpman, 1991). If this cannot be changed, the divergence between industrialized and developing countries will grow.

Finally, one should not forget to mention the *Senior Expert Service* (SES), a non-profit-making enterprise that increasingly promotes vocational and further training of skilled and managerial staff at home and abroad with retired experts. Thus far the focal points have been on technical and economic areas. The specialized fields in which these senior experts work include water and waste management, energy supply and, very generally, environmental protection. These experts are able to convey valuable ideas through a combination of experience, motivation and working on site. It would be useful to determine how the SES can be used even better for technology transfer in the environmental field.

2.2.2
Assessment of Institutional Approaches

The necessity of global coordination for the application of existing knowledge and technical potential in order to solve global environmental problems is formally met by providing for such coordination in environmental protection agreements to a greater extent (*see, for example*, AGENDA 21, Chapter 31). Consideration is also given to the fact that the differing initial conditions in the individual states require adaptation of the available know-how (*see, for example*, Art. 3 of the *Convention to Combat Desertification*). These global aims do not involve a direct obligation, however, because they have not been linked to specific provisions for the planning, financing, implementation or monitoring of transfer measures. Therefore, measures for controlling the transfer of know-how are preferably organized by individual states or within the framework of bilateral agreements. Institutional regulations that might contribute the exchange of know-how and technology transfer can be found in the broader sense in environmental, development and industrial policy.

Many environmental policy measures are taken to extend know-how to include the state of the environment. They include the setting up and expansion of monitoring systems. Substantial differences remain with regard to assessing the state of the environment and its effects, however. They are mainly due to diverging individual preferences worldwide concerning environmental quality, as demonstrated by the example of the differing evaluations of noise in Germany and in Italy.

Large research institutes (e.g. in the climate field) are combined into international bodies in order to determine the cause-and-effect relationships involved in global environmental changes and these bodies stipulate the direction of research. It is questionable here, however, whether this constellation can give sufficient consideration to the worldwide diversity and complexity of social development. Greater efforts are being made to record the knowledge of polluters about the substances and processes they use as well as consumed resources and residual substances that accumulate (e.g. World Bank or World Resources Institute) in order to develop specifications on further behavior on that basis. This may result in a central steering of existing knowledge in a certain direction and thus a restriction of the room for maneuver in carrying out further research, but initially this collection and processing of available know-how is an important prerequisite for a global exchange of know-how.

The significance of development policy applies, in particular, to the transfer of knowledge about avoidance, reduction and elimination of global pollution from countries having a high per capita income to those with a low per capita income. This transfer takes place within the framework of individual projects and is usually carried out by public or non-profit-making organizations - in Germany, for example, by the GTZ. The project basis makes it possible to adapt to the concrete problems connected with the specific situation and to the basic conditions that have to be taken into account. Development cooperation is increasingly oriented to small, decentralized projects that provide for better consideration of local preferences via participation of those concerned. This also leads to a development of approaches on the basis of reciprocal transfer of the existing knowledge in each case and thus to a more efficient use of existing information. Incentives for permanent further development of the projects are created through greater acceptance and commitment on the part of those involved, a key aspect since the continuity of such cooperation is especially important.

Due to the existing organizational structures in the development institutions, inefficiency in the exchange of know-how may result if state funds are awarded on a central basis (World Bank, 1991). Thus it is questionable whether information on the most efficient ways of solving individual global environmental problems is actually available in the case of central appropriation of state funds. A special threat is posed by an orientation to political opportunities. This political orientation may manifest itself in the volume of funds that is adjusted to the general budgetary situation. On the other hand, concrete priorities can be stipulated on the basis of preferential promotion of individual economic sectors in the respective country and spatial priorities can be set on the basis of combining development aid with more far-reaching political objectives through the fund allocation structure. An example of the combination of development aid with sectoral promotion of exports is the elaboration of so-called "FZ joint financing concepts" of the German federal government, in which low-interest loans with federal guarantees are offered for infrastructure and industrial projects as a supplement to direct financing from the budget of the *Ministry for Economic Cooperation and Development* in order to create advantages for German suppliers of corresponding services on international markets (anonymous, 1994). Furthermore, the opportunities for influencing individual interest groups in connection with politically motivated development cooperation must be taken into account with regard to the incentive effects in the target country (Amelung, 1987; Frey, 1985).

Basic industrial and trade policy conditions play an important role due to the great significance of private enterprises for the development and transfer of relevant know-how in dealing with global environmental changes. Both global regulations and provisions in potential countries of origin and target nations involved in the transfer of know-how must be looked at in this connection. Greater worldwide coordination for the development of a framework for transferring information on individual economies should be effected through implementation of the measures for the protection of intellectual property rights within the GATT/WTO regime by the year 2006. The related legal security increases the incentives for enterprises to make increased use of their know-how internationally, without having to fear that this know-how will be made accessible to additional competitors. However, one must keep in mind the long period of introduction between 1996 and 2006 and the question of how to handle conflicts between the protective interests of enterprises in industrialized countries and the owners or users of traditional know-how, particularly in the countries with a low per capita income. These conflicts may lead to a situation in which existing know-how oriented to the local circumstances is no longer used to solve global environmental problems and there is less acceptance of information from industrialized countries.

The trade and industrial policy in the individual countries is usually characterized by an attempt to control the transfer of know-how by private enterprises in the national interest. In addition to promotion of basic research in the industrialized countries, for example, measures supporting individual sectors are taken which are expected to lead to the development of innovative processes and products which are in demand because of, among other things, their contribution to solving global environmental problems, e.g. subsidization of biotechnological processes. This is intended to increase promotion of the competitiveness of individual countries (Lieschke, 1985). The justification put forward for such measures is that economic advantages are provided on the basis of a broadly dispersed applicability of central technological development lines (Krugman, 1993; Romer, 1990b). Problems result because research activities are directed by the individual states (Stolpe, 1993; Raaflaub, 1994). This form of support can only lead to efficient results if the staff at the state institutions have information as to which process and product know-how can contribute the most to solving global environmental problems in the future. If support is concentrated on making use of the advantages of individual countries based on know-how, there is a danger that the potential which is related to further development of this know-how and which would first emerge during an international exchange and during testing for various forms of application could not be put to use at all. As a consequence, one can expect parallel developments and a failure to take advantage of opportunities to add to existing knowledge.

Besides limited willingness to transfer know-how in the individual states, there are other institutional restrictions regarding the acquisition of information that must be taken into account. Numerous multinational enterprises, for example, first begin by opening up national markets as locations for sales activities or for the production of standardized processes before higher-value corporate segments are relocated on a long-term basis (Merten, 1987). The predominant interest in the target countries, however, is to establish export-oriented corporate structures so as to achieve positive foreign trade effects. This is reflected in restrictions on import and on foreign exchange as well as in requirements of local content (such as in the form of an obligation to purchase domestic primary products, to employ local labor and to establish joint ventures with domestic enterprises) with the aim of disseminating know-how in the target country. The consequences of these requirements include fewer incentives for a commitment on the part of private enterprises in and thus for the transfer of know-how to the respective countries. Isolation from international developments induces, in turn, deficits in the countries that impose requirements with regard to adaptation of globally available know-how to the specific environmental problems existing in these states.

Furthermore, the interest of the target countries in an increase in their economic production potential and in an improvement in their foreign trade balance causes the attraction of private enterprises to be less oriented to their contribution to long-term expansion of the knowledge related to solving global environmental problems than to their short-term contribution to increasing the national income. This may enable multinational corporations to transfer less developed know-how for application in countries with a lower per capita income with the argument of creating jobs, and the economic importance of these investments may induce governments there to make concessions regarding the cause of negative environmental effects. This development cannot be evaluated negatively in that it merely represents a consideration of positive and negative effects of the establishment of enterprises in a specific country (*see Section C 7*). However, there is a need for action with respect to the inclusion of any global effects and the introduction of a competitive framework for dealing with globally operating enterprises.

Among the institutional questions that will become more important in the future is the issue of in-tellectual property rights to products produced by means of bioengineering and the genetic material of wild animals and plants. There is still controversy over the question of whether patenting of plant and animal variants should be fundamentally possible and, if this should be the case, which country may act as the beneficiary of a product innovation resulting from the manipulation of genetic material: the countries that provide the genetic material or the enterprise that carried out the manipulation? (Dexel, 1995) The developing countries which represent the predominant source of genetic material are afraid that in the second case it would be more difficult to obtain access to the growth in biotechnological know-how, which primarily takes place in the industrialized nations, especially in the USA, and they would therefore like to at least have a share in the profit derived from use of the new "products" as "owners" of the genetic resources (*see Section C 5*).

In summary, an analysis of the existing basic institutional conditions regarding organization of an exchange of know-how for solving global environmental problems shows no uniform features. As a rule, global objectives exist only in the form of general declarations of intent in multinational agreements that are inadequately linked to concrete statements on planning, financing, implementation and monitoring of transfer activities so that these objectives are, for all practical purposes, not of a binding nature. Decisions on the specific formulation are usually made in the individual states whereas implementation takes place either within the framework of bilateral agreements or at the nongovernmental level. The transfer of know-how is influenced directly by means of state instruments provided for in environmental, development, industrial and trade policy, though there are hardly any signs of coordination between these political segments. In addition, there are a number of open questions regarding future regulation of intellectual property rights to manipulation of plants and animals that are primarily native to developing countries via genetic engineering.

Environmental policy demonstrates increased interest in central determination of available information in order to create a basis for political decisions. This particularly applies to knowledge on the state of the environment, causes and effects whereas globally relevant avoidance, reduction and elimination methods have been tuned to the individual situation up to now. A greater decentralization and concentration on smaller projects can be observed in development policy, though the decision regarding the allocation of funds by government institutions may hinder this orientation. In industrial and trade policy there is a great emphasis on the advantages of individual countries regarding the development of know-how in the industrialized nations and in economies with a low per capita income. This represents an obstacle to transfer and mutual exchange of know-how. Moreover, support is usually provided selectively in favor of individual sectors or research institutions, thus neglecting the necessity of decentralized and diversified approaches to the exchange of know-how and technology transfer.

2.3
Summary

The statements above demonstrate that there are certain requirements for effecting successful exchange of know-how and technology transfer which can be divided into two large groups. The first group comprises measures that can be regarded as state priorities and sectoral recommendations and - in the language of economics - are more of a process policy nature. The second group is focused on the specification of basic institutional conditions that can be interpreted as an improvement of search processes involved in the expansion of the range or exchange of know-how - and are thus, in economic language, more of a regulatory nature.

PROCESS POLICY MEASURES

The first group is predominantly derived from the store of experience of those developing countries (newly industrializing economies) that have been able to shake off the long-term binds of an absolute or relative lag in development. This implied the closing of a technological gap that, at the same time, opened up room for environmental policy action. Most studies involving an analysis of these economically successful states (such as Theierl, 1989) draw the conclusion that the first step must be to meet requirements, including (*as shown in Section B 2.1 with regard to environmental concerns*):

- education and training,
- setting up the country's own scientific and technological research infrastructure,
- development of the transport and telecommunication infrastructure,
- exchange of scientists,
- setting up an environmental consulting infrastructure,
- increasing the productivity of agriculture through environmentally sound use,
- provision of market economy requirements,
- provision of location-related requirements for acquisition of direct foreign investments,
- greater export orientation of the economy,
- orientation to need for environmental policy action in the developing countries and

- environmental policy action taking greater consideration of environmental concerns.

The importance of the first three points has been stated above. They favor the accumulation of human capital in the educational system as well as in research institutions. A controversial question here is still the extent to which the success observed in the newly industrializing countries in Asia has resulted from cultural influences, particularly the Confucian system of values (Kahn, 1979), and the extent to which other cultural conditions possibly act as an "obstacle" to a certain (Western or European) development model, thus requiring a search for other models. There is, however, general acceptance of the thesis that the Confucian basis of the societies in East Asia has at least favored a high work and saving ethic, discipline as well as the willingness to make great efforts for education (Pascha, 1990; Draguhn, 1991). It is also evident that training and technological development were given a decisive boost by the state in most newly industrializing countries of Asia, a fact that has been reflected in increasing proportions of the gross national product being spent on the educational system as well as on research and development (Hofheinz et al., 1982). Numerous research institutes were established and at the same time there was an attempt to get foreign enterprises to locate there (Henke, 1990) or to introduce and further develop programs for systematic vocational training in certain sectors (such as in the information technology field in Singapore) in cooperation with such enterprises (World Bank, 1993a). In this way it was possible to meet the prerequisites for facilitating the handling of complex high technology. Experience shows that the implementation of modern technology in the developing countries frequently ends in failure for the sole reason that the personnel lacks the qualitative requirements (operation, maintenance, care, etc.). This also applies to the use of complex environmental technology since the latter can hardly be distinguished from other forms of technology and in the end it is not techniques but environmental (technology) systems that are transferred.

A distinction can be made here between sectors in developing countries oriented to the domestic market and those oriented to the world market. The former include, in particular, agriculture, which is responsible for the food supply and establishes local ties in the population. Both aspects are important from a global point of view because migration from rural areas to the cities can be prevented or at least checked. This problem has been a focal point of developing country research for some time now and led to the search for the "suitable" technology to create agricultural and nonagricultural employment opportunities in rural areas. The idea of a "medium-scale" or appropriate technology emerged very quickly, especially after Schumacher's book "Small is Beautiful" (Schumacher, 1973), because of the lack of infrastructure, primarily in connection with energy and water supply, the shortage of capital, the deficiencies in qualifications and the specific natural features of these regions. The island-like structure of the energy supply, water supply as well as sewage and waste disposal systems in thinly settled areas, for example, requires in itself different technical solutions from those common in the densely settled industrialized countries (such as for the construction of treatment plants). Simple or appropriate technologies are required in such cases so that either a downgrading of the technology applied in industrialized countries takes place, an upgrading of traditional techniques used in the developing countries themselves is carried out or new methods are developed - even if this involves only a transitional solution (Theierl, 1989). To take this step, the developing countries must have their own scientific and technological infrastructure. The following criteria should be taken into account here (Theierl, 1989):

- orientation to the need for environmental policy action in the developing countries,
- adaptation to training level of population (reduction of need for care, increasing operator convenience and ruggedness of equipment),
- minimization of dependence on industrialized nations (for example, regarding supply of spare parts, repair, consulting, etc.),
- primary orientation to the use of local resources,
- consideration of the initial sociocultural situation,
- optimum adaptation to the local environmental conditions and
- high degree of flexibility in order to ensure a transition to more complex technologies, if necessary.

In the case of export-oriented production sectors that are usually based in centers, on the other hand, a rapid adaptation to the technologies of the industrialized countries is necessary if a developing country wishes to enter into trade within the global division of labor. The problem of technical dualism in the developing countries arises here. Transfer of know-how is carried out very quickly and efficiently via direct foreign investments, as is shown by the examples of Hong Kong, Singapore and, to a certain extent, Taiwan (Galenson, 1985). Other countries, such as South Korea, preferred the acquisition of licenses, though sectoral differences can be seen (Yoo Soo Hong, 1994). However, both methods require a guarantee of market economy principles, the safeguarding of property rights to know-how, and a lowering of investment risks. In many cases in Asia it was political systems of the bureaucratic, authoritarian type that created such conditions, with the support of a specific

national consciousness which also included Confucian values, and accorded the state the role of a development agency, such as the Economic Planning Board in South Korea.

Thus far there has been a lack of more comprehensive studies on the transfer of environmental technology from the industrialized countries to the developing countries. This is due, in part, to the problems connected with the determination of environmental technology goods (RWI, 1994). The greater the role played by integrated environmental protection, the more difficult it is to characterize a unit or plant as a typical environmental facility. The environmental protection function emerges most clearly in the classic retention technologies, disposal technologies or landfill and sewage treatment techniques. The techniques developed in the industrialized countries can be applied to these fields in the developing countries only to a limited extent, however. Often not only hardware but an entire environmental system (including financing) has to be transferred.

In the case of direct investments made by multinational corporations that are increasingly faced with an environmentally sensitive public, it can be assumed, as a rule, that they transfer entire environmental systems today and adopt or comply with the standards of the industrialized countries. Frequently it is the state-owned enterprises that display shortcomings in environmental technology, as in Brazil, for example. In some cases an adaptation to local conditions takes place in conjunction with direct foreign investments. Otherwise, however, one can assume that implementation of imported environmental protection techniques is extensively the reflection of environmental policy requirements and thus of the environmental awareness in the developing countries themselves. For this reason specialized agencies for analyzing the environmental situation or setting priorities, an enforcement system (administrative agency) as well as a consulting industry are important. The federal government can provide valuable assistance in setting up specialized agencies.

Many organizations (such as the IUCN and, to some extent, the World Bank) are currently attempting to create a local "environmental way of thinking", i.e. to improve environmental perception and environmental knowledge. In particular NGOs achieve a successful international "transfer of convictions" and represent a kind of "awareness industry". The technological competence of local consultants is also growing so that fewer and fewer foreign consulting companies are required for general environmental questions. If there are deficiencies, then they are deficits in know-how regarding conservation. These cannot be eliminated so much via know-how transfer as via financing assistance for conservation research in the developing countries.

REGULATORY REQUIREMENTS

These process policy measures are supplemented by activities for the creation of regulatory requirements, i.e. the establishment of an institutional framework at the global level whose structure enables individuals (enterprises, households, organizations, local authorities) to look for problem-solving approaches independently and, at the same time, provides incentives for using the resulting know-how efficiently. Efficiency orientation here refers to the employment of know-how in cases where the resulting advantages for all those concerned exceeds all related costs. To achieve such an efficiency orientation, the scarcity of the available global environmental functions must be illustrated to those concerned when institutional regulations are worked up. Worldwide acceptance of this objective will be more easily attained, the more clearly the advantages - such as strengthening of the negotiating position at international environmental conferences and obtaining revenues from higher acceptance of the exported goods - can be seen when this framework is specifically defined in the individual states. For the industrialized countries this means dispensing with direct taxation of the performance of know-how transfer activities so as to give individual persons greater freedom of action and development.

The formulation of such regulatory provisions is intended to provide for an efficient exchange of know-how. On the one hand, this applies to knowledge based on natural science and concerning key determinants of the state of the environment as well as findings from the field of instrumentation and control engineering. On the other hand, an assessment of whether a certain state of the environment indicates the need for action or not can only be carried out on the basis of the preferences of those concerned. These preferences diverge worldwide to a great extent, however, and thus cannot be adequately determined on a central basis. Consequently, it is advisable to draw up regulations that motivate individuals to disclose their preferences since this is the only way to ensure successful coordination. As has been emphasized several times, very clearly defined rights of action and disposal regarding environmental functions are suitable for this purpose because individuals negotiate with each other in exercising these rights and, therefore, they have to make their preferences known. In view of the differing conditions in the individual countries, the specific features of the respective natural and human environment must be taken into account in the definition of the rights - for example, the importance of common property traditions.

The complexity which is involved in the application of coordination methods in a global context and which is connected with the heterogeneity of the structural elements illustrates the need for strict limitation to environmental problems that are, in fact, globally relevant, while keeping in mind the principle of fiscal equivalence, i.e. spatial accordance between the beneficiary and cost-bearer of a decision on utilization of environmental resources (Klemmer et al., 1993). The form in which the preferences ascertained on the basis of coordination procedures of individual states are represented - democratic representatives, representatives of corporate forms of organization and the like - depends on the situation and must be determined while taking into account the legal traditions of the individual countries.

Knowledge about the interrelationships between causes and effects is subject to constant expansion and modification due to scientific progress. At present a great concentration on large research institutions and on certain paradigms that are extensively recognized in the respective discipline can be observed in the existing scientific structure of industrialized countries. If one accepts the fact that a wide variety of different approaches can perform a stimulating function in the formation, dissemination and adaptation of know-how, then greater spatial and content-related diversification must be promoted so that competitive elements can also play a role in this context (Karl, 1994). Such approaches might involve increased establishment of research institutions in the developing countries, in which case networking with other institutions would initially have to be ensured by setting up a telecommunications infrastructure. Furthermore, experts should be selected from different fields and represent different interests so that they can be induced to present their respective arguments.

When information on causes and effects of global environmental problems is transferred, incentives are necessary to get individuals to disclose their preferences as extensively as possible. An attempt to record this knowledge via a central agency and to use it as the basis for specifying certain modes of behavior will only result in a situation in which those possessing the information are motivated to make strategic use of their advantage in order to influence decisions on how to react to the findings obtained on causes and effects. For example, they may submit claims for high damage sums as victims or deliberately play down the extent of their contribution to an environmental change or refer to inadequate substitution options as polluters. Checking the credibility of such statements, in turn, induces costs.

Those concerned tend to develop their own interest in substantiating the credibility of their statements if this substantiation is linked to financial advantages. This might apply, for example, to the introduction of liability regulations (Karl, 1992 and 1994). Elements for the implementation of an international liability law already exist and should be expanded. If those concerned do not have to reckon with losses in the event of faulty or unsubstantiable information, they will undertake measures to strengthen and/or justify the basis of their argumentation. If, on the other hand, sales advantages or improved public relations may result for private enterprises, for example, potential polluters could be induced to cooperate in the determination of the causes (Kaas, 1993). The large number of individual parties involved as well as heterogeneous initial conditions give rise to problems regarding the application of such incentive-based instruments in a global context. Therefore, appropriate measures must first be taken in the individual states (*see* Jost, 1993 for an evaluation of the existing law in Germany from a theoretical point of view) whereas greater application of the liability principle between the individual countries should be striven for in a global context (*see also* Erichsen, 1993).

At present there is a wide range of environmental, development, industrial and trade policy measures which are taken to bring about and control transfers of know-how for the avoidance, reduction and elimination of global environmental problems but which also restrict individuals in their freedom to decide on the development of research approaches. It is therefore necessary to reduce the existing restrictions on such exchanges of know-how at the level of individual economies. Moreover, greater efforts should be made to include the consequences of global environmental changes in the decisions on utilization of global environmental resources, such as by implementing liability regulations or with the help of international markets for rights to environmental assets (*see* WBGU, 1993 regarding certificate approaches and Becker-Soest and Wink, forthcoming, regarding markets for global soil use rights). In addition, a binding legal framework should be introduced for the exchange of such know-how via institutional channels. This framework is to provide for property rights to immaterial goods, a reduction in restrictions regarding the transfer of know-how, provisions of local content and the introduction of procedures for competitive protection. The stability of such a legal framework would make it possible for those concerned to make their decisions on a longer-term basis and would thus promote incentives for investment in the production and transfer of know-how, leading to economic advantages only in the long run (see Brennan and Buchanan, 1993 regarding the determinants of long-term orientation of individuals).

Complementary measures also appear necessary, in spite of such general regulations. As a result of the differing ecological, economic and sociocultural conditions of the information sources and target groups, for example, problems arise that require adaptation of the relevant knowledge and of the transfer procedures. This can be done by means of intergovernmental "clearing agencies", via "company networks" (Lahaye and Llerena, 1994) or on the basis of self-managed institutions of nongovernmental organizations (Chambers of Commerce, development organizations or the like). The importance of the appropriate information for implementing exchange of know-how and technology transfer can be seen in the payments made for the services offered by these institutions.

2.4
Recommendations for Research and Action

In general, there is still a lack of comprehensive studies regarding the transfer of know-how and technology, though it is evident that the exchange of know-how is largely determined by market and competitive processes. Greater integration of the developing countries in the international division of labor, therefore, strengthens the exchange of know-how. At the same time multinational corporations and their global investment and location-related decisions play a special role in this connection. Analyses of these decision-making processes are still extremely inadequate and should be pushed forward.

Special attention should be devoted to the patent system and/or to the allocation of rights of disposal of intellectual property. This applies in particular to property rights to genetic engineering products and to genetic material from wild animals and plants living and growing in the wild. In addition, a closer look has to be taken at the question of whether patenting of plant and animal variants should be fundamentally possible and who the beneficiary of product innovations should be, i.e., either the countries which provide the initial genetic material or the companies that carry out the manipulation.

The extent to which this approach can be used to improve technology and capital transfer should be examined in the light of the *Berlin Climate Conference*, which led to an upgrading of the concept of Joint Implementation. In this way projects involving a high technology transfer could receive a special status in the framework agreements.

The question as to which institutions can contribute to a departure from determination of the exchange of know-how according to market factors remains unanswered. This applies in particular to the international organizations involved in basic research, which still concentrate on the highly developed industrialized nations and their problems.

Many organizations (such as the IUCN and, to some extent, the World Bank) are currently attempting to create a local "environmental way of thinking", i.e. to improve environmental perception and environmental knowledge. In particular NGOs achieve a successful international "transfer of convictions". The technological competence of local consultants is also growing so that fewer and fewer foreign consulting companies are required for general environmental questions. If there are deficiencies, then they are deficits in know-how regarding conservation. These cannot be eliminated so much via know-how transfer from the industrialized nations to the developing countries as via financing assistance for conservation research in the developing countries. This support should be increased.

It has been shown that strategic action may cause distortions of information in international exchange of know-how. These distortions can be countered through a more stringent liability law. However, there is still a significant need for research in this connection, particularly regarding jurisprudence.

It was also pointed out that the exchange of know-how is plagued with substantial adaptation problems. Research in this field unfortunately displays numerous shortcomings that have yet to be eliminated.

3 Institutions and Organizations

3.1
Problems Related to Institutional Innovation of Global Environmental Policy

Since the first UN Environment Conference in Stockholm in 1972, more than 60 multilateral environmental agreements have been signed and ratified – from protection of the stratospheric ozone layer and regional seas to monitoring of trade with hazardous wastes (UNEP, 1991; UN, 1991, 1992 and 1993). Despite these many years of experience, there is still little knowledge as to whether and how such institutional arrangements contribute to an effective global environmental policy. This lack of knowledge is primarily due to the fact that the social and behavioral sciences have underestimated the institutional dimensions of international and global environmental policy. Emphasis has been put on policy contents, with the policy process itself being rather neglected. Little attention has been given to the changing institutional framework governing this process. This section will therefore focus in greater detail on institutional innovations of global environmental policy. The Council views this section as a first step towards a systematic and continuous examination of issues related to global environmental policy, an analysis that shall be continued in subsequent annual reports (*see also Sections C 3 and C 7*).

Social science generally uses the term "institutions" to describe rules that influence both public and private action, and which thus perform a frameworking and controlling function (Caldwell, 1990; Höll, 1994; Kay and Jacobsen, 1983). The term institutions refers both to organizations as bureaucratic structures and to control systems (or regimes) (*for details see Box 16*). This section focuses on institutions that have been or could be agreed on by countries as a means of addressing global environmental problems.

The various functions of institutions may be classified as follows (Prittwitz, 1994a):

1 Direct control aims at steering the target groups in a certain direction through "command and control" mechanisms (like bans and rules, require-ments, limits) as well as through economic and informational incentives and sanctions specifically geared to the aims in question.

2 *Indirect control* influences target groups without specifying their behavior in detail. Process, resource and organizational control are characteristic features of this type.

3 Direct and indirect control are generated, as *vertical control* of certain target groups, through superordinate control, but may also take various forms of *horizontal self-coordination*. In this case a conscious act of self-control on the part of political agencies is engendered, with a substantial degree of congruence between the controlling agency and the target groups themselves.

The lack of superordinate control levels (such as a "world government") means that global environmental policy institutions primarily exist within the regulatory or organizational patterns of horizontal, national self-coordination (type 3), though instruments of direct (type 1) and indirect (type 2) control are utilized.

The topic of this chapter is the extent to which institutional innovations in global environmental policy can address global environmental problems. *Section B 3.2* outlines the development of global, environmentally relevant institutions, prior to their assessment and the derivation of recommendations for policy action and research in *Section B 3.3*.

3.2
Institutionalization of Global Environmental Policy

3.2.1
The Formation of Global Environmental Policy

In the course of the early 1970s, a global perspective on environmental problems developed. The reports to the Club of Rome were generally based upon the "One World" concept. At first this global di-

Institutions are defined as persistent, inter-linked rules and practices that prescribe behavior, demarcate activities and shape expectations (Haas et al., 1993). They are thus a designation for internalized behavioral patterns which have a regulative function (Krasner, 1983), are quasi fixed, and influence both public and private beha-viour. In the social science debate on international regimes (Haas et al., 1993; Rittberger, 1993) and institutional economies (March and Olsen, 1984; Shepsle, 1992 and 1989), institutions are under-stood as goal-oriented arrangements. They have a framework-setting effect, though they themselves can be rearranged, redesigned and even dissolved (Scharpf, 1991). Institutions can take on the form of *organizations* (bureaucratic structures) and of *regimes* (regulatory mechanisms that do not nec-essarily have an organizational foundation); the latter may explicitly have a contractual basis or be implicitly based on mutual agreement.

Regulatory mechanisms and organizations may be either linked to each other or imple-mented independently of one another. Far-reach-ing agreements have been made concerning "stratospheric ozone" and "acid rain", for exam-ple, without substantial organizational substruc-ture, while several organizations operating with-out standardized rules were created in connection with "population".

---

mension in the public debate had few or no conse-quences for environmental policy. Although several international environmental regimes were set up, particularly in the field of species protection and pro-tection of the seas, the United Nations Environmen-tal Programme (UNEP) was established in 1973 and environmental protection activities were initiated by other international organizations (Kilian, 1987), the focus of environmental policy in the following years was almost entirely restricted to national or trans-boundary environmental issues. Even in the second half of the 1970s, the OECD published several re-ports in which international environmental policy was essentially confined to the transboundary prob-lems of adjacent countries (OECD, 1974, 1976, 1977 and 1979).

After this phase of domestic environmental pol-icy, a new start was made toward international envi-ronmental policy at the beginning of the 1980s. In contrast to the beginning of the 1970s, when environ-mental policy expectations mainly focused on inter-national organizations (such as the European Coun-cil and the ECE), this development was based on the active commitment of a larger number of nation sta-tes. Environmental policy was gradually becoming a component of national foreign policy or intergovern-mental policy (environmental foreign policy). It was now influenced to a greater extent by political inter-ests, but at the same time the prospects for practical implementation were growing (Prittwitz, 1984). A further phase of internationalization of policy came about during the mid-1980s. This was due to a large extent to the development of a common environ-mental policy by the European Community (Höll, 1994). On the other hand, the global dimension of en-vironmental policy was reinstituted, although now underpinned by more or less effective national insti-tutions and sound approaches to transnational envi-ronmental policy (Sand, 1994a; Young, 1991).

Milestones in this internationalization of environ-mental policy included the formulation and imple-mentation of a regime for the protection of the stratospheric ozone layer (1985 Vienna Convention and 1987 Montreal Protocol), the report of the World Commission on Environment and Development (1987), and a revival of debate over the need for glo-bal climate protection (Grubb, 1990; Sebenius, 1991; Sessions, 1992). This development culminated in June 1992 with UNCED, the "Earth Summit" in Rio de Ja-neiro, in the course of which framework conventions on climate and biodiversity were adopted, as well as the Rio Declaration, the principles on the use of fo-rests and AGENDA 21.

3.2.2
Approaches to the Innovation of Global Institutions

The current framework of *international environ-mental policy* still restricts in many ways the options for effective global action. In accordance with the ba-sic principle of national sovereignty, environmental policy depends on the approval of national represen-tatives in each individual case. Accordingly, the deci-sion-making process takes place within specific nego-tiation systems (for the theory *see* Benz et al., 1992). Without the "shadow of hierarchy" (Scharpf, 1992) that promotes decisions in *national* environmental policy, international environmental policy requires voluntary agreement in order to develop problem-related control mechanisms (*international regimes*). Decision making is thus shaped by the differing interests of individual states and is often a difficult and lengthy process (Sand, 1990; Sebenius, 1992). Often problems are addressed too late and the specific provisions agreed upon remain weak (Susskind, 1994).

Also, the actual implementation of action programs is difficult to survey and in most cases can only be monitored on the basis of national reports by the member states (Strübel, 1991). If violations against such arrangements are actually identified, compliance may only be enforced under very specific circumstances (Chayes and Chayes, 1993).

In addition, there is a significant need for coordination due to the variety of interaction between the different environmental policy areas. On the one hand, duplication of work on environmental problems in different bodies sometimes means a waste of resources; on the other hand, agreements in one field may ultimately block agreement in another. For example, supply-side priorities operated by the Multilateral Development Banks in the energy sector may counteract efforts to develop a (demand-oriented) climate policy. The Convention on Climate Change and the Convention on Biological Diversity are not necessarily compatible, either (WBGU, 1995). As the number of international environmental agreements has grown over the past 25 years, so, too, has the need for coordination. UNEP, which inter alia has been assigned the task of coordinating the environmental activities of international institutions, has not been in a position to perform this task adequately to date (Kilian, 1987).

However, a variety of institutional innovations have been initiated since the mid-1980s, as environmental policy has become increasingly internationalized. These involve the establishment of institutions for finance and technology transfer from North to South, as a form of *direct control* (*Section 3.2.2.1*), as well as changes with respect to process, resource and organizational control, as forms of *indirect control* (*Section 3.2.2.2*).

3.2.2.1
Transfers of Technology and Financial Resources

New institutions in the form of funds have been set up to coordinate transfers of finance and technology. Article 5 (4) of the Montreal Protocol and Article 4 (7) of the Convention on Climate Change, for example, explicitly link environmental commitments on the part of developing countries to legally binding transfers on the part of the industrialized countries.

These developments represent an attempt at direct control since transfers are largely dependent on the implementation of environmental measures. At the same time they stand for two different approaches to transfer payments. The Multilateral Fund set up within the framework of the Montreal Protocol in 1990 exclusively provides for transfers related to the protection of stratospheric ozone and thus represents a problem-specific way of organizing compensation payments (Wood, 1993; Round, 1992). The GEF, by contrast, is a central institution not limited to one specific problem area, but was set up to finance measures for protecting the ozone layer, international waters and biological diversity, and for combating the greenhouse effect (WBGU, 1994). Whereas the advantages of joint action were obvious in the case of the ozone fund, the volume of transfer funding administered by the GEF has remained marginal compared to the level that is really needed to finance global environmental policies (Helland-Hanse, 1991; Reed, 1993).

Economic incentives for implementing environmental measures have not been confined to such funds, however. Since the second half of the 1980s, so-called debt-for-nature swaps have become an important instrument of financing environmental projects in developing countries (by the WWF, for example). Within these debt-for-nature swaps, foreign debts of developing countries are bought on the market for executory titles, in most cases far below their nominal value. If such executory titles are transferred in local currency within the scope of an agreement with the debtor country, the funds acquired can be invested in specific local environmental projects, while the respective developing country does not have to expend foreign currency to repay its foreign debts. Usually nongovernmental agencies purchase these debts. Over 20 such transfers with a total volume of more than 100 million US dollars have been carried out since 1987 (UN, 1994a). Although this is still only a marginal sum, this approach should be pursued further.

3.2.2.2
Process, Resource and Organizational Control

There has been a certain change in the operation of traditional environmental policy, as seen in the fact that forms of indirect process, resource and organizational control are being increasingly favored.

CAPACITY BUILDING
In this respect capacity building plays a central role. Every international environmental program and project now contains a capacity-building component. This programmatic reorientation (Kimball,

1992) is based on the realization that inadequate implementation of environmental policy is often due to the lack of the respective nation's ability to take action (Young, 1992; Rosenau and Czempiel, 1992). There is a specific chapter on this topic in AGENDA 21 (*Chapter 37*).

Capacity building refers primarily to generating and strengthening personnel capacities and administrative structures in developing countries. Measures include developing national administration, education and training, financing research for determining causes and potential environmental damages, and improving the infrastructure for the exchange of information provided by international institutions (Haas et al., 1993). Whether this programmatic reorientation to capacity building effectively leads to innovation or whether it is just a redeclaration of traditionally implemented projects, however, requires careful analysis in each case.

This focus on capacity building is also reflected in the activities of the newly-created institutions for finance and technology transfer. The Multilateral Fund of the Montreal Protocol has made institutional strengthening of recipient countries a major effort by supporting research institutes and national administration units. The direct aim here is not to reduce the production and use of CFCs directly, but to provide the institutional prerequisites for achieving this objective (UNEP, 1993).

A similar increase in activities can be noted in connection with "swaps", where NGOs are involved substantially. In accordance with the concept of sustainable development, "debt-for-nature swaps" have been expanded to "debt-for-sustainable-development swaps" in recent years. Consequently, not only environmental protection in the debtor country is being financed via the purchase of executory titles, but also other projects aimed at promoting sustainable development. Swaps have financed training programs and health projects, for example (UN, 1994a). Nongovernmental agencies no longer address purely local or national environmental problems, but, more comprehensively, the question of sustainable global development and the framework conditions of global environmental problems (GERMANWATCH is an example here).

Flexibilization of Institutional Regulations

Another field of global environmental policy in which institutional innovations have come about can be summarized as flexibilization (Young, 1989). This refers both to the functional capacity of existing institutions and to the decision-making and voting procedures for national implementation of international programs.

In particular, a functional *differentiation of communication* can be observed in international environmental regimes. Panels of experts or subcommittees debate on scientific, technical, economic, financial and implementation issues to an increasing extent, separately from the political negotiation process. Within the framework of the Montreal Protocol, for example, working groups have been set up to assess the state of knowledge, and committees have been established to manage the Multilateral Fund and to deal with the difficulties involved in national implementation (Gehring and Oberthür, 1993).

This differentiation of the communication process has proved to be conducive to the reconciliation of interests and to decision making, thus enhancing the effectiveness of environmental regimes. Political negotiations themselves are relieved of certain tasks, politically uncontentious issues can be clarified more quickly and the willingness to reach consensus on the part of individual countries grows. Whereas the highest-level political negotiating body still keeps the right to make the final decision, consensus formation and hence "joint learning" are promoted by committees and working groups. In the case of the Montreal Protocol, the rapid and successful reconciliation of interests can be attributed to such advisory bodies (Benedick, 1992; Haas, 1992; Parson, 1993). These discursive elements have also created room for debate and thus for coming to a consensus within the Commission for Sustainable Development (CSD, UN, 1994a). Normally, this process permits a representative limitation in the number of participants within international institutions that often facilitates the communication process and decision-making (see Haas, 1990 on the role of *epistemic communities*). The Implementation Committee of the Montreal Protocol, for instance, has only 10 members, the committee managing the Multilateral Fund 14, while the executive body of the restructured GEF has 32 members (WBGU, 1994). The optimal level varies, of course, from case to case and from institution to institution. In the view of the Council, there is need for more research on these questions.

The flexibilization of global environmental policy also applies to *decision-making and implementation*. The tradition that international resolutions have to be adopted unanimously often leads to a situation where only the lowest common denominator with respect to specific measures can be decided on. At the Climate Conference in Berlin it became evident that individual Parties or small interest groups gain a minority status allowing them to block efforts to achieve unanimity. Such a mechanism can make decision-making a very difficult and slow process. A deadline, at least, should therefore be imposed on adjourning vetoes.

Improved voting procedures have now been instituted, however. An authorized majority may add stricter measures to agreements that, in some cases, do not require ratification, and all parties to the agreement are bound by these provisions (*binding majority decision*). The Montreal Protocol, for example, calls for a *qualified two-thirds majority*, and a *double weighted majority* has been introduced for the GEF (WBGU, 1994). In the GEF case, a 60% majority of votes, representing at the same time 60% of the membership fees, is required for a decision by the executive body, which consists of equal representation from developing and industrialized countries. Thus the main donor countries and all developing countries as a group can exercise a veto, whereas individual states by themselves are unable to block collective decisions. Even if such decision-making rules are not, in fact, applied, they discipline the negotiating parties since resistance against broadly supported resolutions would be pointless.

Linked to the recognition of such decision-making rules is a partial surrender of national sovereignty, in that noncompliance with collective decisions is rendered virtually impossible (Soroos, 1986). The directly binding effect of such decisions can be weakened in other cases in the form of an "opting-out" procedure (Sand, 1990), where opposing states have to declare explicitly that they will not comply with the decision so as not to be formally bound to it.

Besides innovations within agreements binding under international law, various forms of so-called "*soft law*" have been used in past years. These legally nonbinding regulations are another way of shortening the, often, long period between adoption of international resolutions and national implementation (reaction times) (Weiss, 1992; Chayes et al., 1992). The provisional establishment of the Multilateral Fund of the Montreal Protocol, for example, was based on a resolution that does not fall under formal international law (Ott, 1991). Resolutions like those are regularly adopted at the annual meeting of the Parties to the Montreal Protocol (UNEP, 1993).

INVOLVEMENT OF NONGOVERNMENTAL AGENCIES
Innovations in process or organizational control have also resulted through the *enlarged participation of nongovernmental agencies*, especially environmental NGOs. Gradually these agencies have implicitly or explicitly received greater rights of participation in various fields, thus contributing to the partial elimination of traditional nontransparency of intergovernmental policy. In most international environmental institutions NGOs now have the right to observe and to obtain information on the negotiations. More far-reaching rights of participation have been implemented in the CSD, in which NGOs have the

right to speak not only at official meetings, but also at the discussions of the informal working groups (Martens, 1993). The question of optimum participation rights of nongovernmental agencies at such negotiations is still open, however, or must be answered differently from case to case.

Trends towards extended participation for NGOs can also be noted in the *implementation and evaluation of global environmental policy*. This has happened implicitly within the framework of the Montreal Protocol since 1992. Here, nongovernmental agencies can contribute information on implementation problems of individual countries within the adopted *noncompliance procedure* by notifying the Secretariat. Within the CSD the NGOs are allowed to include their own reports in the work of the Commission and thus identify shortcomings or new possibilities. Little use has been made of this participation option to date, however, a situation that indicates the often limited resources of the NGOs, especially in developing countries.

COORDINATION OF ENVIRONMENTAL ACTIVITIES
The establishment of the CSD represents a major global environmental policy innovation of the indirect control type. It is subordinate to the *Economic and Social Council of the United Nations* (ECOSOC), is supposed to monitor the implementation of AGENDA 21 and other global environmental agreements, and ensure coordination of international activities. In addition, the High Level Advisory Board was founded, an advisory body to the CSD which can draw attention to new environmental threats (UN, 1994c).

A new committee has been established to improve the coordination of the environmental activities of the various UN organizations, for which purpose a separate *Department for Environment and Development* has been set up in the course of restructuring the UN Secretariat. Almost all international organizations are now carrying out internal audits and restructuring with the aim of integrating aspects of sustainable development in all relevant activities.

Besides these new intergovernmental institutions, innovations in the organization of the nongovernmental sector have led to greater concentration and effectiveness of activities. These innovations include the *Business Council for Sustainable Development*, consisting of representatives from environmentally sensitive business organizations (Schmidheiny, 1992), as well as the formation of the Planet Earth Council, a group of environmentally committed personalities (Haas et al., 1992), the *European Business Council for a Sustainable Energy Future*, the *International Green Cross* and the *International Council for Local Environmental Initiatives* (ICLEI).

3.2.3
Reforming Global Environmental Institutions

The following proposals relate, firstly, to basic restructuring of the institutional network of global environmental policy, and, secondly, to practical innovative approaches for extending and reforming existing institutions. Thirdly, it is proposed that the present institutional network be modified and expanded.

3.2.3.1
Basic Restructuring

Proposals for basic restructuring of the existing network of institutions for global environmental policy experienced a boom prior to the 1992 "World Summit". For example, consideration was given to granting the *United Nations Environment Programme* (UNEP) the status of a special organization and extending its mandate, as well as to forming a special organization by combining UNEP and UNDP (UN, 1991). These ideas have not been definitively put aside. Another proposal is to establish a *"Global Environmental Council"* or a *"UN Environmental Trusteeship Council"*, with a status equivalent to that of the UN Security Council.

UN ENVIRONMENTAL TRUSTEESHIP COUNCIL/
GLOBAL ENVIRONMENT AGENCY
Under Article 87 of the UN Charter, the Trusteeship Council, operating under the authority of the General Assembly, supervises the trust territories with the aim of promoting their development towards self-governance and independence. Since the end of 1975 it consisted entirely of permanent members of the Security Council. On November 1, 1994 its work was discontinued when the last trust territory, Palau-Palau, became independent, but it has not been dissolved. Consideration is therefore being given to either extending the responsibilities of the UN Security Council to embrace "environmental security" or converting the UN Trusteeship Council into an "Environmental Trusteeship Council" (Dolzer, 1992).

A similar proposal calls for establishment of a *Global Environmental Organization* (GEO), modeled after GATT/WTO as the most important world trade institution, providing a similar forum for the formulation and implementation of international environmental policy (Esty, 1993 and 1994). The basic idea behind GEO is that it should not only bring together the existing international environmental regimes directed at specific problem areas, but also become the central institution for finance and technol-

ogy transfer in support of sustainable development in the developing countries. The fact that the EU has established a *European Environment Agency* indicates that a GEO is by no means a utopian idea. Institutionalizing supranational environmental policy is a feasible option, at least in comparable cases of regional integration – e.g. in ASEAN; regional environmental policy cooperation has already been strengthened within NAFTA through the *North American Commission on the Environment* (NACE).

AN INTER-GENERATIONAL ENVIRONMENTAL
CONTRACT
Existing as well as new institutions can help solve environmental problems, but in themselves are not sufficient to secure environmental quality for future generations. Sustainable development, by definition, is an intergenerational undertaking. Although economic concepts such as cost internalization and discount rates also take the interests of future generations into consideration, long-term costs are not fully evaluated and included, and the implications of irreversible damages are neglected as well. Also international law provides certain norms for intergenerational equity, but to date there are no appropriate instruments available which could apply the principle of intergenerational equity in practice (Weiss, 1992; Hurrel and Kingsbury, 1991; Gehring, 1991; Ostrom, 1990; Oye, 1986). Therefore, a number of new and innovative proposals have been presented in the literature and in political discussions such as the "Treaty of Intergenerational Equity" (Weiss, 1993), "Planetary legacy" and "Fiduciary trust" (Weiss, 1989), "Ecological Rights of Future Generations" (Cousteau-Society, 1991), "Second Chamber in Addition to Parliament" (von Hayek, 1969) and the "Chamber of the Generations" (von Lersner, 1994). These are very different concepts but they all have a common concern – that new institutional arrangements are essential to ensure intergenerational equity.

However, action and research is needed to define and stipulate the rights of and obligations toward future generations. As a consequence, every policy should be examined as to whether and to what extent it complies with the principle of intergenerational equity. The formulation and implementation of a new ethics embracing fairness between generations was a special concern of the World Commission an Environment and Development and is a recurrent theme in AGENDA 21.

3.2.3.2
Extension and Reform

Innovative approaches relating to communication and decision-making processes at international level basically rely on elements that have proven to be either beneficial or obstructive to the reconciliation of interests. Of particular importance in this context are the voting and coordination procedures by means of which such reconciliation of interests can be achieved.

MAJORITY DECISIONS IN PLACE OF UNANIMITY

Often reference is made to the unwieldiness of the procedure based on *unanimous decisions* (Bächler et al., 1993), as opposed to the positive effects of majority decisions. The Council's 1994 report discussed the new arrangements governing the GEF (WBGU, 1994). Consideration has been given in other contexts (Jäger and Loske, 1994) to the adoption of binding *majority decisions* under the Montreal Protocol – an issue that was repeatedly iterated throughout the Climate Conference in Berlin. At the same time, however, when considering the demand for *majority decisions* one must keep in mind the possibility of group veto rights on the part of the industrialized or developing countries, as in the case of the Montreal Protocol. The question is as to how the minority can be brought to accept the agreement. Trade restrictions (as provided for under Article 4 of the Montreal Protocol), embargos or the involvement of the UN Security Council because of a threat to "international security" may also determine the future of environmental policy. International law has already started to treat large-scale pollution as an "international crime" prohibited by binding international law (*ius cogens*). In 1992, as a precautionary step, the UN Security Council declared itself responsible for environmental matters, where these might involve a threat to security.

ORGANIZATION OF CONSENSUAL KNOWLEDGE

In order to make global environmental institutions more dynamic, stringent objectives, deadlines and measures could become immediately binding for all Parties, thus circumventing the long process of national ratification. *Consensus-building communication processes* (Benedick, 1992) and *organization of consensual knowledge* (IAE, 1995) which make joint learning and searching for solutions possible, are underlined as important elements for facilitating and accelerating international decision-making (*see also* Heinze and Kaiser, 1994 on the "ecology dialogue"). Members of the CSD, for example, are seeking further extension of the discursive elements that have

already been introduced in the form of podium discussions (CSD, 1994).

PARTICIPATION RIGHTS

Further opening of international institutions to NGOs has been repeatedly called for, given that many international organizations have not yet granted participation rights similar to those within the CSD. Environmental organizations, for instance, are striving to obtain greater *participation and information rights* within the new WTO (Cameron et al., 1994). These efforts are in concord with the principle of transparent international institutions specified in AGENDA 21 (Chapter 38). Barriers to extended participation of these agencies, however, are not restricted to procedural rules; at the Climate Convention negotiations, for example, complaints were raised about the constraints on participation of developing country NGOs due to lack of funding.

A central issue in the debate on institutional innovations is the actual implementation of global environmental policy. Increasing reference is made here to the option of applying *"soft" instruments of international law* in order to initiate implementation prior to generally accepted rules or even without them being translated into *"hard"* international law (Sand, 1990; Chayes et al., 1992). In view of the difficulties in obtaining rapid agreement on a formal climate protocol, there have been efforts to bring the international climate convention a step forward, based on "soft" but politically binding international law (Jäger and Loske, 1994; Oberthür, 1994; Ott, 1994).

ECONOMIC INSTRUMENTS

New forms of implementation centering on greater use of *market forces* for the protection of the global commons need to be addressed here as well. Economic instruments for implementing global environmental policy, such as *taxes* and *charges* for utilization or consumption of environmental resources and internationally tradeable permits (entitlements) are currently under discussion (Ewringmann and Schafhausen, 1985; Bonus, 1991). The latter involves establishing a market where none exists so far (Simonis, 1994). The experience with such instruments at the national level has been documented (Howe, 1994; UN, 1994a; Hansjürgens and Fromm, 1994). Under certain conditions such market-oriented mechanisms represent a way of effecting a positive North-South transfer of financial resources beyond centrally or intergovernmentally organized payments.

The intensive (albeit controversial) discussion over the *Joint Implementation* of climate policy measures as provided for under Article 3 (3) of the Climate Convention is also an expression of the rethinking process regarding the institutional framework for implementing environmental policy at the international level (for more details *see* WBGU, 1994). While the industrialized countries tend to favor this mechanism, most of the developing countries have resisted thus far, though with varying intensity (Loske and Oberthür, 1994; Oberthür, 1994). The first Conference of the Parties to the Climate Convention achieved a modest breakthrough by agreeing on a pilot phase. The main question now is not whether, but how *Joint Implementation* shall be used in the future.

In addition to *tradeable permits* and *Joint Implementation*, *environmental taxes* and *charges* also feature largely in the international debate over innovative instruments. At its second meeting in 1994, the CSD also discussed proposals on the introduction of internationally coordinated taxes and charges (UN, 1994d). With regard to climate policy, the primary issue to date, proposals have been put forward for the introduction of an international, OECD-wide or at least Europe-wide CO_2 *charge/energy tax* as well as a *nitrogen tax* (Jäger and Loske, 1994; UN, 1994a). A tax on international flight tickets or aircraft fuel (kerosene) has also been urged, in accordance with the 1980 Brandt report, in order to mobilize additional financial resources for global environmental policy (UN, 1994a). This would undoubtedly be a way of mobilizing substantial funds for environmental protection.

Environmental Funds

There is also discussion about the extension and reform of the existing institutions involved in the transfer of finance and technology to low-capacity developing countries. This refers, firstly, to greater use of national *environmental funds or debt-for-sustainable-development swaps* (UN, 1994a; Klinger, 1994). Secondly, several OECD member states have put forward the idea of transferring the Multilateral Fund of the Montreal Protocol to the GEF (Rowlands, 1993). In addition, the introduction of finance and technology transfer aimed at specific problem areas has been urged in other fields of global environmental policy within the GEF – as anticipated by the Council in connection with the Desertification Convention (WBGU, 1994). *See also Section C 3* regarding the setting up of a Blue Fund for protection of the seas.

Sustainable Development as a Cross-sectoral Task

A cross-sectoral dimension of the debate over institutional innovations going beyond the above proposals concerns the integration of the concept of sustainable development into the activities of all existing international organizations. The current focus here is on trade and finance institutions, with the *en-*

vironmental reform of GATT/WTO being the key agenda topic (Helm, 1995). Proposals for reforming GATT/WTO include the integration of principles specifying minimum environmental standards for products as well as for production processes. This discussion is closely connected to the question of finance and technology transfers (*see Section B 2*), in that the exchange of goods and services via the world market is of far greater importance for environment-related transfers of resources than any agreed or planned intergovernmental transfers, including the GEF (Eglin, 1994).

Development Assistance

The integrating of the *concept of sustainable development* and consideration of global environmental problems raises a problem in connection with *multilateral and bilateral development assistance*. Projects organized by multilateral development banks, for example, may make implementation of global environmental conventions more difficult. In response to this dilemma, environmental organizations have proposed that the executive bodies of the Climate, Biological Diversity and Desertification Conventions should draw up criteria for how to utilize development assistance funds (Eco, 1994).

The ideas on how the Commission on Sustainable Development should continue its work in the future are also aimed at greater integration or coordination of other policy fields with environmental policy (AGENDA 21: Chapter 38). The Council therefore welcomes the efforts of Germany's Federal Ministry of the Environment to involve not only the environment ministers of national governments, but also the ministers for other domains in the work of the CSD so as to promote the dialogue on sustainable development beyond environmental and development policy boundaries.

3.2.3.3
Modification and Expansion

Environmental Audit for Individual States

Recommendations of various kinds have been made not only for extending and reforming existing institutions for global environmental policy, but also for *expanding the network of such institutions*. Following the experience with public audits of services and subsequent assessment of national behavior within the framework of the *International Labor Organization* (ILO), there has been an upsurge of interest in applying this procedure to environmental issues and producing *"environmental audits"* for the individual states (Sand, 1990) (*Box 17*). This idea originated in the field of business management (Sietz and Saldern, 1993), where official environmental audits serve as a special incentive for enterprises to meet or indeed exceed given environmental requirements. After an assessment of environmentally relevant performance by external auditors, the participating enterprises receive a certificate that they can then use for public relations purposes. In the course of such audits, previously undetected environmental protection potential is frequently identified and can be made use of at no additional cost; in fact, they may even reveal economic savings potentials, thus representing genuine *"win-win situations"*.

Monitoring environmental policy, programs and management systems of states and international organizations has been urged for reasons similar to those for private sector environmental audits (Kimball, 1992) (*see Section B 3.3.2* regarding structure and assessment of this approach). The OECD "Environmental Performance Review" Project, launched in 1991, is an important precursor in this respect. Such reviews involve a comprehensive assessment of the environmental policy of all (25) OECD member states using a standardized procedure, although the sequence of states surveyed is irregular (*see* OECD, 1993).

Since there is no legitimized authority which could enforce compliance with commitments at the international level, it is important to ensure that compliant and noncompliant behavior is detected, i.e. that the behavior of the agencies is kept *transparent* (Young, 1979; Göhler, 1987; Mitchell, 1994). The CSD has meanwhile been assigned the function of carrying out public monitoring of environmental policies, programs and management systems on the part of states and international organizations. However, it is still unclear whether this body can guarantee an independent audit and whether its work will concentrate on comprehensive assessment of national environmental policies using specific criteria and the environmental audit technique.

ESTABLISHMENT OF AN INTERNATIONAL COURT FOR THE ENVIRONMENT

The proposal for establishing an *International Court for the Environment* also focuses on expanding the existing system of institutions (Rest, 1994) in order to improve the management of environmental conflicts and create an additional instrument for outlawing environmental crimes. States that violate international environmental agreements would then have to reckon with court proceedings and, in the extreme case, with public condemnation. In contrast to environmental audits, which are aimed at giving positive incentives to the relevant agencies, the effect hoped for with such a court is to have tougher sanctions against environmentally harmful behavior (International Court for the Environment, 1994). However, this court would have to be endowed with jurisdictional authority vis-à-vis nation states.

A milder version of this proposal calls for activation of the International Court of Justice in Den Hague in matters concerning environmental law (*Chamber of Environment*). The states would then be subject to its jurisdiction within the framework of the respective environmental agreements, as only Finland, Sweden, Norway and the Netherlands have done up to now with regard to the ozone regime.

On the basis of these considerations, there is need for action on the part of the German federal government, which has not yet come forward with specific concepts on mandatory international environmental jurisdiction.

3.3
Assessment of Global Environmental Institutions: Recommendations for Action and Research

The developments and proposals outlined in *Section B 3.2* give rise to a multitude of ideas for future institutional innovations in global environmental policy. In the sections that follow, some of these proposed innovations with respect to *capacity building, environmental audits and flexibilization* will be subjected to further analysis, from which appropriate recommendations for action and research are then derived. Other proposals will have to be evaluated in future Council reports.

3.3.1
Capacity Building

As mentioned above (*see Section B 3.2.2.2*), the programmatic concept of capacity building is a response to the fact that implementation of environmental policy measures often fails because governments lack the relevant capacities. Capacity building primarily means the building up of personnel capacities and administrative structures, especially in the developing countries; in general, however, capacity building is aimed at further development of the "human, scientific, technological, organizational and financial potential" of all states (AGENDA 21: Chapter 37.1). A closer look shows that this concept requires further discussion.

3.3.1.1
Qualified Capacity Building

Capacity building is a linkage between direct and indirect control: the primary aim is to enable certain target groups (particularly, but not only, the developing countries) to implement international regulations. Strengthening of these target groups, however, requires a certain degree of autonomy and thus independence. The result is a potential dilemma: in keeping with global environmental objectives, it would be reasonable, from the point of view of the donors, to tie the funds for capacity building (human resources, organizations, programs) strictly to the specific purpose. If agencies who have not displayed environmentally sound behavior are strengthened in their autonomy, there is the danger that allocated funds will be used indifferently or even contrary to the desired environmental objectives of the donors. If "capacity building" of recipient countries involves an environmentally harmful expansion of production, consumption and infrastructure, this would even increase the risk of a global environmental collapse.

On the other hand, it is politically unfeasible to tie all funds, because a certain independence on the part of the recipient countries is needed if environmental targets are to be achieved by the countries themselves. Instead of resource transfer tied to certain conditions (*"green conditionality"*), reinforcing capacity (or deepening capacity building) can prove more effective. This could extend to improved terms of trade in favor of the developing countries. The capacity building concept, as called for in AGENDA 21, in fact addresses the wider issue of the future development of economic and political structures.

3.3.1.2
Responding to Differing Capacities for Action

Given their respective economic situation, it is obvious that poorer countries cannot implement costly environmental policies and that, accordingly, environmental policy progress in these countries can only be stimulated through external financial compensation. The common but differentiated responsibility for international environmental policy (as demanded

---

BOX 17

"Environmental Audit" at EU level

Council Regulation (EEC) No. 1836/93 "concerning the voluntary participation in a Community system for environmental management and audits" (Environmental Audit or EMAS Regulation, Official Journal 1993) provides a uniform framework for external auditing of environmental management systems. It is analogous to industrial auditing in that it is aimed at external validation of internal organizational procedures and is based on the structure of the Quality Assurance Standard ISO 9001, and it went into effect on July 13, 1993. Commercial and industrial enterprises in the European Union can take part in the system voluntarily. The scheme can also be extended to the service sector. It is the government's task to set up a system for authorizing and controlling independent environmental auditors. The relevant groups involved in Germany have reached agreement on a model that serves as the basis for the legislative process launched and its implementation.

The aim of the Directive is to constantly improve environmental protection through specification and implementation of location-related *en-*

vironmental policy, programs and management systems in the individual enterprises and by encouraging self-monitoring. This is to be achieved by means of a systematic, regular assessment of the extent to which the measures taken conform with the environmental declaration in the scheme, and by providing the public with information on industrial environmental protection. Granting of the declaration of participation requires that the rules, conditions and work flows are carried out in accordance with the EMAS Directive. Before a location can be registered, the enterprise must specify an industrial environmental policy which is formulated according to defined criteria and which at least provides for compliance with all relevant environmental regulations and also contains an obligation to ensure "appropriate, continuous improvement of environmental protection". The environmental impacts of an enterprise have to be reduced to the extent possible with an "economically feasible application of the best available technology".

The certified environmental declaration (*audit*) is transmitted to the responsible agency of the Member State and published after entry in the list of enterprises certified in accordance with the Directive, possibly supplemented by a *seal of approval* (Heuvels, 1993).

by the Climate Convention) consists of an environmental responsibility of the polluter (on the part of both donor and recipient countries) and, at the same time, an economic responsibility (of the donor countries) for existing capacity deficits (of the recipient countries).

Although this interpretation of the "imperative of responsibility" (Jonas, 1984) is of great importance in political practice and can be regarded as the *conditio sine qua non* for the success of global environmental policy, a closer analysis shows that it requires some differentiation (Jänicke, 1993). Remarkable differences in environmental (policy) behavior exist both in countries with large economic, technical and administrative capacities (see, for example, the climate policy of the USA and Denmark) and in countries with low such capacities (Malaysia and Brazil, for instance).

It is evident, therefore, that basing the assessment of environmental policy capacity solely on economic and technical variables ("donor/recipient countries") is inadequate and that the notion of a linear capacity gap between countries is quite obviously a misleading one (Oberthür, 1993). Environmentally benign behavior can be founded not only in societies with "post-materialistic" values (Inglehart, 1977; Scherhorn, 1994) and the widespread application of sophisticated technology, but also in (modest) forms of economy and society that are better adapted to scarcity and the actual carrying capacity of ecological systems.

Therefore, in addition to economic determinants of institutional capacities, there are also specific *sociocultural capacities* that either favor or hinder effective environmental policy. The idea that capacity building is nothing but a simple transfer of resources from donor to recipient countries is qualified further by the fact that environmental policy decisions do not necessarily lead to a significant improvement in the capacities of the respective recipient country. Here again, the fundamental problem of implementing adopted programs arises. Another, final, point is that environmentally motivated transfers can be offset by unfavourable development of the terms of trade.

Given these considerations, it is necessary to relate capacity building not only to the transfer of information and technology, or the organizational support of administrations, but also to the possible influence on the underlying socioeconomic structures. Environmentally harmful consumption and production patterns must be overcome, both in the North and in the South, and new structures have to be shaped in an environmentally sound manner in both hemispheres. Various chapters of AGENDA 21 contain important pointers in this connection. On the other hand, of course, lessons must be drawn from the deficient implementation of development assistance in the past. It is crucially important here that the allocation of funds to national authorities be supplemented by subnational institutionalization, i.e. by *decentralized global environmental policy*.

3.3.2
Introduction of International Environmental Audits

In the future, the environmental audit procedure presented above (*see Section B 3.2.3.3*) could become a decentralized incentive for fulfilling or implementing international environmental policy, programs and management systems. Making environmental policy more transparent can generate incentives to implement environmental measures effectively, by identifying and utilizing additional potentials on the part of the target groups. However, if such audits are to form an integral part of global environmental policy, a number of aspects must be taken into account.

Transparency is an institutional element of modern democracy. Establishment of it within the global framework is initially faced with contrary cultural concepts and authoritarian political structures. To this extent, *internationally organized transparency* may conflict with national sovereignty since secrecy has always been taken for granted as an essential feature of foreign relations. The participation of NGOs raises the same problem to a certain extent, in that it runs counter to what national authorities traditionally view as their domestic sovereignty. For these two reasons, open, or at least latent, resistance against international auditing of national environmental policies can be expected.

The audit principles of decentralization, voluntary participation and environmental statement are useful arguments against such misgivings. It is obvious, however, that international environmental audits may take the form of a social, possibly even *political and legal norm for participation*. This again may result in circumvention and deception of that instrument.

Environmentally sound behavior can play a significant role as an enhancer of competitiveness. One illustration of this is the controversial debate currently being waged at international level over the issue of *eco-labeling*. As a rule, in a market economy technical innovation takes place only if it is possible to keep new products or processes secret until they are ready to be put on the market or patented. International environmental audits therefore should be prevented from inducing a complex interplay between efforts to maintain secrecy and the final dissemination of technical standards. The more that environmental audits are geared to management processes rather than purely technological aspects, the more effective they will be at reducing obstacles to innovation.

Of course, the value of even gradual progress in implementing environmental audits should not be underestimated. The greater the extent to which environmental audits are internationally institutionalized (i.e. acceptance is gained for standardized criteria), the greater the competitive pressure generated in the field of environmental policy. Furthermore, environmental audits can be an instrument for informing the public about new forms of environmentally sound behavior. This *competitive or public pressure* could play an important role in countries with little political transparency and extensive bureaucratization.

In view of these problems and potential benefits of international audits, it is recommended that special emphasis be placed on voluntary participation and on the positive impact of environmental audits (incentive effect). They can and should be conducted with the active participation of regional and national organizations. However, comprehensive global audit structures must also be striven for parallel to the development of global regimes. As such audits may come into conflict with prevailing values and interests, creating a *global public forum* is an important condition for their successful implementation. The following issues may be crucial:

- international panels of experts and professional publications should act as forums for controversial global environmental debate;
- news agencies and mass media should be encouraged to point out differences in values and interests as they relate to environmental policy issues at the global level, as well as to unorthodox approaches;
- environmentally relevant cooperation should be further developed toward two-way communication, in contrast to the prevailing "one-way" flow of information from the rich to the poor countries.

The design of international environmental audits is thus closely linked to the general development of environmental communication. Global environmental discourse, which has usually resulted only in response to particular environmental catastrophes, must be radically extended to embrace environmental risks and their management by a professional global public (with enhanced competence to take action). The expansion of such environmental communication could be located within future global environmental regimes (i.e. in addition to the Ozone Protocol, within the framework of the future Climate, Biological Diversity and Desertification Protocols). As these regimes address only some of the existing environmental problems (soils, water and forests are not

dealt with, for example), all *global communication institutions* having an implicit environmental audit character should be strengthened. This would mean, among other things:

- greater support for national and private global information media,
- supporting conferences and projects through global information,
- facilitating international exchange of scientific know-how and experts in the field of environmental policy,
- ensuring the right to free access to information (environmental information), particularly for NGOs, within global networks,
- inputting the results of international environmental audits to regional and local action programs (*see also* the recommendations for research in WBGU, 1993 and 1994).

3.3.3
Flexibilization
of Global Environmental Institutions

As described above (*see Section B 3.2.2.2*), *flexibilization* refers to the functional differentiation of environmental communication and negotiation processes, the development of functional coordination mechanisms, and to enabling international environmental agreements to enter into force more easily. The relevant proposals for innovation are assessed below.

3.3.3.1
Legitimation of Action

An assessment of the problems related to institutional flexibilization should focus on the specific requirements for a *democratic legitimation of action*:

- One argument against a more flexible enactment of international agreements is that the democratically legitimate sovereignty of national (and subnational) decisions must be guaranteed. This objection has recently been raised against the international competencies of the European Union, e.g. by the German Constitutional Court in 1992.
- One objection to the participation of nongovernmental agencies in negotiation processes is that the latter do not possess the same degree of legitimation as democratically elected bodies or the administrative institutions legitimated by the latter. An emphatic, normative view of representative democracy (Fraenkel, 1964) will similarly question the right of lobbies to participate actively in international negotiation processes.

– Asymmetric coordination mechanisms, such as double weighting of agencies within environmental regimes, contradict the principle of equality in modern democracy. Weighting of votes may occasionally be viewed as a requirement for political progress in cases where major disparities in capacity exist; in general, however, asymmetries should be eliminated rather than new ones being created.

In the light of these (and similar) objections, all processes for flexibilization of global environmental policy that are demanded or already in progress should be handled with a deliberate situational focus or in a sequence of phases. Where clearly defined donor-recipient relations exist between those involved in global environmental policy (with the World Bank, for example), asymmetric coordination mechanisms cannot be done away with completely, since otherwise the donor side would have little motivation to become involved. The new GEF arrangements (*see* WBGU, 1994) are an interesting solution, but only one of many variants. For example, *minority rights* (from rights to a hearing to veto positions), *resolution thresholds* associated with them (qualified majorities) and *weighting of votes* (by a factor between 1 and n) can be combined with each other depending on the situation; and *majority decisions* in combination with regional *group veto rights* (similar to those in the ozone regime or the GEF) may point the way. They make it possible for "mavericks", such as Norway and Japan on the issue of whaling, to be outvoted, while creating sufficient international consensus for effective environmental action.

The establishment and further development of international environmental regimes make most sense when carried out in the following order: 1) search for areas in which consensus is possible, 2) negotiation and 3) voting. The results of this process should then be defined as the future basis, even in the absence of a full consensus. In this way the equality of all those involved can be ensured and the opportunity for reaching agreement can be exploited, while at the same time increasing the pressure (the "shadow of imminent coordination") to show more willingness to compromise. Such a sequence of institutional procedural rules could (should) also be adopted by bodies having a less strict donor-recipient structure. In all general questions of environmental policy, but particularly in matters in which the recipient states have considerable scope for action, veto potential and autonomy, unweighted voting procedures should have priority in view of the global interest in establishing democratic institutions.

3.3.3.2
Examples of Flexibilization

In addition to the proposals regarding flexibilization of the negotiation process described above (*see Section B 3.2.2.2*), there are other options for making environmental negotiations more flexible:

1 The progressive global interrelations between the economic and environmental domains are opening up an increasing number of compensation options. New specific and agency-related *package* deals can be developed, addressing the cross-sectoral demands of environmental policy, especially. Virtually all areas of infrastructure development (education, social security, transport, etc.), for example, are being affected by international environmental requirements. As in the case of "debt-for-sustainable-development swaps", problem-solving approaches can be implemented in a larger reference framework. "No-regrets solutions" or synergistic *"win-win situations"* often exist – and will be important issues in global environmental policy in the future.

2 The development of *environmental discourses* is another way in which negotiation processes can be made more flexible, since this helps to reduce distributional conflicts. For this reason the training of professionals should be promoted, especially within the context of global environmental negotiations (on the issue of *policy dialogue*, see Benedick, 1992). Prerequisites for such discourse (besides full respect for all involved) are the comprehensibility of oral and written information, adequate translation aids, and careful institutionalization of the communication process in order to facilitate consensus formation. Reaching consensus is not only the essential basis for specific solutions, but also of indirect importance for creating the methodological capacity for discourse.

3 Large-scale singular negotiations ("environmental summits") intended as the starting point for a continuous negotiation process may play an important role by virtue of the fact that they enable productive and integrative formation of will at the international level. However, this aspect may be offset by the special negotiation pressures at such big events. In consequence a deliberate sequence of large-scale conferences and (relatively) continuous small-scale negotiations seems advisable, a concept already pursued within the framework of the ozone, climate and biodiversity regimes. However, improvements are possible in their timetabling, professional preparation and systematic follow-up.

3.3.3.3
Lean Management
and Global Environmental Management

The demand for greater flexibility of global environmental policy relates not only to institutional procedures. What are also needed are efforts to increase administrative and organizational flexibility. An essential requirement here is that the action of all relevant agencies be geared to *maximum effectiveness*. This approach has much in common with the philosophy of lean management in the field of business administration. The latter concept has been much in vogue in recent years, and can be summarized as follows (for more detail, *see* Metzen, 1994; Naschold, 1993): flattening hierarchies; reduction of bureaucratic fragmentation; cross-functional teams instead of "process-oriented" bureaucracy; result-oriented assessment; decentralized budgeting.

Modern management techniques like these have already been applied or tried in various international organizations and environmental regimes. Continuous monitoring of the structure and efficiency of international environmental management should become established practice, in any case, so that structures adequate to the complexity and dynamics of the problems can develop. The UNEP, GEF, UNESCO and CSD, and perhaps the European Environment Agency, are possible subjects of future evaluations by the Council.

3.4
Special German Contributions to Institutional Innovation of Global Environmental Policy

The issue of "institutional innovation of global environmental policy" gives rise to ideas that have already been the subject of discussion for some time or even implemented in the German-speaking countries, though under different terminology. One such idea is that of *"regulatory policy"* ("Ordnungspolitik"); others include *"co-determination"*, *"social contract between generations"*, *"trusteeship"*, and *"social market economy"*. Some of the lessons that have been drawn from these experiences could be applied to future global environmental policy, and thus contribute to a structural "environmentalization" (Jänicke, 1993) of international relations.

One should not forget, however, that Germany has played an outstanding role in the formulation of global environmental policy in only a few specific areas until now; indeed, many an opportunity to exert stronger influence has been wasted (e.g. energy tax/CO_2 charge). Nevertheless, Germany possesses a significant potential for influencing the further de-

velopment of global environmental policy by virtue of its economic and technological strength, its grown political importance, especially in the European Union, as well as the high degree of environmental awareness and broad support from the population. In some global agreements, Germany has shown a relatively strong financial commitment – in the GEF, for example.

Given this background, the Council sees an active role on the part of Germany in the institutional innovation of global environmental policy as having considerable prospects of success. Appropriate initiatives should be based on the discussion outlined and assessed above (*in Sections B 3.2. and 3.3, respectively*), so that specific aspects and needs can be included in the international political discourse. Section C 3 points out the special role that Germany could play in the (short-term) establishment of a *Blue Fund* and in the (long-term) implementation of a *Convention on the Protection of the Sea*. In its 1994 annual report the Council demanded an active German role in the formulation of a global *Soil Convention*.

In addition to the respective government policy options, various *subnational* and *transnational agencies*, especially environmental organizations, employers' associations and standards committees, have substantial scope for action. Global policy of this kind can be supported by governments to a certain degree through access to information and allocation of funds, thus creating an interactive system for policy formulation. This includes the promotion of globally oriented environmental research projects and program proposals (*see also* the research recommendations in WBGU, 1993 and 1994).

Important *domestic requirements* for the development of an original German position favoring institutional innovation in global environmental policy, i.e. proactively influencing the innovation approaches presented in *Sections B 3.2 and 3.3*, include:
- increased political perception and public discussion of global environmental problems and the institutional conditions under which they occur,
- creation of sufficient or, in most cases, expanded capacities for the diagnosis and therapy of global environmental problems,
- raising the commitment and strengthening the competence of German representatives in international institutions that are directly or indirectly related to environmental issues.

The following selected examples show quite clearly the degree to which Germany's scope for exerting influence is still restricted with regard to the aspects listed above:
- global environmental policy research (such as institutional research) still plays a minor role in overall German research,

- the Federal Office of the Environment as the top-level agency under the Federal Ministry of the Environment still has little capacity for the formulation and implementation of projects and programs related to sustainable development,
- German nongovernmental organizations have to restrict their participation in international meetings considerably because they lack sufficient funds for participation, evaluation and implementation of results, etc.

Another aspect that may be just as important as infrastructural requirements and political will for effective involvement at the global level concerns the *credibility of institutional reforms* at the national level, meaning that innovations in the field of global environmental policy have to be consistent with foreign and domestic policy projects – with regard to, for example:

- the coordination and/or compatibility of initiatives with other German foreign policy activities, such as ecological reform of the GATT/WTO regime or further development of the concept of collective (and hence environmental) security,
- the structural development of domestic institutions; for instance, economic instruments and liability law can only be internationally propagated in a convincing manner if appropriate methods of direct and indirect control are applied and made more effective within German national environmental policy.

Action is particularly needed in the latter area, as shown by the continued dominance of environmental command and control policy over economic instruments (such as *environmental taxes, tradeable permits, forms of compensation*). Appropriate national (and European) reform initiatives would undoubtedly enhance the credibility of stronger German commitment in favor of institutional innovation in global environmental policy, i.e. create a stronger emphasis on legal and market-based solutions to problems of sustainable development.

The growth and distribution of the world's population are decisive factors for the solution of global environmental and development problems (WBGU, 1993). The basic demographic conditions for coping with global environmental problems are characterized by

– an annual increase of approx. 95 million people,
– increasing urbanization, particularly in developing countries, and
– the growing pressure of international migrations to Europe and North America.

Because of the radical nature of the pending issues, the most precise knowledge possible regarding the expected developments is necessary. Even slight changes in trends of population growth, urbanization and migration result in significant forecast shifts and can have a substantial influence on the political room for maneuver within the framework of global environmental and development strategies.

4.1
Current Trends

4.1.1
Population Growth

There is a problem connected with statistical studies of demographic development in that reliable data can only be published after a two-year delay. The 1995 annual reports of the *World Bank* and UNPD, for example, are based on data from 1993. For these reasons evaluations of statements made on population development should be conducted carefully.

The population forecasts of the *World Bank* for the year 2025 have constantly been modified in recent years (World Bank, 1992, 1993 and 1994). Whereas the world's population in the year 2025 was estimated at 8.3 billion in 1992, the 1993 forecast assumed a figure of 8.24 billion and in 1994 a population of "only" 8.12 billion people was forecast. In 1992 the *World Resources Institute* (WRI) assumed a

population of 8.5 billion for the year 2025, but revised this figure in its 1994/95 annual report to 8.47 billion. The population forecast of the 1994 Human Development Report, on the other hand, assumes an increase in population growth; the forecast of 6.33 billion people for the year 2000 was higher than the figure calculated the year before, namely 6.22 billion.

In addition to the absolute forecast figures, the growth rates that have been provided every year by the World Bank, WRI and UNDP since 1992 are useful for an assessment of demographic development. The estimates of this growth rate have generally been modified downwards. The continents selected for a comparison of figures, i.e. Africa, Asia and Latin America, were those characterized by high population growth. They account together for nearly 97% of the global population growth (*Tab. 3*).

Almost all figures in *Tab. 3* show lower growth rates for 1994/95 than in the year of the *UN Conference on Environment and Development* (1992). The most significant correction was made by the WRI for Latin America: whereas the 1992/93 annual report showed a growth rate of 1.7%, in 1994/95 the institute assumes a growth of 1.3% up to the year 2000. The World Bank's growth forecast for Asia is an exception. Starting with an estimate of 1.8% in 1992, the 1994 World Bank report shows a slightly increased growth figure of 1.9%. From a global point of view, however, the World Bank also assumes a lower population growth

The exceptional case of Asia may point to a slight shift in the growth poles, as already described the WBGU's 1994 annual report. Nevertheless, while mortality rates remain constant, the fertility rates are lower than assumed in 1992, especially in Latin America. For the period from 1990 to 1995, for instance, the WRI estimated a growth rate of 3.2% in 1992, but corrected the figure to 2.9% in 1994.

Depending on the modified growth rates, corresponding changes result for the forecasts of the absolute population. On the basis of the estimated worldwide growth rate of 1.5%, which is 0.2% lower than the calculation basis in 1992, the estimated figure for the world's population in the year 2050 already devi-

Annual Report	Africa [%]	Asia [%]	Latin America [%]	Developing Countries [%]	World [%]
World Bank 1992	3.0	1.8	1.8	-	1.6
World Bank 1993	3.0	1.9	1.6	-	1.6
World Bank 1994	2.8	1.9	1.6	-	1.5
UNDP 1992	-	-	-	2.0	1.7
UNDP 1993	-	-	-	1.9	1.6
UNDP 1994	-	-	-	1.9	1.6
WRI 1992/93	3.0	1.7	1.7	-	1.6
WRI 1994/95	2.7	1.4	1.3	-	1.4

Table 3
Modification of estimated world population growth rates since 1992.
Sources: World Bank 1992, 1993 and 1994; WRI 1993 and 1995; UNDP 1992, 1993 and 1994

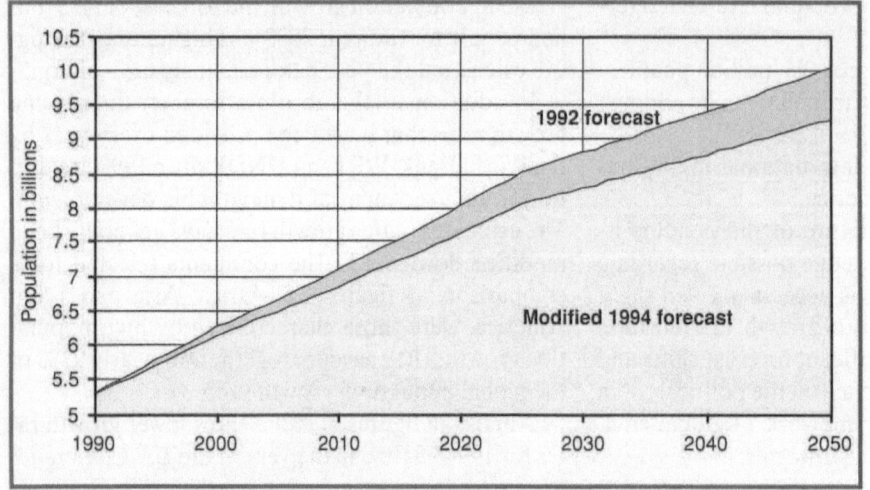

Figure 8
Modified population forecasts up to 2050.
Source: WBGU based on the 1992 World Population Report

ates from the 1992 forecast by 733 million people. The WBGU assumes here that the growth rates from the years 2000 and 2025 onward, relative to the current correction, can also be reduced (progressive extrapolation). In comparison to the present estimate, the growth rate from the year 2000 onward would then be 1.08% (instead of the 1.23% figure calculated by the World Bank) and 0.57% from the year 2025 onward (instead of the World Bank's 0.65% figure). If the growth rates from the year 2000 onward should not drop more than expected in 1992, a world population figure of 9.8 billion people (instead of 10.0 billion) can be assumed for the year 2050 (linear extrapolation).

The 1992 forecast and the modified, more optimistic estimate up to the year 2050 are compared in *Fig. 8.*

In view of these "optimistic" forecasts, there is definitely hope that the extent of demographically related global environmental changes can be modified in the long run. However, one should be warned against making a too optimistic assessment of the population development or drawing false conclusions from the so-called revolution in reproduction behavior. Birg (1994), for example, showed that the decline in the fertility rate has slowed down from decade to decade. Between the second half of the 60s and the first half of the 90s the fertility rate dropped from 4.98 children per woman to 3.26, but this strong decrease should not blind us to the fact that it fell by 1.05 in the first decade but only by 0.41 in the second. Moreover, Birg examined how long it will take for the fertility rate, on world average, to drop to the reproduction level (replacement level) of 2.13 children per woman. He obtained results different from those of the World Bank, which expects the end of the population increase roughly in the year 2050. Without giving a specific point in time – the calculations vary with respect to when the reproduction level will be reached – Birg is skeptical about the year 2050 indicated by the World Bank. He further assumes that as the fertility level drops, it will become increasingly difficult to achieve an additional decline, particularly due to the lack of social security, and he fears that the

Definition of Demographic Transition

Demographic transition, so-called, takes place in five phases. High birth and death rates are characteristic of the first phase whereas the process ends with much lower rates that hardly vary in the short term. Between these two phases there is initially a period in which considerable improvement in the chances of survival is coupled with a constantly high birth rate that even increases at times. This results in a widening gap between the curves of the birth and death rates so that the population grows rapidly. Only at a much later stage can an adjustment of generative behavior to the altered death rates be noted. In this phase the birth rates decline faster than the death rates and the gap begins to narrow again (Bähr, 1992).

A decisive factor is that the length of this demographic transition has shortened over the course of time. Whereas the transition in England began around 1740 and lasted for roughly 200 years, Japan required only 40 years, beginning in 1920, to complete the transformation (*Fig. 9*).

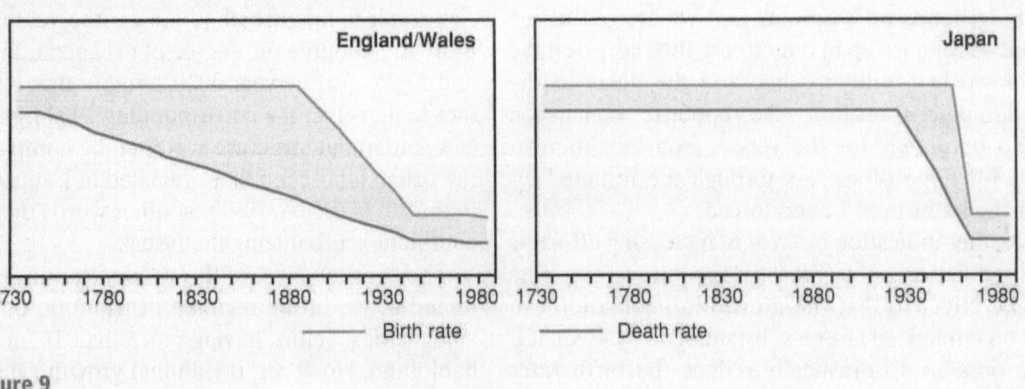

Figure 9
The demographic transition.
Source: Bähr, 1992

development and family planning policy will not be able to keep up with the increasing needs.

Another important question is how this fertility decline will take place; not only are the target year and the desired replacement level decisive factors, but also the progression of decreasing fertility over time. A linear decline in the birth rate to a desired level leads to higher absolute population figures than those in the case of an exponentially (at the beginning rapidly) decreasing fertility rate with the same target level. The following sample calculation illustrates how urgently necessary rapid success is with regard to fertility reduction: if achievement of the goal to reduce global fertility from the current figure of 3.3 children per woman to 2.13 children were to be delayed by only 10 years, the growth in the number of people worldwide up to the middle of the next century would increase by approx. 500 million. The figure of 10.1 billion people calculated by the World Bank for the year 2050 or 12.1 billion for the year 2150 will be exceeded if it is not possible to lower the average rate of live births per woman in the developing countries from 3.6 at present to 2.3 by 2030 at the latest (Birg, 1994).

In view of the slowdown feared in the decline of fertility, Birg forecasts a rise in the world' population to roughly 14 billion people, while referring to the collapsing aid systems due to social change. Another indication of this is the perceptible delay in the "demographic transition" in the developing countries (*Box 18*).

By contrast, one can observe in the developing countries today a substantial delay in the demographic transition that is favored, as on the Philippines, by massive emigration of qualified labor. The emigration of medical personnel, who play a key role for the general health care of the population and for the provision of basic health services, is a special factor here. If access to such basic care continues to be denied, the mortality rate in the region concerned will fall very slowly. This, in turn, may delay a change in reproductive behavior and thus in the demographic transition.

A large number of physicians and nursing staff have migrated across national borders in recent decades. The WHO assumes that 14,000 nurses left their native countries in the early 70s and in the year 1972 alone over 140,000 doctors were working outside of the countries where they were trained or born (World Bank, 1993). This trend continues even today. The consequences of this development are ambivalent in that, on the one hand, labor market bottlenecks in the immigration target countries (mostly countries with high incomes) can be eliminated through additional nursing personnel; on the other hand, the migration flows lead to the problems of short supply in the emigration countries, as described above.

The conclusion that can be drawn concerning demographic development, therefore, is that long-term forecasts with regard to population increase have been modified downward slightly in recent years. Nevertheless, this very small correction should not be grounds for assuming that we are out of the woods or for a let-up in our efforts. Instead, given the slowdown in fertility decline and the delay in the demographic transition, the opposite conclusion should be drawn: for the very reason that there is cause for hope of success through the initiated efforts, the latter must be reinforced.

Another indication in favor of increasing efforts is that population policy measures require a long time to take effect so that the growth in population can only be influenced after a substantial delay. Even if it were possible to drastically reduce the birth rates worldwide from the current figure of 2.6% to the current mortality rate of 1.5%, the world's population would increase by the year 2050 due to the unfavorable age structure (42% of the population is younger than 16) (WBGU, 1993). Furthermore, the natural environment will be subject to considerable pressures because of the long period between implementation of the respective measures and their taking effect.

4.1.2
Urbanization

A total of approx. 83% of the worldwide population growth is accounted for by urban regions, i.e., the urban population will increase by 75 million people annually over the next decade according to the modified population forecast. Existing settlement systems will be subjected to considerable additional pressure or even overloaded in many cases as a result of the natural population growth and inmmigration. The rapid expansion of urban systems will cause immense social and ecological costs as well as environmental damage. If the international community is not able to put a stop to the degeneration of settlement systems, cities, as the focal point of population growth, will "collapse" and air, soil and water in the cities will be increasingly polluted.

This threatening development is progressing at a different rate depending on the region, however, with the developing countries being increasingly affected by uncontrolled urbanization processes. The African continent deserves special attention in this connection. Up to the beginning of the 90s Africa was considered the "rural continent" with the village as the dominant form of settlement. According to a study conducted by the *United Nations Center for Human Settlement*, this assumption no longer applied in the early 80s: while the proportion of the total population living in urban areas in 1960, 18.3%, was still just under the corresponding figure for Asia, it rose to 27.8% by 1980, reaching roughly the same level (DGVN, 1992). The United Nations forecasts for Africa result in a figure of 40.7% for the year 2000, i.e. more than double the degree of urbanization in 1960, and 53.9% for the year 2020, a figure that is three times higher than the corresponding 1960 figure. Africa's settlement structure will then be comparable to the urban landscape that prevailed in Latin America in the 70s (DIESA, 1991). In other words the poorest continent is urbanizing the fastest.

The consequences of this process cannot be determined as yet in the regional distribution of existing "megacities" (cities having more than 10 million inhabitants). However, the annual growth rates of the urban population in Africa are currently around 5% and will not fall below 3% until far into the next century (DIESA, 1991). The populations of the big cities in the developing countries are growing more rapidly than the urban population overall. According to the forecasts cited above, by the year 2025 approx. 4.4 billion people, i.e. almost half the total population of the developing countries, will be living in urban agglomerations, a large portion of them in Africa. The few cities on this continent that had over a million inhabitants up to several years ago will already have grown into megacities by the year 2000 (DGVN, 1992). Many other towns numbering among the intermediate cities today will then have reached the million mark. Africa's urbanization is accompanied by growing economic problems. Ten years ago only half of the urban population were supplied with water and sanitary infrastructure, and the households in many big cities lived in squatter camps (constructed without town planning support) (DGVN, 1992). Only a minority has a regular income.

- The town of Bangui in the Central African Republic had a population of 500,000 in 1985, the majority of whom lived in squatter camps. These 500,000 inhabitants were still using a sewage system that had been built for 26,000 people in 1946.
- Of Cairo's 7.7 million inhabitants in 1985, over 1 million lived in the city's cemeteries.
- Nouakchott, Mauritania's capital, which was still a

small town of 5,000 inhabitants in 1965, already had a population of 135,000 in 1977, a figure that had doubled by 1985. More than two-thirds of the population suffers from water shortage (DGVN, 1992).

These examples bear witness to the fact that there is still no such thing as resource management according to the principle of sustainable development in the rapidly growing cities. Although more recent data is not available, it can be assumed that the problems have aggravated substantially. The trend towards increasing urbanization is unchecked and will continue to confront the existing settlement systems with environmental effects on a global scale. Nevertheless, the living conditions in the cities are frequently better than in rural regions, resulting in increasing migration to the cities.

Fig. 10 points out the nations in which urban development is threatened as a consequence of the rapid population growth.

4.1.3
Migration

International migration, in particular the streams of refugees, increased significantly at the beginning of the 90s. According to estimates of the *United Nations*, roughly 50 million people (i.e. 1% of the world's population) lived outside of their native country in 1989. The World Bank estimated the total number of cross-border migrants for the year 1992 at 100 million (World Bank, 1993).

Most of the currently estimated 17 million refugees can be found in Africa, Asia and Latin America and their numbers are rising rapidly. Another 3.5 to 4 million people live in a refugee-like situation (World Bank, 1993).

Cross-border migration has not yet attained the quantitative dimensions of urbanization. Nonetheless, the importance of migration movements must not be underestimated since their effects bear no relation at all to the number of people involved. In many countries "immigration" has become a highly explosive political issue. It has become the subject of negotiations at the highest level in connection with the G7 summit meeting, the OECD *Council of Ministers and the Summit Conference of the European Union* in the past three years. The Council emphatically pointed out the causes of migrations and their significance for global environmental changes in its 1993 and 1994 annual reports and made reference to international migration towards the industrialized countries of the North. These migrations can be primarily attributed to the fact that the expectations of many intranational migrants (rural exodus) are not

fulfilled in the target regions (cities). As a rule, the cities cannot cope with the influx of immigrants and thus themselves become sources for international migration in the end.

Due to strongly diverging data bases, changes in known development trends are difficult to determine. Rising population figures, growing poverty, environmental disasters occurring at shorter and shorter intervals (Wöhlke, 1992), creeping environmental destruction (e.g. soil degradation) as well as armed conflicts like in Rwanda (*see Box 19*) indicate that the number of migrants continues to increase. Ignoring hot spots such as in Rwanda, *Population Action International* (PAI) estimated the number of refugees worldwide at 19 million in 1994. This figure did not include the 2.6 million Palestinian migrants (PAI, 1994).

4.2
Conferences and International Agreements

4.2.1
UNCED 1992: AGENDA 21

The AGENDA 21 action program, whose aims include a tolerable population growth and promotion of sustainable settlement structures, and the so-called *Rio Declaration* were adopted during the 1992 *Conference on Environment and Development in Rio de Janeiro*.

In view of the complex issues and global significance of the Rio Declaration, it is not surprising that the principles it contains are formulated in a both reserved and nonbinding fashion. For instance, human beings are entitled to a healthy and productive life in harmony with nature (Principle 1), developing countries should receive preferential treatment (Principle 2), states shall cooperate in the spirit of global partnership (Principle 7) and an internalization of external costs should be striven for at national level (Principle 16). In this form the principles can be regarded more as ideal solutions lacking any force. Only the third principle of the Rio Declaration stands out from the others by virtue of its demand that the right to development be fulfilled so that the developmental and environmental needs of present and future generations can be equitably met (Principle 3).

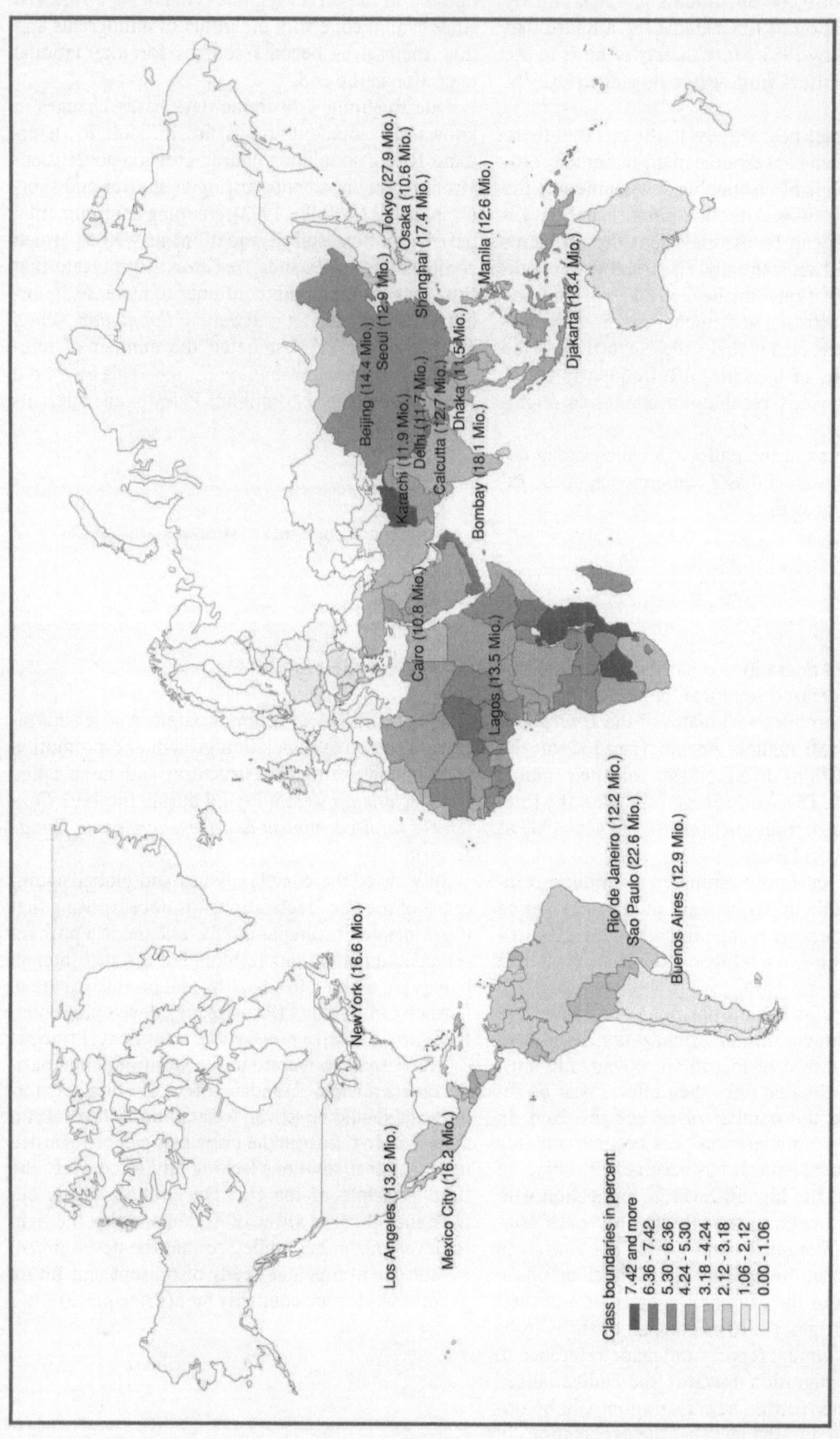

Figure 10
Urbanization rates 1990-1995 and largest cities in the year 2000.
Source: DGVN, 1992 and own calculations

BOX 19

Case Study of Rwanda

International news reports usually describe the civil war in Rwanda as an escalation of ethnic conflicts between the Bahutu and Batutsi tribes. The tribalistic argument does, in fact, appear plausible in view of the various ethnic groups in Rwanda. In addition to the Bahutu and Batutsi, Rwanda's population consists of five other tribes who, however, have been drastically reduced in number in some cases by the systematic mass murders of recent months and years. A historical perspective also nurtures the presumption that ethno-nationalistic claims and unreconciled resentments alone triggered the military conflict.

The centuries of suppression of the Bahutu (roughly 85% of the population in Rwanda) by the Batutsi did not end until the revolution in 1959, which forced hundreds of thousands to flee to Uganda in the north. A second revolution in 1973 then appeared to cement the dominance of the Bahutu. Since then the regime under President Habyarimana has ruled over the fate of the central African state, which has the highest population density on the continent, 285 inhabitants per km^2. Nepotism, corruption, regionalism, persecution and suppression of opposition forces have become part of daily politics since that time, thus inevitably provoking resistance movements on the part of minorities. A civil war started smoldering in Rwanda again in October 1990. Troops from neighboring Uganda invaded northern Rwanda. The Inkotanyi, as the army of invaders is called, fought to repatriate refugees who had to leave the country as victims of the 1959 revolution and later conflicts in the name of the "Front Patriotique Rwandais" (FPR, the political arm of this rebel movement).

The invading troops were, indeed, recruited from the refugees mentioned above, some of whom had been living in Uganda for decades and a large number of whom had been trained in the Ugandan army. However, it would be wrong to claim that the attackers were only Batutsis and that none were Bahutus (Mayr, 1991).

To this extent, the view in favor of a purely tribalistic conflict cannot be upheld. The rivaling parties possibly use the "ethnic gambit" only to distract people from other causes of the strife. In fact, the Bahutu and Batutsi speak the same language and there are no linguistic indications of an alienation of the two tribes. The Bahutu and Batutsi share the same culture, such as concerning the upbringing of children, construction of houses, livestock breeding, farming or religious beliefs (Mayr, 1991). Furthermore, they share the same settlement region so that there is no geographic-ethnic demarcation. It must also be pointed out that besides pure Batutsi and Bahutu clans, who by the way have common ancestors, mixed clans composed of both Bahutus and Batutsis also exist. A third of the two ethnic groups are not able to verify their tribal descent today. Assuming that tribalism is only a secondary cause of the military clash, there remains the question as to the primary factors of a conflict that turned every second Rwandan into a migrant. Rwanda, Africa's third smallest country, has an area of 26,338 km^2, corresponding to that of the German state of Hessen. Food supply for its rapidly growing population (growth rate of 3.4%) is proving to be the dominant development problem. Constant overexploitation of the soil by the cultivation of subsistence products and basic foodstuffs as well as extensive livestock breeding combined with the clearing of forests to obtain fuel wood in recent years have led to increasing soil erosion and a dramatic deterioration of soil fertility. The geographically isolated location is also a disadvantage for further development of this inland country. All trade in commodities must be carried out at high transport costs via the ports of Dar es Salaam (Tanzania) and Mombasa (Kenya). The subsistence sector is characterized by low area productivity and old production methods. The demand for meat and dairy products cannot be met through domestic production. A high illiteracy rate, a high mortality rate for children, a low life expectancy and extremely poor basic medical care are indications of the precarious social situation in Rwanda.

The interplay of the ecological conditions described above can be interpreted as the actual source of social tensions. Those who lose their settlement areas as a result of woodland being cleared in Gishwati, for example, are forced to secure their livelihood somewhere else in the country. Such a situation may finally develop into a conflict that escalates under the guise of tribalism and, as happened in Rwanda, induce flows of refugees of a magnitude of 3 million people, thus threatening sustainable development both within and outside of the country's borders.

The key statistical data concerning the present and future scope and orientation of the migration movements vary substantially, depending on the source of the data. According to nearly all surveys, the total number of refugees is estimated to be half the population of the central African state. However, statements regarding the intranational distribution and international direction of migrations differ significantly from one another.

Fig. 11 is based on an overall view of the migration flows. It makes no claim to being complete,

but it does illustrate the considerable magnitude of these migrations. In view of the fact that the basic data for this figure diverges greatly in some cases, it is necessary to have a set of instruments that permit the calculation of precise migration statistics and forecasts. Both the Rio Declaration and AGENDA 21 revealed the substantial need for clarity in this area.

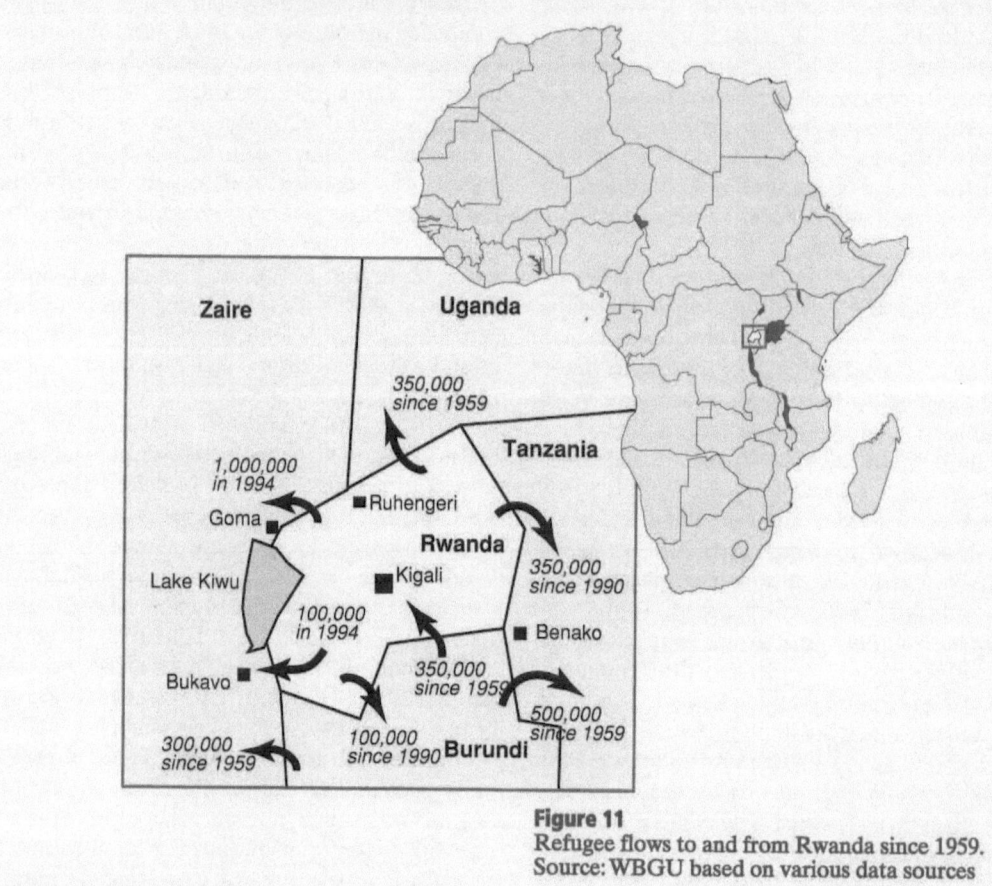

Figure 11
Refugee flows to and from Rwanda since 1959.
Source: WBGU based on various data sources

4.2.1.1
Population Development

At first it is surprising that only one of the 27 principles in the Rio Declaration (UNCED: Principle 8) focuses on demographic development as a key problem for sustainable development (WBGU, 1993). This principle calls for appropriate demographic policies as a basic prerequisite for achieving sustainable development and a higher quality of life. Otherwise there are no references to demographic development in the Rio Declaration.

In AGENDA 21, by contrast, an entire chapter is devoted to this issue. One of the major aims of this chapter, entitled "Demographic dynamics and sustainability", is to show ways of researching and disseminating specialized knowledge on the interrelationships between population growth and sustainable development. Moreover, the international community wishes to harmonize and expand population policy so that population growth can be checked and a rapid demographic transition can be carried out on this basis.

According to AGENDA 21, the key elements of a successful population policy include combating poverty, providing adequate medical care, guaranteeing basic needs, improving the social position of women and developing educational systems (AGENDA 21, 5). However, AGENDA 21 does not state how these – undoubtedly justified – demands are to be implemented. Similar weaknesses are revealed, particularly in the third focal point of Chapter 5, which calls for an implementation of population policy measures within the framework of other political areas, such as health, science, research and economics.

Thus realization of the demographic transition remains the most clearly stated demand in AGENDA 21 with regard to population development. The synergies of this issue in connection with environmental concerns and the necessity of a rapid decline in the birth rate are underlined by linking demographic aspects with ecological demands such as: "They should combine environmental concerns and population issues within a holistic view of development" (AGENDA 21, 5.16), and the formulation of aims: "Population programmes should be implemented along with natural resource management and development programmes at the local level that will ensure sustainable use of natural resources, improve the quality of life of the people and enhance environmental quality" (AGENDA 21, 5.43).

In summary, an evaluation of the success of the UNCED three years later, particularly from the point of view of population, cannot provide reliable findings. In some segments the demographic trends and trend corrections are progressing in accordance with the Rio Declaration and AGENDA 21; however, a direct connection between UNCED and the actual development cannot be demonstrated on that basis. There are many indications that development programs launched prior to 1992 are showing initial success today.

4.2.1.2
Urbanization

Cities are highly complex organizational structures of human coexistence. They are the result and basis for developments with a wide variety of tasks and social patterns that reflect the diversity of the individual cultural regions. In addition, cities are the locations of industry, business, institutions and service enterprises that supply the surrounding area. "Healthy" urban growth, however, requires that all municipal activities, such as culture, schools and transportation, housing, energy and water supply as well as disposal (sewage, waste), are coordinated with each other. In view of the continuing process of urbanization, it is questionable whether larger and larger cities can remain functional and it is unclear

whether there is an optimum or maximum magnitude of urban development. In some cases expansion is limited by the available settlement space, the transportation infrastructure and water supply as well as by other supply problems (Vöppel, 1970).

AGENDA 21 takes a position on "urbanization" in Chapter 7, entitled "Promoting sustainable human settlement development". A distinction is made between industrialized and developing countries with respect to global environmental changes. Whereas the metropolises of the industrialized world are primarily characterized by a high and thus environmentally harmful level of consumption, more and more raw materials, energy and economic growth are required for cities in the developing countries in order to cope with the elementary economic and social problems of urban regions there.

One reason why the standard of living in the cities of most developing countries is declining is a lack of necessary investments in such areas as infrastructure because crises have to be averted in other sectors. Government expenditures for housing and social security in low income countries, for instance, were roughly 5.6% of the total budget on an annual average (UNCED, 1992; the corresponding figure for the OECD states was 39.3% on average; UNDP, 1991). In addition, there is little financial support on the part of international organizations for cities in developing countries: only approx. 1% of the UN expenditure in 1988 was earmarked for settlement purposes.

On the other hand, the existing cooperation regarding settlement systems has triggered substantial public and private follow-up investment. Every US$ spent on technical cooperation by the UNDP in 1988 led to follow-up investments amounting to about 120 US$: a result that was not achieved in any other UNDP sector (UNCED, 1992).

On the basis of these positive incentive effects that were triggered by technical cooperation with the UNDP, the main objective of AGENDA 21 is to further promote public-private partnerships (cooperation between public and private actors) so as to achieve progress in the areas of settlement management, regional planning, infrastructure planning and industrialization.

The hope here is that the urban regions in which 60% of the world's gross national product is earned will be in a position, given appropriate management, to expand capacity and increase productivity so as to improve the standard of living of the population in accordance with sustainable development (AGENDA 21, 7.15).

The Council, however, is very skeptical about this hope regarding the development opportunities of urban agglomerations because of the uncontrolled nature of urban growth. The Council examined in detail the problems related to current urban developments in its 1994 annual report. The "São Paulo Syndrome", for instance, pointed out the threat to urban structures and indicated the possible collapse of many megacities.

The dangers stemming from the rapid population increase are presumably underestimated in AGENDA 21. This may be due to a lack of the necessary information, especially since the following demand was expressed: "Sociodemographic information should be developed in a suitable format for interfacing with physical, biological and socioeconomic data. Compatible spatial and temporal scales, cross-country and time-series information, as well as global behavioral indicators should be developed, learning from local communities' perceptions and attitudes." (AGENDA 21, 5.10). Moreover, "better modeling capabilities should be developed, identifying the range of possible outcomes of current human activities, especially the interrelated impact of demographic trends and factors, per capita resource use and wealth distribution, as well as the major migration flows that may be expected with increasing climatic events and cumulative environmental change that may destroy people's local livelihoods." (AGENDA 21, 5.9)

As long as the highly inadequate state of research into interrelationships continues, i.e. as a systematic analysis of global human-environment relationships, the risk of succumbing to radical misjudgments or playing down key problems related to global environmental changes will increase.

Urbanization in Africa is primarily a consequence of increasing poverty and less the result of development. If even the satisfaction of basic human needs cannot be guaranteed in Africa's cities today, then it is impossible to postulate (as was done in AGENDA 21) that the conditions for sustainable development have been met. Even the promotion of intermediate cities (AGENDA 21, 7.19) or public-private partnerships (AGENDA 21, 7.21), as aimed at in the action program, does not take into account the real development trends and the extent of the global environmental changes they cause. This situation points to information deficits that have to be eliminated by prompting research into the specific interrelationships in the global human-environment system.

4.2.1.3
Migration

The United Nations described the interlinkage between migration and global environmental changes in its 1993 population report. The latter states that "the gradual destruction of the environment is the main cause of population movements" (DGVN, 1993). Astonishingly, neither intranational nor international migration are focuses of attention in either the Rio Declaration or in AGENDA 21. One can only guess at the reasons why this issue is ignored in both documents. One possible reason is that migration is frequently viewed as a barometer for changing social, economic and political conditions or as the result of individual or family-related decisions and was thus not an immediate subject of discussion at the *Conference on Environment and Development*. It is also conceivable that the international community deliberately fixed its range of competence in Rio de Janeiro to a list of issues that ruled out overlaps with migration-related fields of research of other institutions. From the council's point of view, in any case, this situation obviously represents a significant shortcoming which both points to deficits with regard to a systematic, analytical understanding of the global environment and demonstrates the substantial need for research.

4.2.2
State, NGOs and Churches

After a relatively long phase of discrete treatment the causes and effects of the rapid population growth have again become the subject of intensive and open public discussion. This increasing attention to questions of population development may provide a significant contribution to the shaping of public awareness and, in the end, bring about a change in generative behavior. The following will take a look at the basic attitudes and opportunities offered by nongovernmental organizations (NGOs), churches and government bodies in this connection.

There is now extensive agreement among governmental organizations, nongovernmental organizations and churches at the international and national level that development and population policy are not separate elements but have to be closely interlinked. The *Council of the Protestant Church* in Germany, for instance, demanded in 1984 that family planning be integrated into economic and social development and that it must take into consideration people's religious, cultural and social characteristics (EKD, 1993). The *Catholic Church* in Germany has also discussed population problems and their consequences in detail (Kommission Weltkirche, 1993).

The *German Council of Women* emphasizes in this connection that women's and family policy is a task sui generis and warns against using these political fields as demographically oriented instruments. The Council of Women criticizes the fact that the contribution of the German federal government to the

International Population Conference (see below) does not contain a separate "approach to women's issues". Rather, women were only indirectly mentioned as "mothers and pregnant women", as "labor market reserve" and as "recipients of development aid". The Council of Women condemns the fact that development aid is expressly tied to population programs so that the one-sided aim of support to women was for women to have fewer or no children. Population policy should not "refer and be restricted to women as objects of the reproductory function, [...] but must require the joint and holistic responsibility of men and women" (Deutsche Stiftung Weltbevölkerung, 1994). German nongovernmental organizations unanimously agree that major importance should be attached to improvement of the social position of women (*Box 20*).

Seven (German) NGOs, including the *German Foundation for World Population*, the *German World Famine Relief Organization* and the *German Society for the United Nations*, will be organized under a single umbrella organization in the future so as to lend greater force to their demands with respect to political bodies and make their public relations work more efficient (Deutsche Stiftung Weltbevölkerung, 1994). Among other things, the organizations demand binding transfer payments on the part of the North, improved world economic conditions for the countries of the South, free access to family planning options for all people as well as measures in the areas of education, health and old age insurance. Special focus is to be placed on the role of women in the development process and equal access to education and occupation is to be promoted (DGVN, 1994). The German NGOs thus share the views of comparable institutions at the European level, such as Eurostep (1994).

Germany, as the third largest donor worldwide for population programs, has more than doubled its appropriation of funds for family planning in the period from 1990 (approx. 74 million DM) to 1993 (approx. 160 million DM) (DGVN, 1994). The German federal government supports the aims of the *World Population Action Plan* (*see Section B 4.2.3*), about which there is general consensus.

The federal government criticizes the fact that the lack of coordination on the part of the donors leads to inefficiency in international cooperation. The donor states often lose sight of the necessity for a coordinated procedure as a result of their efforts to implement their own concepts. The federal government provides increasing support to the population fund of the United Nations in order to implement the 1984 Mexico recommendations and has made population policy a focus of its development cooperation for the 90s (BMI, 1994). The *Federal Minister for Economic*

Cooperation and Development underlines the dual strategy of improving the economic situation of the population and supporting family planning. Provision of information to groups that have been neglected up to now and support for women form the focal points of population policy cooperation (BMI, 1994).

The *European Union* also regards measures to check population growth and to combat mass poverty as important instruments for reducing migration pressure. Population growth is seen as an obstacle to the development of the economy, income and employment. In addition, the EU Commission gives consideration to armed conflicts and failure to comply with human rights which bring about an increase in refugees (Commission of the European Communities, 1994).

Even if it is not possible in the end to quantify the contribution of the increasing discussion about population-related questions in almost all areas of public life – both at the national and international level – one can presume that the implementation of population-regulating measures will be indirectly facilitated by virtue of the great importance now attached to population dynamics on the part of NGOs, churches and political bodies.

4.2.3
The 1994 International Population Conference

A central topic of the 1994 *International Conference on Population and Development* (IPCD, *Box 20*) in Cairo was empowerment of women. This issue is considered to be the most significant point in connection with reproductive rights (human right to family planning) and reproductive health (access to basic health services) in the action program of the International Population Conference. At the same time the "women's issue" led to a debate over abortion and access to contraceptives.

One of the main causes of the high population growth in the estimation of the ICPD is the weak social, economic and political position of women (*see also Box 20*). Two-thirds of the slightly less than one billion illiterate persons worldwide are women according to UN studies. As long as women continue to be denied rights to self-determination, cannot decide for themselves how many children they have and how long the intervals are between births and are hindered from gaining access to education and social recognition outside of the household, there will be no decisive changes in reproductive behavior.

Even though the conference participants fundamentally agreed that abortion should not be promoted as a means of family planning, it is not ruled

BOX20

Women's Education and Birth Rate

Women are still subject to sociocultural discrimination in most societies today. An improvement of the position of women in all areas of life, therefore, is a value in itself and an ethical requirement. A central element in improving the social position of women is their education and training. In addition, there is a close connection between the level of education and the reproductive behavior of a society, a fact that is of major importance regarding the question of population growth.

The success of women's education programs depends to a great extent on the cultural and social conditions in a society. Numerous studies have proved empirically that education for women generally induces a decline in birth rates. A UNFPA study, for instance, showed that women in Brazil without any schooling have 4 children more on average than women who have attended school. In India the proportion of women who had completed primary school increased slightly from 1989 to 1991, resulting in a decline in the birth rate from 4.2% to 4.0% in the years thereafter.

The education and training of women have an effect on the birth rate through several factors. First of all, education is a basic requirement for women to be able to improve their economic situation. Studies conducted by the World Bank have shown that children in families where the mother earned her own income grew up in a more healthy environment than in families where the father alone provided the household income. Women who have their own income spend a larger portion of it on goods necessary for the existence of the family. Two examples: The expenses necessary to provide children in Guatemala with better nutri-

tion are fifteen times higher on average when the income is earned by the father and not by the mother. In the Ivory Coast a doubling of the income managed by women reduced the share of the household budget spent on alcohol by 26% and the amount spent on cigarettes by 14% (World Bank, 1993). In addition to the positive effect on children's health, the income of women is a significant factor with respect to population. First of all, an improvement in the health situation of children means a lower mortality rate for children. As the mortality rate for children declines, the birth rate also drops. Moreover, when women have their own income, they are less dependent on their husbands and therefore on giving birth to children as security for their old age (Sadik, 1994).

Another important aspect is the link between women's education and the use of contraceptives. Other factors besides women's education also play a role here, in particular the existence of family planning services and the level of education of men. In a study conducted by the World Bank in 15 African countries, however, a significant positive correlation was found in most cases (12 countries) between the number of years women had attended school and the use of contraceptives. In fact, longer school attendance had a more than proportional effect on contraceptive use (Ainsworth, 1994).

The education that women receive at school also influences the kind of education they want for their children. Since children who go to school usually produce costs or at least opportunity costs, parents often decide to have fewer children. This means that better education for women can reduce the birth rate due to this aspect as well. This positive correlation was verified in two studies, in Ghana and in the Ivory Coast (Montgomery and Kouamé, 1994; Oliver, 1994).

out as ultima ratio because it is allowed under certain conditions in almost all UN states. This issue was the subject of broad discussion, not least of all due to the intervention of the Catholic Church. The Vatican does not contradict a strengthening of the role of women in this connection. The Pope himself even pointed out that there "may be objective reasons for restricting [...] births" (Kommission Weltkirche, 1993). In June 1994 the *Pontifical Academy of the Sciences* financed by the Vatican stated that it is necessary "to curb birth figures so as to avoid the creation of insoluble problems" (DGVN, 1994).

However, the Vatican immediately dissociated itself from the Academy's study. 114 of the 139 Catholic cardinals published an appeal in which they came out against the "imperialism" of the *Cairo Conference* that in their view leads to abortion on request, sexual promiscuity and a distorted understanding of the family. The Catholic Church would like to avoid a situation in which "the change in reproductive behavior in the developing countries drives people [...] into the dead end of human poverty and the loss of human values" (Kommission Weltkirche, 1993). The Vatican fears that the "human right to family planning" (proclaimed at the *International Human Rights Conference* in Teheran) will be misunderstood as a "human right to abortion". Its well-known position ("natural" contraception) was reinforced in a position paper of the *Popal Council for the Family* that was also published on the occasion of the International Population Conference. This paper repeatedly calls for a commitment against world population and family planning organizations and appeals against artificial methods of contraception (Kommission Weltkirche, 1993).

An analysis of the demands of national and international NGOs, including churches, the German federal government, the European Union and the United Nations, shows – apart from certain differences primarily regarding the ethical assessment of specific measures of birth control – that there was extensive agreement on the following objectives within the framework of the International Population Conference:

- ensuring fundamental preventive health care and aftercare, particularly to reduce infant and mother mortality; this is coupled with improvement of hygiene and basic medical care,
- introduction of adapted, small-scale systems of financing basic social security and health care,
- improvement of economic situation and legal position of women,
- comprehensive educational programs, especially for women,
- extensive advisory networks to provide women, men and young people with access to sex education and all recognized contraceptives.

Implementation of population-related measures should be carried out without state compulsion and while guaranteeing human rights to family planning; exclusive financing by the patients themselves is categorically rejected. Furthermore, health programs must not be restricted to family planning and the interlinkage between development and population policy measures should be kept in mind.

The International Population Conference in Cairo has opened up a significant opportunity for setting the course to solutions for population and development problems. The course set will decide, for example, whether the world's population will level off at 9 billion or at 13 to 14 billion. It must be pointed out, however, that even if the right course is set, the success that results from Cairo will only take effect in the long run.

The increase in armed conflicts illustrates the explosive nature of population growth and of the resulting aggravation of economic and ecological bottlenecks. Such conflicts are frequently attributed to ethnic causes, but the triggering factors can presumably also be found in increasingly restricted living space and worsening future prospects (*Box 21*).

This interrelationship underlines once again the necessity of pursuing the paths pointed out by the 1994 ICPD consistently and with the greatest urgency.

4.3
Recommendations for Action and Research

The 1994 International Conference on Population and Development in Cairo made the international public aware of the fact that the consequences of population growth affect all nations of the earth, even though the industrialized countries currently have very low birth rates. The high migration pressure in the nations of Africa, Asia and Latin America, which will presumably continue to grow, primarily has an effect on the countries of the European Union and North America. This possible migration will cause the already high level of resource consumption to increase even more in the potential immigration regions. Thus the international community is faced with significant environmental and developmental challenges that in most cases require a jointly supported solution concept in view of the extensive interlinkage of problems. An alarming sign in this connection is the fact that the necessity for rapid and far-sighted action is still frequently underestimated. In conformity with the Rio Declaration and AGENDA 21, the Council (WBGU, 1993 and 1994) still views the following as the most important objectives with regard to population growth and distribution:

- long-term stabilization of the population by
  - combating poverty (old age insurance),
  - providing equality for women,
  - recognizing the right to family planning as an individual human right,
  - improving family planning opportunities,
  - reducing child mortality,
  - improving education and training.
- prevention and reduction of forced migration through
  - international cooperation coping with international migration movements,
  - further efforts to intensify awareness of the consequences of uncontrolled migration movements and urbanization processes because they make the development of integrated problem-solving approaches more difficult.
- creation of functional urban structures through
  - international cooperation in the field of regional planning policy,
  - specification of regional planning models that

BOX 21

The 1994 UN International Population Conference in Cairo

The International Conference on Population and Development (ICPD) took place in Cairo from September 5 - 13, 1994 and was chaired by Nafis Sadik, director of the *UN Population Fund*. The Conference was organizationally and physically separated from the meeting of experts of the nongovernmental organizations, the 1994 NGO Forum. The draft for the UN action program of the International Population Conference, comprising 16 chapters, was drawn up at the preparatory conferences (PREPCOM I-III) prior to the ICPD, and the nongovernmental organizations left a noticeable mark on the structure and formulation of the program. The purpose of the ICPD was to establish an international consensus regarding controversial passages of the program draft.

The participants of the ICPD consisted of approx. 3500 members of the national delegations that, in turn, were composed of governmental representatives, representatives from semigovernmental organizations and in a few cases NGOs (the German delegation, for instance). Sudan and Saudi Arabia were among the few states that did not take part in the Conference.

The center of the NGO Forum was Cairo's Indoor Stadium, which contained 120 stands. Opportunities for exchanging opinions and maintaining contacts were offered at approx. 70 daily events. The purpose of the NGO Forum with approx. 5,000 registered participants from roughly 1,500 organizations was a general exchange of information and ideas with regard to population de-velopment and policy.

The focus of discussion at the ICPD was placed on "Reproductive Rights", "Reproductive Health" and "Strengthening the Position of Women". The great importance of strengthening the role of women for population policy and sustainable development was recognized for the first time within an international consensus - even by representatives of nations in which women traditionally have a weak social role. The emphasis on "Reproductive Health" is less a reorientation of population policy than a health policy recognition of the fact that approx. 500,000 mothers annually die as a result of pregnancy. However, no concrete agreement was reached with regard to the question of improperly performed abortions, which account for 25% to 40% of all cases of mortality involving pregnant women. The basic ethical positions on the legality or illegality of abortion did not permit such a consensus.

The ICPD action program represented the first time that far-reaching, detailed statements on reproductive rights were made in a conference document, including stipulation of the right to deciding on the number of children and the right to a healthy and safe sexual life. Different forms of families and their right to reproduction and family planning were recognized, though emphasis was also placed on the importance of the family as the core of society.

The long debate over population policy prevented the conference from dealing more intensively with the second part of its title, "Development". An essential discussion regarding the distribution of resource consumption and economic disparities between North and South did not take place. Thus the contents of the action program do not go beyond AGENDA 21 in this respect.

allow for a harmonization of "environment and development" (e.g. through a balanced mixture of land-use structures within the city, see Chapter 7 of AGENDA 21),
– creation of polycentric instead of monocentric structures of land use (see Chapter 7.19-7.22 of AGENDA 21),
– technology transfer (avoidance or reduction of emissions and wastes in urban agglomerations).
The Council places special emphasis on the demand for recognition of the individual right to family planning, with particular focus on information on and access to methods and means of individual family planning. The Council recommends the most rapid

possible provision of an institutional basis for population policy in the states concerned and feels that it is additionally necessary to reinforce efforts in support of women within the framework of the development cooperation.

These objectives should be supported by binding agreements and long-term financing programs (*see* Chapter 2 of AGENDA 21). At the same time the respective countries must be given the opportunity of implementing these programs on their own responsibility, while taking into account the respective conditions and sensitivities (culture, religion).

An in-depth study of the "man-society-environment" system is essential in the research field of "population growth, migration and urbanization". Such research requires well-supported statements on the carrying capacity of the Earth, the people-related consumption of finite and renewable resources as well as forecasts on the quantitative and qualitative magnitude of local, regional and global pollution.

Another requirement is an analysis of the political structures, including individual and social behavioral patterns and their effects on sustainable development. The development of innovative analytical and forecast methods is imperative for collecting and evaluating environmentally relevant data in the natural and social sciences so that interrelationships on which the "man-society-environment" system is based can be adequately assessed.

Special importance is also attached to the development of practicable analytical and forecast methods with regard to the research field of "migrations", the focus of which is on the quantification and qualification of push and pull factors.

Carrying capacity determinants have to be identified and assessed for the research field of "urbanization". This would be an important step towards localization, qualification and quantification of the urbanization process with respect to global environmental changes as well as towards determination of functional city sizes, depending on the local and regional conditions.

Global information networks (and coordinated databases) for quick data exchange can facilitate rapid, selective action on the part of the international community as soon as crises are detected. The *Global Resource Information Database* (GRID) of the UNEP is an example of this, though no network nodes have been set up for it in Germany as yet.

International Conventions Aimed at Solving Global Environmental Problems

C

1.1
The Berlin Climate Conference – Results and Assessment

1.1.1
Climate Politics Between Ecological Necessities and Political Restrictions

The signing of the Framework Convention on Climate Change at the *UN Conference on Environment and Development* in Rio de Janeiro in 1992 and its subsequent coming into force were a major step towards a global climate policy. The international community declared through this convention that it is determined to pursue a global policy to protect the climate: "The ultimate objective of this Convention is to achieve stabilization of greenhouse gas concentrations in the atmosphere at a level that would prevent dangerous anthropogenic interference with the climate system. Such a level should be achieved within a time frame sufficient to allow ecosystems to adapt naturally to climate change, to ensure that food production is not threatened and to enable economic development to proceed in a sustainable manner" (Johnson, 1993).

The Convention for protecting the global climate has meanwhile been ratified by more than 120 states (as of April 1995). However, since it is explicitly a Framework Convention, it is still of necessity rather vague (WBGU, 1994; Enquete Commission, 1995b), both with respect to objectives as well as the definition of the instruments for implementing and monitoring the Convention. The significance of the Convention lies in the institutionalization of the climate policy process, the establishment of a legal framework and some general principles; further steps, especially binding targets and schedules for the reduction of greenhouse gases, will have to be specified at subsequent Conferences of the Parties.

It is against this background that the first *Conference of the Parties* (COP-1), held in Berlin between March 28 and April 7, 1995, should be evaluated. The Conference failed to reach agreement on specific measures for reducing CO_2 emissions, despite the urgent need for such action as shown by the alarming scientific evidence for anthropogenic destabilization of the climate (IPCC, 1994; MPI, 1995; Enquete Commission, 1995b; WBGU, 1995). Despite the warnings by scientists about the disturbing trends, there has been constant increase from one year to the next in the use of fossil fuels and hence in the level of CO_2 emissions on a worldwide scale (IEA, 1994). There is no empirical evidence for a change in this trend, nor can any such change be anticipated until now, reasons being, for example, the growth in world population and the expansion of the world economy.

According to calculations made by the Advisory Council, a reduction in greenhouse gases can be accomplished by relatively small steps at first, but subsequently high-level reductions over many decades would have to follow (*see Section C 1.3*). These calculations also show that the time scale remaining for these measures to take effect is extremely short, and for this reason the failure of the Berlin Climate Conference to reach agreement on appropriate reduction targets was very disappointing.

The sessions of the *Intergovernmental Negotiating Committee* (INC) prior to COP-1 had already dampened expectations. Some states had even taken the view that the nonbinding and vague commitments in Article 4.2(a) and (b) of the Climate Convention were adequate. This assessment would have rendered obsolete any further steps toward an active climate protection policy. The Framework Convention on Climate Change itself, and the results of INC I-XI made the following demands on the Conference (Estrada Oyuela, 1995; Merkel, 1995):
- The adequacy of commitments entered into hitherto must be reviewed. If such a review determines that commitments are inadequate, further steps are to be laid down and a protocol drawn up containing binding commitments.
- A substantial mandate for the working out of a reduction protocol is to be agreed upon.

– Agreement must be reached on whether and how the instrument of Joint Implementation referred to in the Climate Convention should be established.

– A decision has to be made on the finance mechanism, and the reports of the industrialized countries on their national climate protection policies are to be subjected to review.

– Agreement is to be reached on the voting procedures for the Conference of the Parties.

– An infrastructure (e.g. a secretariat) must be established for implementing the Framework Convention on Climate Change.

1.1.2
The Results of the First Conference of the Parties – An Overview

Article 4 of the Framework Convention on Climate Change does not set a binding deadline for stabilizing CO_2 emissions at the 1990 level. The target year 2000 is only mentioned in passing, and does not represent any firm undertaking of the Convention, e.g. analogous to Montreal Protocol commitments. Above all, no position is taken on the reduction of greenhouse-gas emissions after the turn of the century (post-2000 target). Nor does Article 4 refer to other greenhouse gases. It was therefore of crucial importance that the Conference determined that previous commitments were essentially inadequate. This resolution paved the way for the Mandate to work out a draft protocol for further reductions in greenhouse gases.

1 The Parties adopted the so-called "Berlin Mandate" (UN Document FCCC/CP/1995/L.14), which establishes a negotiation process with a view to adopting the results in time for the Third Conference of the Parties in 1997 for a protocol or other legal instrument with the following contents:
    - reductions in excess of the stabilization of greenhouse gas emissions already agreed upon,
    - specified time frames for achieving these objectives could be 2005, 2010 and 2020,
    - a concept for concrete measures for reducing greenhouse-gases must be presented,
    - the industrialized countries provide a commitment in relation to their historical and present responsibility as primary polluters for emission reductions,
    - no commitments are envisaged for the developing countries, but instead they are granted the right to sustainable development and hence a reasonable increase in greenhouse-gas emissions.

Even though many wishes remained unfulfilled, these provisions of the Berlin Mandate can be considered an important partial success of the first Conference of the Parties.

2 A pilot phase for climate-protection activities which are to be implemented jointly (Joint Implementation) was launched (UN Document FCCC/CP/1995/L.13). This shall last for a maximum of five years and serve to gather experience on the basis of which binding criteria for the long-term application of this instrument for CO_2 reductions can be developed step by step. Until such time as agreement is reached on application criteria, joint activities during the pilot phase shall be conducted under open conditions. Project organizers could be state bodies, private companies and nongovernmental organizations. Participants can be those states listed in Annex I to the Framework Convention on Climate Change (industrialized countries and countries with economies in transition), on a voluntary basis. Due to the absence of any binding international rules and in order to allay the fears expressed by developing countries in particular that industrialized countries would neglect to achieve reductions in their own countries in favor of Joint Implementation, no credits on national reduction targets can be claimed for reduced emissions achieved in other countries during the pilot phase. The Council expressly welcomes the fact that a consensus – albeit limited – was achieved in this highly controversial area, in that Joint Implementation of climate protection projects offers substantial potential for additional climate protection as well as for additional developmental impulses (*see Section C 1.4.4*).

3 The Parties did not manage to agree in Berlin on Rules of Procedure. The reason was disagreement regarding the voting mechanism to be chosen. However, in view of the conflicts which can be expected, a departure from the "consensus procedure" (unanimous voting) would appear necessary. The Framework Convention on Climate Change already provides for a three-fourths majority vote for adopting amendments, so the logical conclusion is to include majority voting in the Rules of Procedure.

4 The *Global Environmental Facility* (GEF) was maintained for a further four years as the financial mechanism (UN Documents FCCC/CP/1995/L.1 and FCCC/CP/1995/4). Its financial volume is now, therefore, totally inadequate (WBGU, 1994), and urgently requires strength-

ening, in the view of the Council, on account of the major structural adjustments that can be expected.

5 One important instrument for international climate protection under the Framework Convention on Climate Change is technology transfer (Articles 4.1, 4.5, 4.7, 4.9, 9.2 and 11.1 of the Convention; *see also Section B 2*). The industrialized countries have a special responsibility for ensuring that technology and know-how is passed on to developing countries, as required by AGENDA 21 (Chapter 34). To ensure progress in this field, the Secretariat of the Convention shall report on the status of technology transfer before COP-2 is held in 1996. In addition, the requirements and conditions for successful technology transfer in accordance with the climate protection objectives will also be laid down and submitted as results for COP-2 (UN Document FCCC/CP/1995/L.10) (*see also Section B 2*).

6 Another factor that is important for the functioning of international conventions involves checks and controls as to whether or not emission reductions are actually being achieved (verification). What is needed is standardization of the national communications, which have so far often proved of little use, in order to improve their relevance and comparability. It was therefore essential that the Conference of the Parties agreed on commonly accepted methodologies for preparing national communications. The Conference was able to reach agreement in this regard, refering to the proposals submitted by the IPCC (*Intergovernmental Panel on Climate Change*), for example, on common measurement and assessment issues.

7 The choice of Bonn as the location of the Permanent Secretariat of the Convention is linked to the hopes and expectations being placed in German policies for climate protection. This is the background for the national commitment that the Chancellor stated at the Berlin Conference. This target is so high that its instrumental implementability must now be demanded to an even greater degree than before.

8 Some of the other activities directed at climate protection which occurred in Berlin during the Conference are also of importance. Two initiatives deserve special mention: the establishment of a *European Business Council for a Sustainable Energy Future* (*Box 22*), aimed at strengthening initiatives within the business community, which had hitherto been articulated at national level only, and the combination of local government efforts to protect the climate in the *International Council for Local Environmental Initiatives* (*Box 23*).

BOX 22

European Business Council for a Sustainable Energy Future

The objective of this organization founded in Berlin is to strengthen, through an alliance of industry representatives with environmental groups, administrators and politicians, a form of economy that is tolerable for the climate, and to strengthen climate protection policies which are geared to sustainable development. The Business Council wishes to show that companies can respond to the demands of climate protection with the most modern, internationally competitive technologies, so that environmental protection considerations can improve economic success. The earlier businesses recognize the trend toward environmentally benign technologies, the better equipped they will be to survive the international competition in energy efficiency.

The Business Council for a Sustainable Energy Future, formed by companies and associations in the fields of electricity, gas, renewable energies and energy efficiency, was established in the USA as early as 1992. It takes action for rapid market introduction of low-CO_2 and CO_2-free (regenerative) energy resources, and for increasing the energy efficiency in industrialized and developing countries. This involves technologies and measures such as automobile engines that do not burn mineral oils, renewable energy sources in the power supply industry, greater use of combined heat and power generation and district heating, and, last but not least, a rethinking of research priorities in the energy field.

The Business Council pointed out at the Berlin Conference that climate protection measures should not be seen as a cost factor only, but that they can lead to economic benefits in the long term. The Council therefore advocates the introduction of energy taxes, greater support for renewable energy sources and a departure from environmentally damaging subsidies.

The Council's summary assessment of the Berlin Conference is that it failed to fulfil the hopes placed in it at Rio de Janeiro: firstly, no protocol was established, and secondly, the wording of the Berlin Mandate gives rise to worries that the substance of the protocol to be adopted in two years might not match up to original expectations. On the other hand, there is no denying that the first Conference of the Parties, by acknowledging the inadequacy of the commitments and adopting the Mandate to draw up a protocol by 1997, has taken the next steps for an effective climate protection policy. What is important now is that existing commitments and targets be maintained and developed further in a determined fashion, in order that the Framework Convention on Climate Change becomes a powerful instrument for international climate protection.

1.2
The Relevance of the Human Factor to Climate

A number of new scientific findings were presented on the occasion of the Berlin Conference, all of which underscore the urgent need of political action to protect the climate. Particularly noteworthy are the model calculations of the Max Planck Institute for Meteorology, Hamburg and the Hadley Centre for Climate Prediction and Research, Bracknell, UK, as well as the interim report of the Intergovernmental Panel on Climate Change, requested by the Intergovernmental Negotiating Committee for the Berlin Conference. The most important findings are summarized below.

1.2.1
Evidence for the Anthropogenic Greenhouse Effect

New model calculations and statistical analyses carried out by the Max Planck Institute for Meteorology in Hamburg have shown that there is a 95% probability that the temperature changes of the last 30 years were not caused by natural fluctuations in climate. Even though the word "proof" was deliberately avoided in the discussion (Hasselmann as cited by Göpfert, 1995), the Council assumes that for such a high probability people are to be seen as the causal agents of global warming: only anthropogenic deposition of trace gases in the atmosphere is able to explain the observed degree of radiative forcing.

In the study produced by von Hegerl et al. (1994), the observed trends of mean surface temperatures in the period since 1860 were tested using the so-called "optimum fingerprint method" to see if they could be explained by natural climate variability. This method uses geographical patterns, whereby observations of regions with relatively low greenhouse-gas signals relative to natural temperature variability are weighted lower than regions with relatively large greenhouse-gas signals relative to natural temperature variability. However, there is insufficient longitudinal data as yet to quantify these findings. The global geo-

---

BOX 23

ICLEI – International Council for Local Environmental Initiatives

ICLEI, the association of communal environmental protection initiatives, showed at the Municipal Leaders' Summit on Climate Change held parallel to the Berlin Conference that there are already numerous successful activities at local level worldwide, alongside the slow progress of international climate protection policies. 20 cities have already committed themselves to a 20% reduction of CO_2 emissions by the year 2005.

Representatives of 160 cities from 65 countries (representing over 250 million people) at the Municipal Leaders' Summit presented a large number of local measures for protecting the climate. These include stronger support for renewable energies and ecological housing construction, as well as improved energy efficiency of public facilities and buildings and greater use of less environmentally damaging transportation in cities. Because the participants, as local decision makers, are closer to individual decisions relating to environmentally sound action than politicians at national or international level, they demand greater involvement in energy-sector decision making and the creation of an appropriate national framework.

In their final communiqué, the participants at the Municipal Leaders' Summit expressed strong support for the AOSIS protocol calling on developed nations to commit themselves to a 20% reduction in CO_2 emissions. Nonindustrialized countries are called on not to follow in the footsteps of the industrialized countries, but instead to decouple economic growth and energy consumption by using renewable energy sources and increasing their energy efficiency.

Figure 12
Development of the
detection variable for
observed data and for the
model predictions driven
by anthropogenic climate
disturbance (observed
between 1935-1985,
forecast on the basis of a
Business-as-Usual
emissions scenario from
1985 onwards; global
coupled atmosphere-ocean
circulation model
ECHAM/LSG).
Source: Hegerl *et al.*, 1994

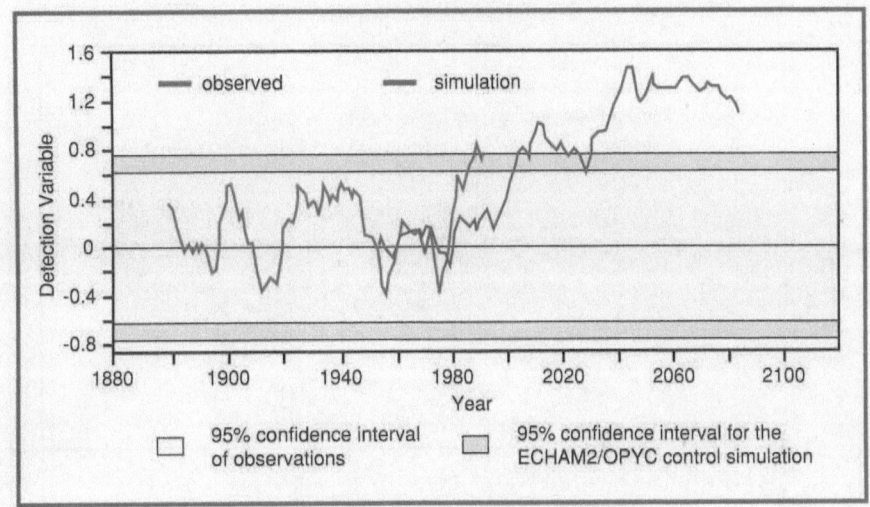

graphical pattern of natural climate variability was
therefore simulated with climate models (global coupled ocean-atmosphere circulation model, simulation without anthropogenic disturbance).

Figure 12 shows that on the basis of an analysis of
the 20-year mean trends since 1935, the variable selected for detection of the anthropogenic climate sig-
nal in 1990 (this value corresponds to the trend for
the period between 1971 and 1990) exceeds the
threshold value equivalent to a statistical probability
of 95% for such detection. The detection variable
represents the geographical pattern of the observed
temperature trend. The regions of the Earth that are
not adequately represented in the observed data due
to inadequate measurements were filtered out during the analysis. The development of the detection
variable, shown as a function of natural climate variability, is tested statistically against two data sets describing climate variability, namely the observations
adjusted to exclude anthropogenic disturbance and
the temperature curves as predicted by the climate
models (excluding anthropogenic disturbance). The
data used to exclude anthropogenic disturbance
from the observed data are simulated by the climate
models with and without anthropogenic greenhouse-
gas concentrations (observations 1935-1985, Business-as-Usual emission scenario from 1985 onwards;
IPCC, 1990).

1.2.2
Accounting for Anthropogenic Sulfate Aerosols in Global Circulation Models

Anthropogenic impacts on climate due to greenhouse-gas emissions are partially offset by increased
release of aerosols, although with considerable regional variations (WBGU, 1993). Aerosols interfere

with the radiation balance. Their strongest anthropogenic sources are located in the industrialized regions
of the northern hemisphere. This partial compensation for global warming, which has not been taken
into account by climate models to date, is considered
to be one reason why model calculations have so far
produced a temperature increase of 0.95 ± 0.35 °C
since the beginning of industrialization, whereas the
actually observed temperature increase is only 0.45
± 0.15 °C.

Efforts are now being made to examine the complex effect of aerosols on the climate using coupled
ocean-atmosphere circulation models. However,
there is still some uncertainty regarding the direct
impact on climate of anthropogenic aerosols: estimates for the globally averaged effect produce a level of radiative forcing since preindustrial times of
between -0.25 and -1 W m^{-2} (*Fig. 14*; Charlson et al.,
1992; Kaufman and Chou, 1993; Kiehl and Briegleb,
1993). It has been virtually impossible to quantify the
indirect aerosol effect: an initial estimate for the global average is around -1.3 W m^{-2} (Jones et al., 1994).

Despite the number of aerosol effects on climate
that have not been integrated into climate models, it
has been possible to improve the congruence between the model results and observed data for globally averaged surface temperature (*Fig. 13*; Hadley
Centre, 1995). In this model, the anthropogenic aerosol is represented by sulfate aerosol alone and its radiative forcing effect exclusively as an increase in al-
bedo. Thus other effects of the aerosol are not represented, e.g. the influence on cloud formation. Anthropogenic sulfate aerosols are generated by combus-
tion processes in which SO$_2$ is emitted. Assuming an
unrestricted increase in emissions (Business-as-
Usual; IPCC, 1992), lower increases in temperature
are expected for the decades up to 2050 compared to
earlier model predictions, namely approx. 0.2 °C/de-

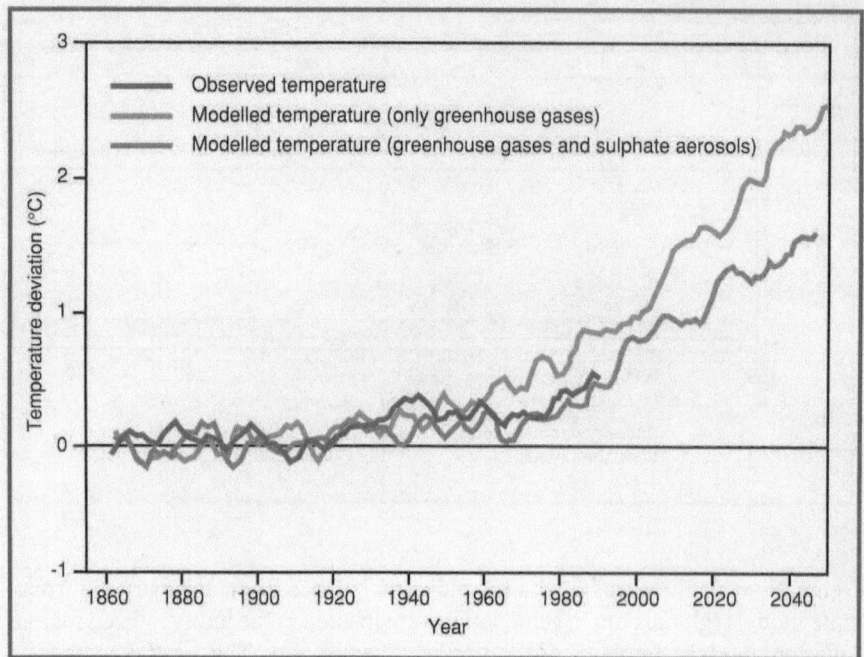

Figure 13
Development of ground-level global mean temperature, observations (1860-1990) and predictions obtained using a global coupled ocean-atmosphere circulation model (1860-2050) with and without aerosol effect. Source: Hadley Centre, 1995

cade (globally averaged temperature). If the aerosol effect is not taken into account, the model predicts a temperature rise of about 0.3 °C/decade.

The forecast for the geographical distribution of temperature change is still imprecise. Maximum reductions in warming through the aerosol effect are expected to occur to the east of the industrialized regions of the northern hemisphere.

1.2.3
Anthropogenic Forcing
of the Earth's Radiation Balance

The anthropogenic increase in the concentration of long-lived greenhouse gases such as carbon dioxide (CO_2), methane (CH_4) and nitrous oxide (N_2O, laughing gas) continued throughout the 1990s, albeit at varying rates. In the case of the fully halogenated chlorofluorocarbons (CFCs), which are exclusively anthropogenic in origin, the rate of increase in atmospheric concentration declined from 5% in the early 1980s to a present figure of less than 2% per annum. This is directly attributable to the enforcement of the restrictions on production imposed in 1989 under the Montreal Protocol and which have been toughened further since then (*see Section C 2*).

It is possible to calculate the radiative forcing caused by changes in the concentrations of greenhouse gases. *Figure 14* shows the scale of human-induced radiative forcing since the onset of industrialization. These perturbations have not yet developed their full impact, since temperature changes are delayed by the thermal inertia of the oceans and the polar regions, which makes it more difficult to identify the anthropogenic contribution to climate change (*see Section C 1.1.1*). The most important conclusions to be drawn here are:

– Radiative forcing by long-lived greenhouse gases, estimated at 2.45 ± 0.3 W m^{-2}, is already equivalent to an increase in solar radiation of more than 1%.

– The increase in the concentration of tropospheric ozone, globally averaged at 0.4 ± 0.2 W m^{-2}, is the second largest contributory factor. The increase in ozone is caused primarily by increasing NO_x deposition, produced mainly by the energy and transport sectors.

– Increased backscattering of solar energy by anthropogenic sulfate aerosol (originating from sulfur dioxide emissions) reduces radiative forcing by somewhere between -0.25 and -0.9 W m^{-2} (*see Section C 1.1.2*).

– Increase in solar radiation due to natural variability is relatively low compared to the sum of the anthropogenic effects.

– There is still considerable uncertainty regarding radiative forcing by increased water vapor in the stratosphere.

If all these factors are combined, however, the result is clearly positive radiative forcing and increased global warming.

Figure 14
Globally averaged radiative forcing due to human activities from pre-industrial times to the present day. An indication of relative confidence in the estimates is given below each bar.
Source: IPCC, 1994a

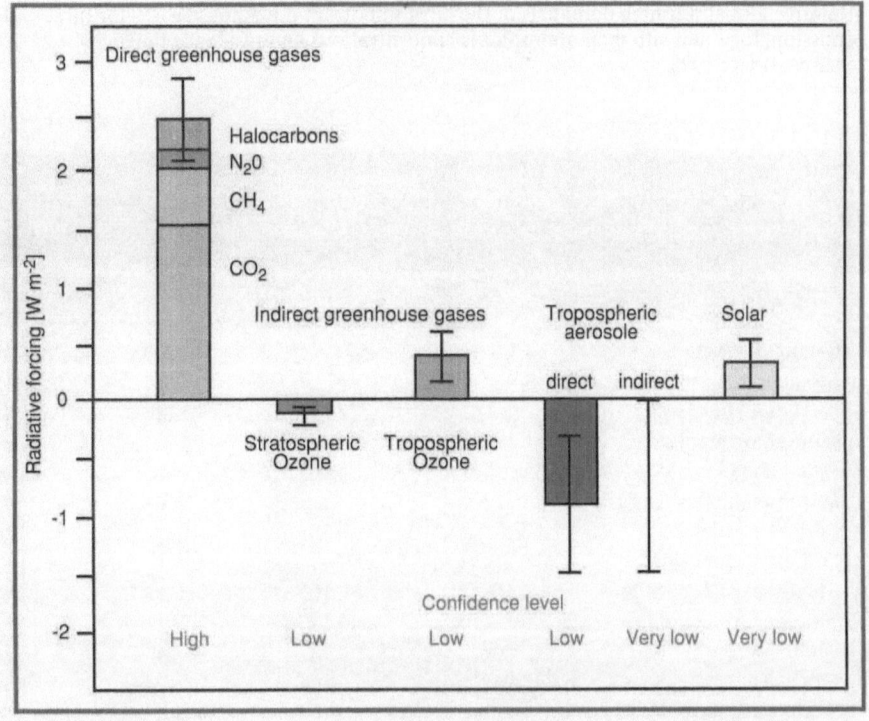

1.2.4
Time-dependent Relative Global Warming Potentials as the Basis for Political Decision Making

The radiative effect of a gas depends on its molecular structure, its concentration and, due to the spectral overlaps and chemical interactions in the atmosphere, to the presence of other gases which absorb radiation. Because CO_2 is the dominant anthropogenic greenhouse gas on account of the high emission levels, it has become common practice to measure the global radiative impact of other gases relative to CO_2. This procedure has been retained despite the problems of determining the residence time of anthropogenic CO_2 in the atmosphere, since several complex processes influence the uptake of CO_2 into the terrestrial biosphere, the oceans and the sediments on the ocean floor.

Important global warming potential figures are shown in *Table 4* (global warming potential per unit mass of an emitted gas relative to the unit mass of emitted CO_2, the so-called relative global warming potential). For example, as little as 0.03 kg N_2O can have the same impact on climate as 1 kg CO_2. The residence time of the gases in the atmosphere (the time taken for concentration to fall to $1/e = 0.37$) displays considerable variation due to very different sinks, so the impact on climate also depends on the time horizon considered after emission. When calcu-

lating the global warming potential of methane (CH_4) in *Table 4*, its indirect impact on climate, namely enhanced ozone formation in the troposphere, increased water vapor concentration in the stratosphere and the formation of a CO_2 molecule were also taken into account. This indirect global warming effect of methane is just as important as the direct effect.

Table 4 reveals the long-term problem that greenhouse-gas emissions are creating. It also shows that the effects of long-lived greenhouse gases with equally high global warming potentials, even if they are only present as minute traces (e.g. carbon tetrafluoride, CF_4), can be equivalent to those of methane and nitrous oxide, taken over a time horizon of a century, even though the current emission rate is very low.

The protocol to the Climate Convention should therefore give consideration not only to CO_2, but also to other greenhouse gases, given these time horizons and their relative global warming potentials.

Table 4
Relative global warming potentials of the most important greenhouse gases for three different time horizons after emission, together with their atmospheric concentrations and residence times.
Source: IPCC, 1995

Species	Chemical Formula	Concentration [ppbv]	Lifetime [yr]	Global Warming Potential Time Horizon 20 years	100 years	500 years
Carbon Dioxide	CO_2	35,500	50-200	1	1	1
Methane*	CH_4	1,714	14	62	24.5	7.5
Nitrous Oxide	N_2O	311	120	290	330	180
Monofluorotrichloro-methane (F11)	$CFCl_3$	0.365	50	5,000	4,000	1,400
Difluorodichloromethane (F12)	CF_2Cl_2	0.503	102	7,900	8,500	4,200
Difluoromonochloro-methane (F22)	CF_2HCl	0.105	13.3	4,300	1,700	520
Carbon Tetrachloride	CCl_4	0.140	42	2,000	1,400	500
Halon-1301	CF_3Br	0.002	65	6,200	5,600	2,200
HFC-134a	CH_2FCF_3	no data	14	3,300	1,300	420
Sulfur Hexafluoride	SF_6	0.003	3,200	16,500	24,900	36,500
Tetrafluoromethane	CF_4	0.070	50,000	4,100	6,300	9,800

*includes the direct effects and those indirect effects due to the production of tropospheric ozone and stratosperic water vapour. The indirect effect due to the production of CO_2 is not included.

1.3
Scenario for Estimating Minimum Targets for Global Emission Reductions

1.3.1
The "Backwards" Mode of the Scenario

Until now, model calculations of the relationship between greenhouse-gas emissions and climate change have been carried out in the forwards mode (IPCC, 1990). Based on various assumptions about demographic and economic development, a set of emission scenarios was predefined, and future climate changes predicted on that basis. Such studies are helpful to an extent in identifying nonacceptable ("nonsustainable") pathways of global environmental change, but do not provide any direct answer regarding the preconditions for acceptable (or "sustainable") pathways.

The backwards mode must be deployed instead if such answers are to be obtained: taking account of the impacts of climate change on human beings and nature, a "window" of tolerable future climate change is defined. The next step is to calculate the global emission profiles which ensure conformity within that window. In this way, the minimum demands for a global reduction strategy can be derived directly.

The most recent analyses already use some elements (IEA, 1993; IPCC, 1994). of this "inverse scenario". The Council pursues this path systematically in an attempt to derive political conclusions within a coherent picture. This approach is illustrated in *Figure 15* (more detailed notes on the "Inverse scenario" and definitions of terms used can be found in the Annex at the end of this book).

In Step 1, a – generously dimensioned – tolerance range is defined for potential stresses caused by climate change. Fixing values in this way, which of course involves specific notions about ecologically and economically desirable conditions, enables the objectives defined in Article 2 of the Climate Convention to be operationalized. In a second step, an assessment is made of the climate developments that lead to tolerable stress levels within the predefined limits. Using simplified models of climate dynamics and the carbon cycle, the permissible global emission profiles for CO_2 are determined in Steps 3 and 4 (other greenhouse gases from human sources are ex-

Figure 15
The "Inverse scenario"
applied by the Council.
Source: WBGU, 1995

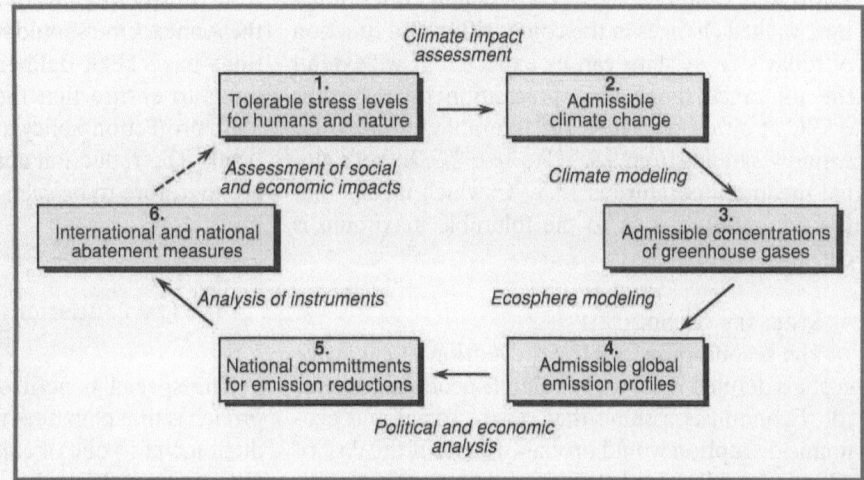

cluded here). It is particularly important here to determine the upper limit for total CO_2 emissions within the selected stabilization period, as well as "optimum" emission profiles that come close to this upper limit.

In Step 5 – subject to the criteria of international environment and development policies – reduction commitments for the individual countries or groups of countries are derived. The sixth and final stage is to analyze which reduction instruments at what locations are able to produce the most effective emission reductions, and which mix of instruments causes the least amount of cost.

Studies featuring a combined focus on all aspects of relevance to the climate problem are appearing at an increasing rate. One current example is a study carried out by Bach (1995) in connection with the German government's Enquete Commission on Climate Protection. However, this analysis, too, is carried out in "forwards mode" (i.e. anticlockwise in *Figure 15*).

In the following analysis, the Council restricts itself in the main to Steps 1 to 4. Furthermore, the permissible minimum reduction function for the Annex 1 countries (Germany included) is determined on the basis of the probable international apportionment index. As far as steps 5 and 6 are concerned, the Council makes a number of points (*Section C 1.4.2*); full clarification of these issues has yet to be achieved, however, and poses a major challenge for research.

The analytical procedure adopted here can be developed further, by assessing the worldwide socioeconomic impact of reduction measures, to produce an integrated model for climate policy (broken arrow between Steps 6 and 1 in *Figure 15*). Within such a model, the costs of adapting to changed climatic conditions can be compared with the costs of preventing climate change. However, the Council dispenses pro-

visionally with this integrated perspective, since it still involves too many uncertainties – the direct and indirect effects of abatement efforts on humans and on nature (e.g. diminishing environmental stress due to reduced traffic levels) are extremely difficult to quantify at present. Taking only direct climatic impacts into account means that the severity of anthropogenic climate forcing is probably underestimated in the scenario.

Within the climate policy process initiated at the Berlin Conference, however, "integrated modeling" obtains greater significance. The Council returns to this topic in a later Section (*C 1.4.2*) and discusses in some detail the function, state of development and perspectives of this scientific instrument.

1.3.2
The Basic Assumptions of the Scenario

In order to arrive at an approximate but well-founded estimate of the possible impacts of climate change, despite the highly complex mechanisms involved, the Council applies the twin principles of
– preservation of Creation
– prevention of excessive costs.
The boundary conditions of the scenario are developed from these two principles.

A TOLERABLE TEMPERATURE WINDOW
The first principle, preservation of Creation, is defined within this scenario in the form of a tolerable "temperature window". This window is derived from the range of fluctuation for the Earth's temperature in the late Quarternary period. This geological epoch has shaped our present-day environment, with the lowest temperatures occurring in the last ice age (10.4 °C) and the highest temperatures during the last interglacial period (16.1 °C) (Schönwiese, 1987).

If this temperature range is exceeded in either direction, radical changes in the composition and function of today's ecosystems can be expected. If we extend the tolerance range as a precaution by a further 0.5 °C at either end, then the tolerable temperature window extends from 9.9 °C to 16.6 °C. Today's global mean temperature is 15.3 °C, which means that the temperature span to the tolerable maximum is currently 1.3 °C.

STRESSES ON SOCIETY

The second principle, the prevention of excessive costs, is defined in terms of a simple economic indicator. Economists assume that severe social and economic disruption would probably ensue if the cost of adapting to climate change, including the cost of repairing damage resulting from climate change, were in the order of 3-5% of Gross Global Product (GGP). In our scenario, a global mean value for the burden on society of 5% of GGP is taken as the utmost tolerable limit (to the extent that this burden can be expressed in monetary terms). One must realize here that, given the uneven spatial distribution of climatic impacts, some states may be affected much more seriously than others (e.g. Bangladesh, island states).

Most estimates of the global annual costs resulting from doubling CO_2 by the end of the next century arrive at 1-2% of global GGP. Doubling CO_2 over that period would lead to a mean temperature increase of 0.2 °C per decade in the various climate models used. However, all of these estimates fail to include either extreme events (droughts, floods, tropical storms), or possible synergies between the various forms of global environmental change. If these events are also included, there is good reason to assume that a temperature change of 0.2 °C per decade would already correspond to an upper limit for adaptation costs of 5% of global GGP. However, much research remains to be done on these questions.

DECLINING ADAPTABILITY

This upper limit for the maximum permissible rate of temperature change is probably only valid for as long as the ecosphere is located in the center of the temperature window. As the upper temperature limit of 16.6 °C is approached, however, adaptability will show constant decline. This means that the tolerable temperature gradient at the upper boundary goes to zero.

With the help of these three basic assumptions, and taking into account a number of other factors (e.g. nonlinear dependencies and irreversibilities), we can now define a two-dimensional climate domain 𝔅 that the climate system should not depart

from. Further details of this analysis can be found in the Annex. One should note that all boundary conditions have been deliberately defined within broad limits to ensure that the resultant demands for climate protection policy are not assessed too pessimistically. The reduction commitments presented below are therefore to be seen as minimum values.

1.3.3
The Key Conclusions of the Scenario

The special benefit of the inverse analytical approach is that climate is not seen as a problem of prediction, but as one of control: the future of the global environment depends to a significant extent on the CO_2 emission profile $E(t)$ of the next centuries, and this profile can be chosen, within certain limits, by humankind.

The model calculations on which this study is based permit the identification and classification of all selectable emission profiles $E(t)$. From the wealth of results obtained, two permissible profiles in particular have been chosen for presentation, each displaying very different characteristics. In view of the major gaps in our knowledge and understanding of the climate system, all of these results are in effect probability statements.

CURRENT SITUATION

The long-term behavior of the climate-carbon model (Hasselmann et. al., 1995) implies that the total volume of all future anthropogenic CO_2 emissions must not be allowed to exceed a certain finite value, if the tolerable climate domain 𝔅 is to be complied with. On the basis of a best estimate of the modeling parameters used here, this value is calculated at almost 1,600 gigatons of carbon (Gt C). If somewhat divergent parameter sets within the existing range of scientific uncertainty are taken, this upper limit could rise to approx. 2,000 Gt C.

The conclusion, however, is the same in all cases, namely that the international community has only a limited "budget" of additional carbon to be deposited in the atmosphere. This budget may well be less than the total amount of carbon stored in the remaining fossil fuel deposits. Of course, there are very different options regarding how this contingent is to be handled: for example, it can be used up as quickly as possible or stretched over as long a time scale as possible.

For physical and chemical reasons, a permanent adjustment of global anthropogenic CO_2 emissions, even at a constantly low level, is not possible, without causing a serious impact on the climate system. Even if fossil fuel reserves were inexhaustible, climate pro-

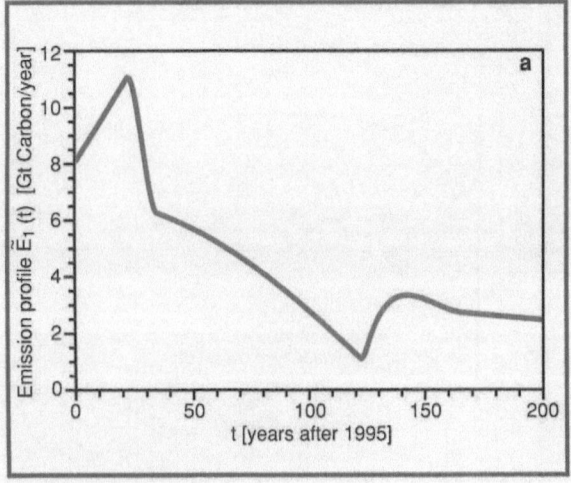

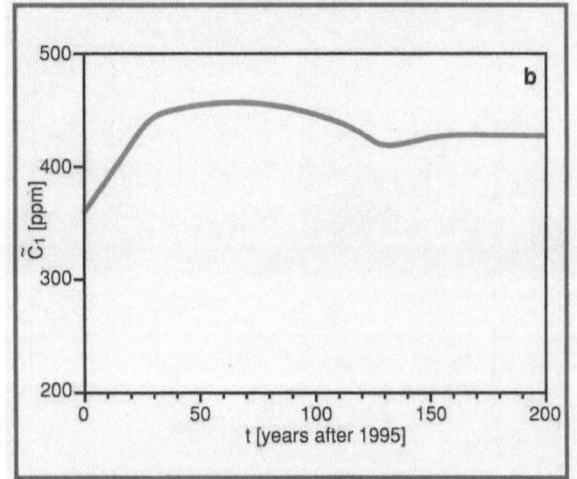

Figure 16 a-c
"Crash-barrier scenario" for initial Business-as-Usual:
a) Global CO_2 emission profile $\tilde{E}_1$
b) Concentration profile $\tilde{C}_1$ resulting from $\tilde{E}_1$
c) Shift in the climate system caused by $\tilde{E}_1$ within the tolerable window. P_0 represents the state of the climate today (X=2.05; Y=0.07).
Source: WBGU

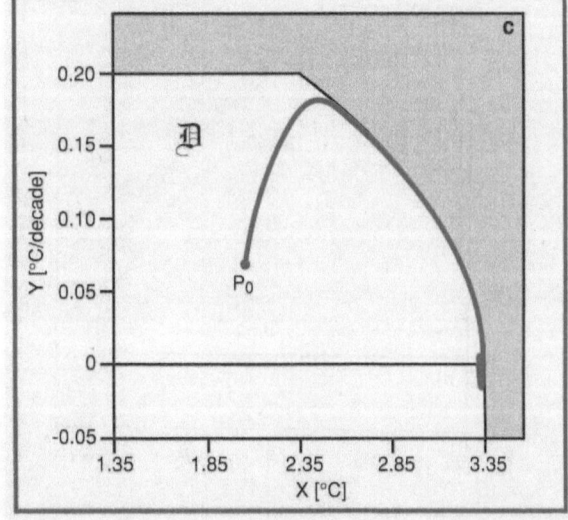

tection considerations would prohibit such a strategy: current scientific knowledge regarding the carbon cycle suggests that a certain percentage of all carbon dioxide emissions remain in the atmosphere for all time, thus leading inevitably to further accumulation (Joos and Sarmiento, 1995). This conclusion would only need to be qualified if there were still some unknown negative (i.e. climate-stabilizing) feedback mechanisms within the Earth System.

The model calculations provide a clear indication of a unique property of the ecosphere, namely that the climate system grants considerable liberties regarding the choice of emission profile E(t), i.e. highly differing distributions of the same emission sum over the next centuries can enable conformity with the climate domain 𝔇. This contradicts somewhat the common assumption that only through immediate and full stoppage of global CO_2 emissions can climate changes be kept within tolerable limits. To that extent there is a certain amount of leeway that permits soci-

ocultural, political and economic criteria to be taken into account when defining the global emission profile. However, this does not mean that future CO_2 emissions can be set arbitrarily: the boundary conditions defined by 𝔇 are of crucial importance in the medium-term planning range especially, as the following business-as-usual example illustrates.

AN UNDESIRABLE "CRASH BARRIER SCENARIO"

Global emissions of CO_2 are currently increasing at a rate of approx. 1.7% per annum relative to the 1994 level. Hypothetical projection of this linear trend is termed "Business-as-Usual". The Council's model computation shows that such emission behavior could lead within less than 30 years to the limits of the tolerable climate domain being reached; the climate system could then be kept within the permissible range only if drastic changes were implemented, i.e. a reduction of emissions by approx. 40% within

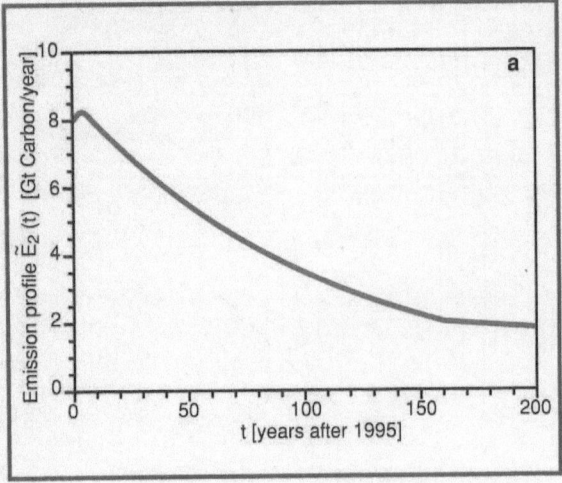

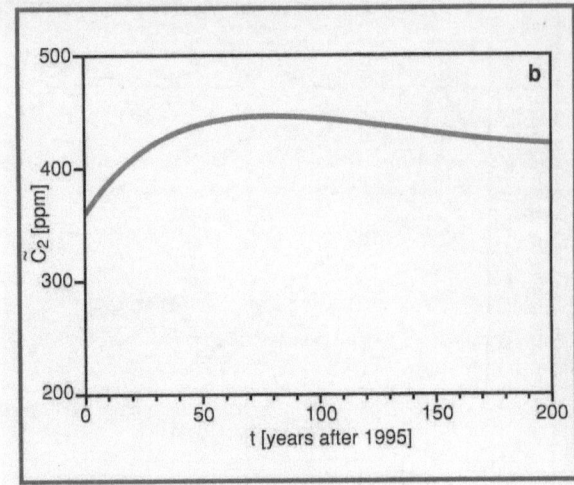

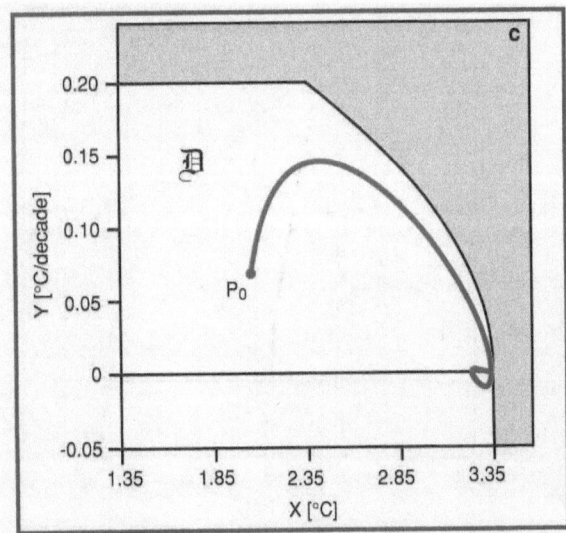

Figure 17 a-c
"Council scenario", with a constant annual percentage reduction rate:
a) Global CO_2 emission profile $\tilde{E}_2$
b) Concentration profile $\tilde{C}_2$ resulting from $\tilde{E}_2$
c) Shift in the climate system caused by $\tilde{E}_2$ within the tolerable window. P_0 represents the state of the climate today (X=2.05; Y=0.07).
Source: WBGU, 1995

only a few years. The relevant emission profile that makes maximum use of the scope defined by 𝔻 has a very irregular shape. If this profile is "smoothed" to account for the actual scope for control (which is limited by technological and socioeconomic factors), the emission function $\tilde{E}_1(t)$ is obtained. *Figure 16a* shows this profile for the first 200 years.

The profile shown is "optimized" only in the sense that the accumulated CO_2 emission level at any point in time is approximated to the maximum realized level acceptable for the climate. This results in annual mean emissions of 4.2 Gt C. As a comparison, the emissions of the industrialized nations (countries listed in Annex 1 of the Climate Convention) are currently about 6.3 Gt , those of the developing countries around 1.6 Gt C. Current global emissions therefore amount to approx. 7.6 Gt annually.

Figure 16b shows the atmospheric CO_2 concentration $\tilde{C}_1(t)$ corresponding to the emission profile $\tilde{E}_1$ for

the next 200 years. $\tilde{C}_1$ leads to a transient peak concentration of 458 ppm and long-term stabilization at approx. 425 ppm. The preindustrial concentration was approx. 270 ppm – well below this hypothetical final level.

$\tilde{C}_1$, on the other hand, produces a shift in the climate system within the phase space (*see Annex*); this permissible climate development for all time is illustrated in *Figure 16c*. It must be noted, however, that this shift occurs largely at the outer periphery of the tolerable domain 𝔻 ("along the crash barrier", so to speak), and therefore represents a high-risk emission strategy.

The Council's conclusion is that emission profile $\tilde{E}_1$, a continuation of Business-as-Usual, is neither feasible nor desirable. The checks and controls on the climate system this would entail would not only dispense with safety margins, but could also mean having to make extreme adjustments. For example, the reduction requirements that would arise after approx. 30 years would exceed the elasticity of the world economic system. After about 125 years, global CO_2 emissions would even have to be reduced temporarily to almost zero in order to make up within a short space of time for the emission reductions that were not made at an earlier stage.

THE COUNCIL'S SCENARIO

In *Figure 17a*, the Council presents the global emission profile $\tilde{E}_2(t)$ which it considers to be the more favorable alternative, both ecologically and economically.

This climate protection strategy would involve, after a 5-year transitional period (initial "bending" of the Business-as-Usual trend), an annual reduction in global CO_2 emissions of 1% until the year 2155, followed by annual reductions of approx. 0.25%. $\tilde{E}_2$ behaves as a compound exponential function and represents, of all curves of this type, the permissible threshold function. All emission profiles of this class that are higher would cause the climate system to depart from ⃞.

Figures 17b and *17c* show the concentration profile $\tilde{C}_2(t)$ corresponding to $\tilde{E}_2$ as well as the phase space trajectory of the climate system this generates. In the view of the Council, the comparison between the two emission strategies $\tilde{E}_1$ and $\tilde{E}_2$ thus favors the latter. $\tilde{E}_2$ offers greater security of planning, but also of control – meeting an annual reduction quota that remains constant over the long term is without doubt a more economically acceptable course to take than drastic course changes in the medium term. Because the climate system under strategy $\tilde{E}_2$ is clearly within the permissible window ⃞, the risk of control errors

and incorrect estimates is much less. In contrast, the level of total emissions associated with $\tilde{E}_1$ over the next 200 years is only about 4.5% more than that for $\tilde{E}_2$: 838 Gt C as opposed to 802 Gt C.

1.3.4
Possible Allocation Formulas

PERCENTAGE REDUCTION PROFILE FOR ANNEX I-COUNTRIES (INCLUDING GERMANY)

In one sense, emission profiles $\tilde{E}_1$ and $\tilde{E}_2$ define global pollution quotas as a function of time. These global quantities can be transformed into national reduction commitments if, for example, a politically negotiated allocation formula is laid down in a CO_2 protocol signed by the Parties when implementing the Climate Convention. The topic of allocation is discussed in greater detail in the Annex. Particular reference is made here to three possible allocation formulas:

a Equal distribution of reduction commitments among the Annex-I states while freezing the emission contingents of the developing countries.

b Allocation of pollution rights exclusively on the basis of current demographic weight of the individual countries.

c Entitlement to a linear increase of emissions over a limited period (doubling of the CO_2 emissions of the developing countries within 50 years), followed by equal distribution of the remaining reduction obligations.

If all these allocation schemes are based on the "optimum" emission strategy $\tilde{E}_2$, this results in per-

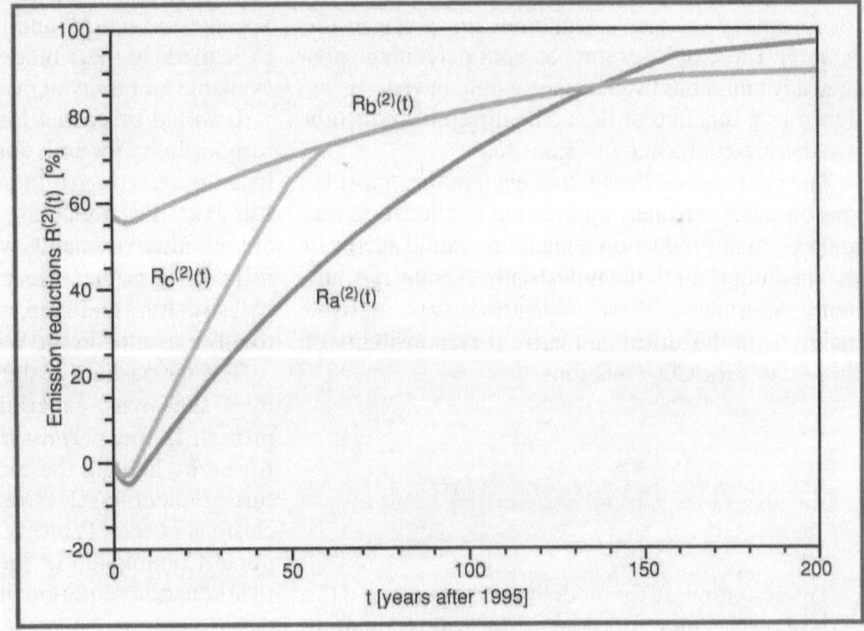

Figure 18
Reduction profile for Germany, based on emission profile $\tilde{E}_2$ and the allocation systems a) - c). Source: WBGU, 1995

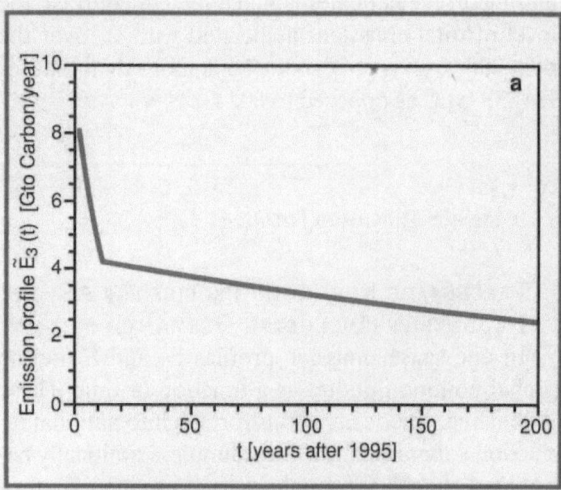

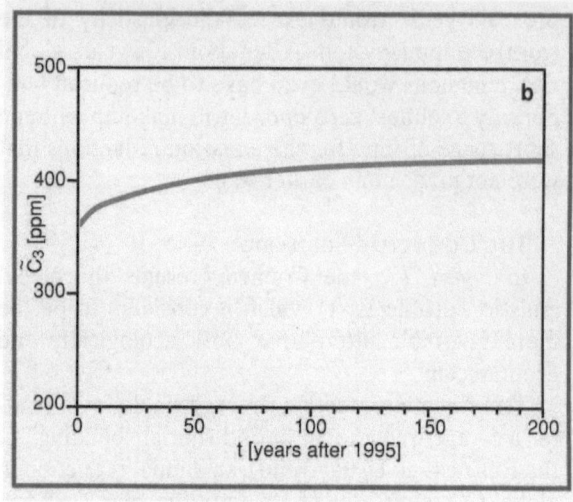

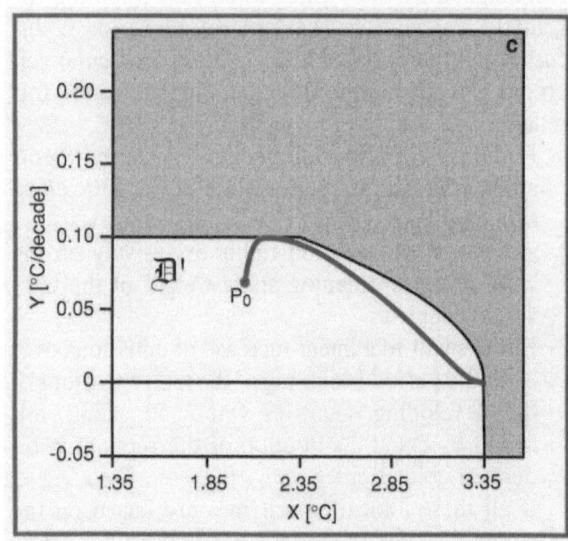

Figure 19 a-c
Reduced climate window
a) "Optimal" global CO_2 emission profile $\tilde{E}_3$
b) Concentration profile $\tilde{C}_3$ resulting from $\tilde{E}_3$
c) Shift in the climate system produced by $\tilde{E}_3$ within the tolerable climate window. P_0 is present state of the climate (X=2.05; Y=0.07).
Source: WBGU

centage reduction profiles $R_a^{(2)}(t)$, $R_b^{(2)}(t)$ and $R_c^{(2)}(t)$ for Germany (precise calculations are given in the Annex). These profiles state at what percentage present-day emissions by Germany would have to be reduced as a function of time. The different reduction profiles are contrasted in *Figure 18*.

The chart shows clearly that even in the "most favorable case" – namely application of allocation formula a – major reduction obligations would accrue in the medium term to the industrialized countries, and hence Germany. These obligations are derived mainly from historical and current responsibility of these states for CO_2 emissions.

1.3.5
Conclusions for Altered Assumptions: A Sensitivity Analysis

The systems used for modeling the climate and the carbon cycle may, of course, be overestimating

anthropogenic impacts on the global climate system. Specific corrections and improvements will continue to be made to these models, but major changes to the available hierarchy of models are unlikely to happen.

It would be wrong for politicians to believe that responsibility for emission reductions can be escaped from by referring to the scientific uncertainties that still exist. The reduction requirements in *Figure 18* are minimum demands within the various allocation formulas. Another aspect is that other greenhouse gases such as methane, nitrous oxide and CFCs are totally excluded in the WBGU scenario.

The seriousness of the situation can be illustrated by a sensitivity experiment using the inverse approach. If the permissible climate window is narrowed by halving the maximum tolerable temperature gradient to 0.1 °C/decade (in line with the conclusions of the "Protection of the Atmosphere" Enquete Commission of the German Bundestag) then the reduction requirements increase in dramatic proportions. If we try to calculate an "optimum CO_2 strategy" for this smaller window which will produce congruence between the desire for security of planning and control with the need for the greatest possible leeway with regard to emissions, we obtain the global profile $\tilde{E}_3(t)$ shown in *Figure 19a*.

If this strategy is applied, the business-as-usual trend must be succeeded within the space of one year by an exponential curve with an annual reduction rate of approx. 6.3%. After ten more years, the transition can be made to a less-steep exponential decline involving an annual reduction rate of approx. 0.3%. The cumulated CO_2 emissions over the next 200 years in the case of $\tilde{E}_3$ would amount to 682 Gt C.

Figures 19b and *19c* show the corresponding concentration profile $\tilde{C}_3(t)$ and the resultant shift in the climate system within the smaller tolerance range $\mathcal{B}'$.

Figure 19c shows how difficult climate control could be under the tighter boundary conditions being considered here. However, even if this control would be achieved, this would still involve enormous short-term reductions for the industrialized countries. This is illustrated in *Figure 20*, which summarizes the reduction profiles $R_a^{(3)}(t)$, $R_b^{(3)}(t)$ and $R_c^{(3)}(t)$ for Germany defined by $\tilde{E}_3$ and the allocation formulas a-c. It should be emphasized that in all cases, reductions of CO_2 emissions of over 50% would have to be made within the space of a decade.

However, the Council wishes to point out that a very pessimistic assessment of the adaptability of ecosystems was taken as the basis for the confinement of the tolerable climate window under discussion, and that nonenvironmental factors were scarcely given consideration. To that extent, the reduction requirements shown in *Figure 20* should not necessarily be seen as fixed general criteria for all future abatement efforts. Having said that, the sensitivity analysis demonstrates how the pressure to implement a global climate protection policy increases disproportionately (nonlinearly) as the tolerance window shrinks.

1.4
Implementation of Reduction Requirements

1.4.1
The Self-commitment Imposed by Germany

In its Cabinet resolution of December 11, 1991, the Federal Government confirmed Germany's 1990 self-commitment to reduce CO_2 emissions. "The Federal Government (...) reasserts its previous decisions of June 13 and November 7, 1990 and shall endeavor to reduce CO_2 emissions by 25-30%, relative to 1987 levels, by the year 2005."(BT Drucksache 12/8557). Total energy-related CO_2 emissions amounted in 1987 to 1,060 million tons (combined figure for the former states of West and East Germany), so a 25% reduction would mean limiting emissions to 795 million tons by the year 2005. A 30% reduction would be equivalent to an upper limit of 742 million tons. A target band was therefore defined for German climate protection efforts.

Figure 20
Reduction profile for Germany and the other Annex I countries based on emission strategy $\tilde{E}_2$ and allocation systems a) - c).
Source: WBGU, 1995

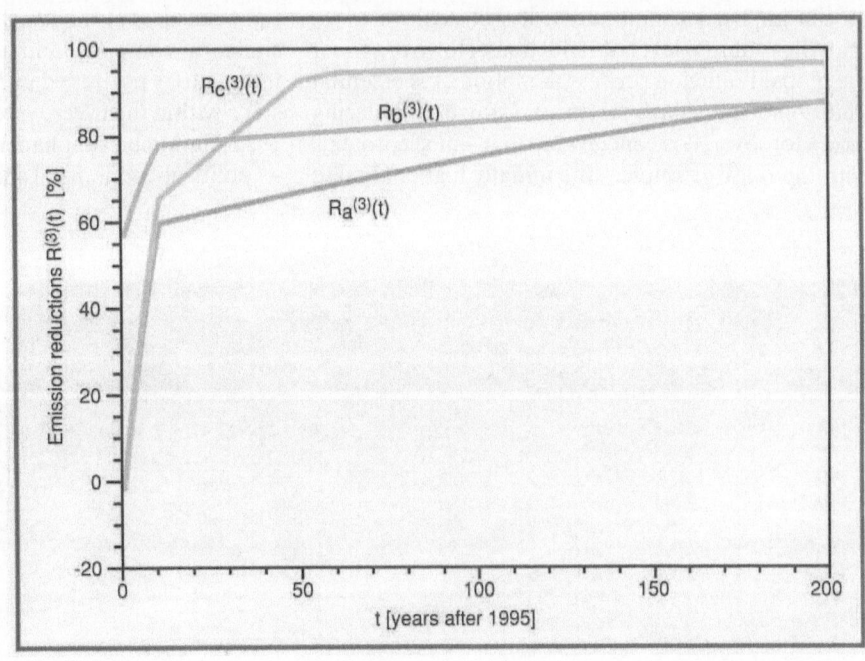

In his speech on April 5 to the Climate Conference in Berlin, the German Chancellor, Dr. Helmut Kohl, specified the above-mentioned target while at the same time relating this reduction to a new base year: "Germany remains committed to its objective to reduce its CO_2 emissions by 25% by the year 2005, relative to 1990 levels." (Presse- und Informationsamt, 1995). Based on emissions in 1987, (figures provided by UBA, 1995 and RWI, 1995), the new reduction target announced by the Chancellor corresponds to a 29% reduction, which is in the upper range of the target band laid down in 1990 (*Table 5*).

The total 1994 energy-related CO_2 emissions in Germany were 892 million tons (UBA, 1995), i.e. a reduction of 11.1% relative to 1990 has already been achieved. The remaining emissions reduction relative to 1990 which still has to be met by the year 2005 is therefore 13.9%.

How should the self-commitment on the part of Germany be assessed against the background of the WBGU's calculations? The answer depends on which permissible climate domains and which international allocation system are chosen as the basis for CO_2 reductions. As far as the latter is concerned, the Council proceeds on the contingents that are taking shape through the actual development in the non-Annex I-states: a medium-term increase in emissions by the developing countries, as reflected in the reduction criteria of class R_c (*see Annex*).

If one assumes the tolerable climate window $\mathbb{D}$ and the constantly falling emission profile $\tilde{E}_2$ favored by the Council, then the corresponding reduction commitment for Germany would be defined by the function $R_c^{(2)}(t)$. This involves a CO_2 abatement of about 10% until the year 2005; so if Germany meets its self-imposed commitment, it will achieve more than the required level of reductions. However, to set the global reduction process in motion, it is essential that some individual states perform a trailblazing role. Moreover, it is conceivable that – in accordance with the relay principle – the initially higher climate

protection accomplishments of leading countries could lead to a reduced commitment at a later date.

If, however, the chosen basis is the confined climate window $\mathbb{D}'$, which reflects a very pessimistic assessment of ecosystem adaptability, then the pressure to achieve reductions is magnified. The reduction requirement $R_c^{(3)}(t)$ which would then operate would mean CO_2 reductions in excess of 60% having to be made by the year 2005. In the view of the Council, reductions of this order are technically, economically and socially unfeasible.

Germany's self-imposed commitment is therefore a meaningful but very ambitious target given the tensions which exist between what is environmentally imperative and what society is willing or able to achieve. Should the economy also display positive growth in the medium-term, then the measures to reduce emissions drawn up by the "CO_2 Reduction" *Interministerial Working Group of the Federal Government* (IMA) will not suffice. The Council therefore recommends that the IMA revise its catalog of measures accordingly and advocates with respect to the protocol negotiations that the German target be made the standard for all Annex I-states. *Table 6* provides an overview of the targets for climate protection that have so far been operating in OECD states.

1.4.2
Cost-benefit Analyses Applied to Climate Protection Strategies

Any responsible emission strategy in the field of international climate protection must take account – if at all possible – of all the consequences that reduction measures may have. Strictly speaking, this necessitates a complete and integrated analysis of the interactions portrayed in *Figure 15*. In particular, the steps within the inverse scenario described in *Section C 1.3* should be supplemented by a

– political-economical analysis to determine reason-

Year	Energy-related CO_2 emissions	Reduction commitment	CO_2 reduction	CO_2 target for 2005
	[Mio. t]	[%]	[Mio. t]	[Mio. t]
1987	UBA: 1,060	25 - 30	265 - 318	742 - 795
1987	RWI: 1,024	25 - 30	251 - 307	717 - 773
1990	UBA: 1,003	25	251	752
1990	RWI: 975	25	244	731
1987*	UBA: 1,060	29	308	752
1987*	RWI: 1,024	29	293	731

*Projection of the target value based on 1990 to the year 1987

Table 5
German self-commitment to reduce CO_2 emissions. Sources: UBA and StBA, 1995; RWI, 1995 and own calculations

Table 6
Status of commitments of OECD countries on Global Climate Change
Source: IEA, 1994

Country	Gases included	Action	Base Year	Commitment Year
Australia	NMP* GHGs*	Stabilisation	1988	2000
		20% reduction	1988	2005
Austria	CO$_2$	20% reduction	1988	2005
Belgium	CO$_2$	5% reduction	1990	2000
Canada	CO$_2$ and other NMP GHGs	Stabilisation	1990	2000
Denmark	CO$_2$	20% reduction	1988	2005
Finland	CO$_2$	Stabilisation	–	End 1990s
France	CO$_2$	Per capita stabilisation	–	2000
Germany	CO$_2$	25% reduction	1990	2005
Greece	CO$_2$	EU Agreement**		
Iceland	All GHGs	Stabilisation	1990	2000
Ireland	CO$_2$	Limitation to 20% growth	1990	2000
Italy	CO$_2$	Stabilisation	1990	2000
Japan	CO$_2$	Per capita stabilisation	1990	2000
Luxembourg	CO$_2$	Stabilisation	1990	2000
		20% reduction	1990	2005
Netherlands	CO$_2$	Stabilisation	1989/90	1994/95
		3-5% reduction	1989/90	2000
	All other GHGs	20-25% reduction	1989/90	2000
New Zealand	CO$_2$	Stabilisation	1990	2000
Norway	CO$_2$	Stabilisation	1989	2000
Portugal	CO$_2$	EU Agreement**		
Spain	CO$_2$	Limitation to 25% growth	1990	2000
Sweden	CO$_2$	Stabilisation	1990	2000
		Reduction	1990	after 2000
Switzerland	CO$_2$	Stabilisation	1990	2000
		Reduction	1990	after 2000
Turkey		No target has been set		
United Kingdom	CO$_2$, methane and other major GHGs	Stabilisation	1990	2000
United States	All GHGs	Stabilisation	1990	2000
EU	CO$_2$	Stabilisation	1990	2000

* NMP = Non-Montreal Protocol (refers to greenhouse gases other than those covered under the 1987 "Montreal Protocol on Substances that Deplete the Ozone Layer" and its subsequent amendments i.e. greenhouse gases other than CFCs, HCFC, halons, carbon tetrachloride, and methyl chloroform)
* GHGs = Greenhouse gases
** "EU Agreement" means the country in question falls under the EU-wide target stated at the end of the table but has not yet developed its own target.

able national reduction commitments (Step 4 → 5),
– analysis of the instruments for identifying effective reduction measures (Step 5 → 6),
– assessment of the social and economic impact of reduction measures (Step 6 → 1)

This is an ambitious program of research that must be tackled vigorously over the next few years. In *Sections C 1.4.3 and 1.4.4* below, the Council provides a number of suggestions regarding the first two items. As far as the third aspect is concerned – "assessing the impact of prevention" – there are already important findings in the literature on the cost-benefit analysis of the climate problem, as is also the case with the evaluation of the socioeconomic impact of climate change (link between Steps 1 → 2 of the scenario). The current status of scientific knowledge is referred to here in brief; the compilations by Kaya et al. (1993) and Nakicenovic et al. (1994) give a more detailed overview.

1.4.2.1
Cost-benefit Assessments

One key element in formal cost-benefit analysis comprises the so-called "damage functions" that express the economic impacts of climate change in monetary form. These damage functions depend on a broad range of geographical, socioeconomic and technological factors, with regional variations. Because these factors change over time, the damage functions themselves will also change – even when the same climate forcing scenario is taken as a basis. Many scientists assume that human activities will be progressively less dependent on climate fluctuations (Ausubel, 1991; Schelling, 1992). Viewed in this way, actual damage by climate change in 50 or 100 years might be significantly less than that predicted by an assessment that applies today's world economy to a future climate.

The pioneering work by Nordhaus (1991b) on systematically quantifying the economic impact of climate change for the USA was the starting point for many similar studies. Numerous writers have criticized Nordhaus's approach as being too narrow and not comprehensive (Cline, 1992; Ayres and Walter, 1991). Other scientists have taken over the basic principles of the study and applied them to other regions of the Earth. Fankhauser (1993a) has even expanded the approach to the global dimension. His estimates for the EU, the USA, the CIS, China, the OECD countries and the world as a whole are similarly based on an increased concentration of greenhouse gases with the equivalent impact of a doubling of CO_2 concentration. Fankhauser's results largely confirm Nordhaus's conclusions regarding the costs of climate change for the industrialized countries (0.25-2% of GNP) and for developing countries (0.5-5% of GNP).

The direct effects of climate change on society are normally determined using a bottom-up method. For example, one assumes the climate sensitivity of certain important cultivated plants in selected smaller regions, aggregates the findings obtained at the level of economic (sub-)sectors and large-scale regions, and scales up the results – provided these are not too fragmentary – to the national economy level (*see, for example,* Parry et al., 1988). Further remarks on climate impact assessment can be found in *Box 24.*

The bottom-up approach, which is largely a static one, provides valuable indications of the rough dimensions of the damage to be anticipated. Nevertheless, the intensive interactions between the development of the climate system and the economic system demands an approach that takes account of the dynamics of that interaction, as well as indirect effects such as intersectoral compensation. Scheraga et al. (1993) were the first to respond to this need: with the help of Jorgensen and Wilcox's dynamic "general equilibrium" type model, they present a top-down analysis of the overall economic impacts for a precisely defined set of climate-induced disturbances (increase in the costs of agricultural production, increase in the costs of electricity, increased expenditures on coastal protection).

Applying general equilibrium models makes it possible to prove that a specific form of climate change can trigger off major intersectoral resource flows. However, even the most advanced macroeconomic models come up against their limits when they attempt to analyze climate change impacts, in that they do not take account of some hitherto important factors that are sensitive to climate change, such as freshwater resources.

As far as avoidance costs are concerned, i.e. the costs to society for climate protection, there are already a wealth of model-based studies (Manne and Richels, 1992; Manne and Rutherford, 1993; Jorgensen and Wilcoxen, 1993). These studies arrive at different cost estimates for achieving the same climate protection targets, because they proceed from different future projections for uncontrolled development of greenhouse gas emissions.

There are considerable differences between different countries and regions of the world with respect to economic structure, energy consumption and the corresponding emissions having an impact on climate. This suggests that a homogeneous worldwide CO_2 reduction might not be the best socioeconomic strategy. Many authors have referred in this connection to the benefits that would arise from a global tradeable permit systems for CO_2 emissions. Edmonds et al. (1993), for example, use a modified version of the Edmonds-Reilly-Barns model and compare the costs of three alternative mechanisms for the implementation of a hypothetical climate protection protocol: uniform taxation, tradeable permit system and individual (regional) reduction commitments. The study comes to the general conclusion that any and every international convention for the abatement of greenhouse gas emissions from fossil fuels has to be "reshaped" due to the permanent change of economic and technological conditions. The cost estimates produced by Edmonds et al. (1993) show that laying down individual (national or regional) reduction targets would give rise to almost double the level of expenditure as an effectively organized and globally operating system combining CO_2 tradeable permits and energy taxes. This topic is dealt with in more detail in *Sections C 1.4.3 and 1.4.4.*

The first attempt to integrate a heterogeneous set of cost-benefit estimates into a closed picture was that of Cline (1992). He identified 16 categories of damage and extracted data from a large number of key studies in order to carry out his own evaluations for each of these categories. As far as avoidance costs are concerned, Cline performed a detailed analysis of six advanced energy-economy models. Avoidance and damage costs were then compared in a cost-benefit model.

The assessment of the direct and indirect consequences of failing to protect the climate remains the weakest link in the argumentational chain of any integrated analysis. The required research lags far behind both pure climate system research as well as the purely economic analysis of avoidance costs. One reason for this is the complexity of climate impacts, but another reason is the absence of, or fluctuation of support: every new doubt leveled at the predictive power of global climate models brings the societal relevance of impact research in question, even though the necessity for a thorough understanding of potential climate effects would be substantiated from the considerable variability of the undisturbed climate system (see results from ice cores in Greenland, GRIP).

Some of the key tasks of climate impact research are the identification and evaluation of the possible effects of global change on natural systems and civilizations as well as the potential protective and mitigation measures with respect to these effects (Schellnhuber and Sterr, 1993). "Avoidance of causes" as a strategic option is deliberately countered with the contrary position advocating "toleration of and/or adaptation to the consequences/effects". Knowledge about adaptation options and costs is, as one would expect, even less than that of regional and sectoral climate sensitivity. The 1995 IPCC report provides an updated summary of the state of significant research in this area.

Work produced in the field of climate impact research can be classified into those which have a "sectoral" and/or "geographical" reference according to the characteristics "empirical", "estimating" and/or "modeling". Some recent studies relating to particular sectors and which received attention at the Berlin Conference focused on the study of climate effects on coastal zones (IPCC, 1994b), the water courses of river systems (Weijers and Vellinga, 1995), biodiversity (Markham, 1995) and the insurance industry (Swiss Re, 1994).

Examples of recent geographically related studies include the impact assessments of climate change for the American Mid-West (Rosenberg, 1993), the Japanese islands (Nishioka, 1993) and the Mackenzie Basin in Canada (Cohen, 1994). A model-based quantification of possible climate change impacts is planned in Great Britain (Parr and Eatherall, 1994). Genuinely empirical studies on the basis of regionally occurring climate anomalies are rare; one exception is the analysis of the hot and dry summer in northern Germany in 1992 (Schellnhuber et al., 1993).

The project entitled "As Climate Changes: International Impacts and Implications" (Strzepek and Smith, 1995) is a current attempt to combine sectorally and geographically related climate change impacts. This study draws largely on the geographical approach as it has been shaped by M. Parry and his co-workers, which also dominates the methodological guidelines used by the IPCC for the assessment of climate change impacts (Carter et al., 1994).

Since the actual effects of global climate change are probably determined to a significant extent by the geography and natural resources of a particular region, as well as by its socioeconomic and political-cultural peculiarities, climate change impact research has to develop almost reciprocally to climate system research: the latter proceeded from the description of fundamental global processes and strives for simulations of the entire spectrum of meteorological phenomena with high geographical resolution. Impact research, on the other hand, can only arrive at tenable conclusions if it concentrates first and foremost on the integrated modeling of discrete regions, in order ultimately to gain a global picture of how regional models are networked.

Cline's most important conclusion from his cost-benefit analysis is that " ... the benefits of an agressive program of abatement warrant the costs of reducing the greenhouse gas emissions if the policy-makers are risk averse, or if one is pessimistic and concentrates on high-damage cases". This view contradicts sharply the earlier cost-benefit assessments of Nordhaus (1991a and b). The latter came to the conclusion that an optimum climate protection policy would have to enforce a drastic cutback of industrial emissions of CFCs and halons. CO_2 emissions, on the other hand, would have to be reduced by only 2%, which could be achieved by a carbon dioxide tax of US\$ 7.33 per ton.

1.4.2.2
Integrated Models

The most compact instruments for determining optimum climate protection strategies are those models which conceive of the globe as a single region, and the world economy as a single producer-consumer pair. Greenhouse gas emissions are calculated in terms of the sector-specific degree of capacity utilization. Both avoidance costs and the costs of climate change impacts are continuously fed into the system in order to include influences affecting total production.

An important contribution in this respect is the DICE model (*Dynamic Integrated Model of Climate and the Economy*) (Nordhaus, 1993a and b). This is a variant of Ramsey's model of optimum growth, extended to take account of direct climate damage and resource shifts through the application of reduction measures. The optimization criterion used by DICE is the maximization of the discounted total benefit of the per capita consumption. The climate-induced development paths of the world economy appear in this analytical framework only as mild deviations from the undisturbed, "ideal" course of growth. Similar models for only one region and with sectoral aggregation have been designed by Fankhauser (1993b) and Maddison (1994).

As already mentioned, individual countries and regions cannot be accorded equal treatment with respect to the climate problem. This means that only geographically explicit models can lead to tenable statements regarding favorable climate protection strategies. One well-known example for this type of model is MERGE (*Model for Evaluating Regional and Global Effects of GHG Reduction Policies*; Manne et al., 1993).

MERGE consists of three components: an improved version of the Global 2100 model (Manne and Richels, 1992) used for calculating the self-organized dynamics of the economy as a whole, a simplified carbon cycle model based on that of Maier-Reimer and Hasselmann (1987) and a module for assessing climate change impacts. MERGE is one of the first integrated models that attempts to take noneconomic climate change impacts into consideration. This is done via the hypothetical willingness of citizens to pay for the preservation of environmental resources such as landscapes or rare species.

One interesting approach with a philosophy similar to that of the Council's inverse scenario was chosen by Richels and Edmonds (1994): they describe the "Economy-Energy-Atmosphere System" using a combination of two global energy models and a carbon cycle model similar to the one used by MERGE. Fixed targets are defined for the stabilization of at-

mospheric CO_2 concentration and the possible emission pathways for reaching these targets are compared. The model provides interesting information about the costs of different short-term avoidance measures and their actual benefits for climate protection.

1.4.3
Reduction Potential and the International Distribution of Responsibility

UNIFORM NATIONAL QUOTAS AND THEIR SIGNIFICANCE FOR INTERNATIONAL INSTRUMENTS

As negotiations to date have shown, the international community appears to have considerable difficulty in reaching agreement on binding national reduction targets. The debate over the fixing of national quotas appears to be heading in the direction that all Annex I-states, with a distinction possibly being made between industrialized countries and economies in transition, will be given the same percentage reduction commitments. To what extent such a procedure will lead to an economically meaningful, i.e. cost-effective solution, can only be assessed in consideration of regionally varying reduction potentials.

OPTIONS FOR INTERNATIONAL DISTRIBUTION OF EMISSION REDUCTION BURDENS

A distribution formula for national reduction quotas should be equitable, i.e. it should give consideration to the respective capabilities of the states and their "common but differentiated responsibilities" for the anthropogenic greenhouse effect. Furthermore, it should lead to an economically efficient reduction of emissions. An economically efficient and at the same time equitable way of distributing burdens among the Parties could be achieved on the basis of the polluter-pays principle. The polluter-pays principle is an internationally recognized maxim (UNCED, 1992; OECD, 1975; UN, 1987) for allocating the responsibilities and obligations to pay for the removal of environmental damage. It also applies for global environmental problems, where the causal agents and injured parties are states (Lass and Schuldt, 1994). No country seriously disputes having a certain national responsibility for external damage, according to the polluter-pays principle. If national environmental policy determines that the emitter is the one that is technically obliged to bear the costs of avoidance measures, then there are various conceivable ways of implementing a polluter-pays system through climate protection policy (Cansier, 1991). One could demand a proportional reduction for all participant countries, i.e. every country should reduce its emissions within a certain time frame by x%. Depending on the reference criteria and the temporal framework, other distribution pattern would then be obtained.

The question that is raised in connection with the time frame is this: should the initial basis be current emissions alone, or should historical emissions generated during the industrialization of today's wealthy nations be also taken into account, which would increase their reduction obligations accordingly? If economic development is included as an objective, the logical consequence would be to increase still further the emission quotas of the developing countries, since this would appear indispensable for economic growth.

Imposing a certain responsibility to pay on a "per nation" basis is generally thought to make sense when determining the international distribution of costs, although often other distribution formulas are used as the amount of costs increase. For CO_2 reduction quotas the per capita emission or emissions per unit of GNP could be used (Simonis, 1993). Quotas that take per capita emissions at least partially into account would reflect the highly differing levels of per capita energy consumption in the world, and, if reduction commitments were determined in proportion to this consumption, there would then be incentives for bringing these more into line. On the other hand, taking emissions per unit of GNP more into account would penalize those states more whose energy efficiency is relatively low, i.e. those countries in which relatively more energy is needed to produce one unit of GNP than in other states. This would provide an incentive to greater development and deployment of energy-efficient technologies. Successes on the environmental front should be achieved at lowest possible costs; it is therefore important that consideration be given to the differing reduction potentials that exist in different countries.

REGIONAL DIFFERENCES IN TECHNICAL REDUCTION POTENTIAL

In order to implement the ecologically undisputed need for action in an economically efficient manner, preference should be given to processes whereby the primary geographical (and also sectoral) focus is where the greatest potentials for emission reductions exist. However, there is no comprehensive derivation of globally existing or presumed technical CO_2 reduction potentials and such should therefore be produced as soon as possible. An approximate picture can be gained from available data, however. A distinction is made here between regions

1 that emit or will emit quantitatively significant amounts of CO_2 in the present or the future,

2 those that are lagging behind countries with ad-vanced technology (i.e. Japan and the states of Western Europe) in the energy efficiency of plant and equipment,

3 those whose economies will most likely have the highest growth rates .

The first criterion, the level of emissions, is important when determining the key geographical aspects. If high CO_2 emissions are regionally linked to high efficiency of power generation plant and/or with low specific energy consumption (second criterion), this is an indication that in such countries there is only marginal scope for significant technical emission reduction in the short term. Experience shows that other reduction measures and/or increases in energy productivity involve very high costs and could exceed the adaptability of the economic system (see the second boundary condition of the scenario). Negative growth rates in the highly developed countries would, as experience in the new Bundesländer shows, reduce emissions but at the same time exacerbate the problem of unemployment in many industrialized nations, have an impact on all world trade and thus ultimately affect the developing countries as well. A sufficiency revolution requires a high level of environmental awareness, however. It can probably only be achieved in the long term, and must therefore be prepared through appropriate educational measures. Major short-term successes in reducing CO_2 emissions will only be attained, as a rule, where energy consumption is high and where there is technical scope for increasing energy productivity. This scope can be all the more easily exploited the higher the anticipated or targeted growth rates are (third criterion), since high efficiency levels and/or a low specific energy consumption are most effectively achieved with new plants and technologies.

If these criteria are applied, a rough estimate of existing technical CO_2 reduction potentials and/or regional prioritizations in the following countries makes the most sense:

– within the OECD: in the USA and Canada,

– among non-OECD states: in the People's Republic of China and in most CIS countries and eastern European states.

These areas accounted in 1990 for almost two thirds of global CO_2 emissions (IEA, 1993b; RWI, 1994; WBGU, 1994) and, in some cases, for the highest levels of CO_2 emissions per capita and year (Table 7). China is the third largest consumer of energy in the world after the USA and Russia, due above all to the high level of industrial demand for energy, which is mainly met with coal (about 72% of industrial demand for energy in 1991). According to calculations by the International Energy Agency, if China continues its present trend and/or attains its planned gro-

Table 7
CO_2 emissions per capita and year, and fuel consumption
per unit of GDP for selected countries and regions.
Sources : IEA, 1993; RWI, 1994a; WBGU, 1994

Region	CO_2 emission per capita and year [t]	Fuel consumption per unit GDP [Gigajoule/$ 1.000 GDP]
North America	20.2	11.9
CIS states	13.6	no data
Germany*	13.0	11.3
Eastern Europe	10.6	28.8
Western Europe	8.1	7.1
Japan	8.1	3.9
Asia (incl. China)	1.5	34.9
India	0.7	13.7

* including the new Bundesländer in eastern Germany

wth, it will emit approx. 5 billion tons of CO_2 (= 1.4
Gt C) in the year 2010, or about 18% of current glo-
bal CO_2 emissions. The fuel consumption per unit of
GDP in the countries and regions listed was far
above that of other important CO_2 emitters in 1990
(*Table 2*).

CONSEQUENCES FOR DEFINING REDUCTION COMMITMENTS

The difficulties outlined above for the international distribution of costs mean that the Parties to
the Framework Convention on Climate Change will
probably agree on equal percentage reduction rates
for the absolute emissions of all Annex I-states.

If one proceeds on this assumption, then it is all
the more important, given the enormous variations in
regional reduction potential, that international strategies permitting these rigid national quotas to be rendered more flexible come into force alongside climate protection policies using purely national instruments. Flexibilization enables very different CO_2 reduction potentials, which already exist among the
Annex I-states and even within the subgroup of industrialized countries not including the countries
with economies in transition (Annex II-states), to be
taken into account. In this way, individual states
could achieve their CO_2 emission quotas at markedly
lower cost. In view of the difficult adaptation process
that implementation of reduction commitments involves for each country, the Council is emphatic in its
support for the exploitation of these opportunities. A
flexibilization of the rigid percentage rule could
prove to be the key to binding agreement in the field
of international climate protection policy and thus
lead to progress being made in combating the anthropogenic greenhouse effect. One important instru-

ment for making national quotas more flexible and
therefore for reducing the costs of adaptation and
prevention is the Joint Implementation of mitigation
and reduction measures, another the introduction of
a global tradeable permit system.

1.4.4
Flexibility Under a System of Uniform National Quotas: Joint Implementation and the Tradeable Permit System

JOINT IMPLEMENTATION

The Climate Convention permits countries to fulfil part of their national reduction commitment by
carrying out mitigation measures in other countries.
The Council welcomes the establishment of a pilot
phase by the Berlin Conference. The Parties should
agree upon generally accepted rules for Joint Implementation as rapidly as possible in addition to institutional integration of such measures (WBGU,
1994). Reference is made here to the importance of
the verification of greenhouse gas reductions. Since
all participants profit from having the highest possible figures documenting successful achievement of
reductions, an independent supranational institution
is required for evaluation and monitoring at project
level and through national bodies. The functions of
this institution should conform to the principle of
subsidiarity, i.e. it should only perform those tasks
that cannot be handled locally by the participant Parties.

In addition to the advantage of reducing avoidance costs, Joint Implementation provides a framework for promoting the transfer of private capital,
technology and know-how, and at the same time for
promoting the capacity building process so essential
for development. This function must not be underestimated by developing countries either, considering
the severe shortage of public funds that are available.

With regard to the crediting of emission reductions achieved through Joint Implementation projects against the reduction commitments of the Parties listed in Annex I, the Council wishes to point out
that if no form of crediting is introduced, an important driving force for Joint Implementation projects
would be lost. Even if crediting must be dispensed
with during the pilot phase in order to achieve some
form of international consensus on Joint Implementation, the concept cannot be put into operation on
any comprehensive scale unless reductions can be
credited against national reduction targets and incentives exist for private sector involvement. The objections raised by some countries and environmental
groups against the crediting of emission reductions
can be removed by committing the Annex I-states to

realizing the major proportion of their reduction obligations – say 70-80% – within that group of states (i.e. including countries undergoing the transition to a market economy). If measures for reducing emissions outside the territory of a specific state are implemented, then the reduction commitment should be raised by a certain specified amount – in other words, emission reductions achieved elsewhere should not be credited in full against national targets.

The prerequisite for crediting emission reductions is the existence of binding national reduction targets (*Box 25*). Against this background, fixing the existing self-imposed commitments on the part of many Parties into the Convention itself would be greatly welcome. This is an additional reason for rapid implementation of binding reduction targets for the period following the pilot phase, in accordance with the Berlin Mandate, for example. Another precondition concerns the enterprise level: participation in Joint Implementation projects must be financially worthwhile for private investors. If national emission targets are implemented by means of command-and-control or economic instruments, then compensatory measures such as exemption from levies or taxes provide the needed incentive; the size of the benefit accruing to the private actor must exceed the costs of the projects carried out abroad. Only in the case of projects financed purely by the state would the provision of such support through the instrument of Joint Implementation not be required (Michaelowa, 1995).

In the long term, protection of the environment will only obtain the desired acceptance and importance at international level – as at national level until now – if the right environmental measures are accompanied by implementation of the right economic policies. On the whole, the Council welcomes the efforts of the Parties to the Climate Convention to integrate economic aspects, in the form of Joint Implementation, as part of the climate protection strategy. This can build confidence in the functional effectiveness of this virtually untested instrument, and thus enhance its acceptance. At the same time, as developments in the USA have shown, this type of compensation model can lead to valuable experience being acquired for later and more comprehensive deployment of additional economic instruments. One of the most important such instruments, especially as a means for making the application of rigid national quotas more flexible, is that of internationally tradeable permits.

AN INTERNATIONAL TRADEABLE PERMIT SYSTEM

The Council expressly advocates the protocol proposed by AOSIS (*Alliance of Small Island States*) and

---

BOX 25

One Perspective: Inclusion of Other Greenhouse Gases (Comprehensive Approach)

The points made in the text, and the Council's inverse scenario relate almost exclusively to CO_2. One major advantage of reduction commitments that embrace all or at least the most important greenhouse gases (the comprehensive approach) is the much lower avoidance costs for the same environmental effectiveness. Each country could then decide which greenhouse gases it will reduce in order to meet its quota in terms of a CO_2 equivalent volume, i.e. choose the most favorable solution in cost terms, e.g. through emission reductions of methane (CH_4) or nitrous oxide (N_2O, laughing gas) (Cansier, 1991). This lower-cost solution at both national and international level would also appear to be more politically feasible than separate conventions for specific gases, because the costs of adaptation to include all gases would be more evenly spread than for exclusive focus on CO_2 emissions.

However, applying the comprehensive approach is barely feasible at present due to the problems associated with measurement and set-offs. There is still insufficient knowledge about the different gases, and the radiative forcing potentials of greenhouse gases have only recently been revised (IPCC, 1995) (*see Section C 1.2.4*). Reducing the various greenhouse gases involves various technical, economic, social and political factors (Simonis, 1994). Due to the long atmospheric lifetime of CO_2, it is necessary for reduction commitments for this greenhouse gas to be established as quickly as possible. Although the Council does not advocate the comprehensive approach at this point in time, it does refer to the necessity of including other gases in reduction commitments and recommends that research resources be channeled more strongly into this concept. A conceivable approach would be to limit an international convention to CO_2 initially, but in subsequent negotiations to permit other gases to be included, step by step, as a way of achieving the reduction target.

the German government's proposals that economic instruments be deployed to a greater extent than hitherto as a means of reducing emissions of greenhouse gases. Climate protection measures are thus made as cost-efficient as possible. At the global level, the introduction of a tradeable permit system would contribute to greater flexibility with respect to national quotas. In such a system, the Parties would be permitted to sell their allocated emissions to other states for given periods in the form of tradeable emission permits. The latter states could then use these emission rights to cover their own emissions until they have fully developed their own emission avoidance technologies, for example. A developing country could also act as a seller of tradeable permits if it is unable to make full use of the emission volume allocated to it, but would first like to obtain income by "renting out" these rights. When the economy has developed further, partly with the income from renting out emission rights, the allocated emission rights could then be made use of.

Such a system would enable targets to be achieved much more cost-efficiently, or greater CO_2 reductions could be achieved with the same budget, regardless of the allocation system (national quotas) initially selected (WBGU, 1993 and 1994). In principle, therefore, such an international tradeable permit system would mean that the Parties could operate a trade in emission permits with each other. What precise measures the individual states deploy in order to comply with their emission ceilings would be a matter of national sovereignty.

The Secretariat of the Framework Convention on Climate Change should be issued with the mandate to investigate and identify the conditions for establishing an international tradeable permit system. This will involve a whole series of definitions and stipulations. The question of initial allocation of national emission quotas can be resolved, in principle – after all, the national quotas that will have to be laid down by the Conference of the Parties in any event will provide a basis for calculating the emission levels that individual countries are implicitly "allowed" to produce. The initial allocation of emission targets will involve implicitly the determination of the financial burdens to be borne by the individual Parties.

Independently of the further development of a global tradeable permit system, the Member States of the European Union should take immediate action to establish such a system within its own group of nations as soon as possible. In the view of the Council, conditions in the EU are particularly conducive to the success of such an endeavor.

1.5
Research Recommendations

Climate system research
- Recording of all sources and sinks of greenhouse gases in a standardized "survey".
- Precise determination of the contributions of individual greenhouse gases to the radiative balance and development of a binding conversion scheme permitting the definition of reduction equivalents.
- Further improvement of climate modeling of the physical climate, with special reference to
  - sub-scale (parameterized) processes,
  - synergies (aerosols, ozone, etc.),
  - regional variants of global climate change.
- Inclusion of atmospheric chemistry aspects in global circulation models.
- Integrated modeling of the physical climate system and biogeochemical cycles, especially through the development of dynamic atmosphere-vegetation modules.
- Systematic comparison of natural climate variability and fluctuation patterns resulting from anthropogenic climate forcing.

Climate impact research
- Analysis of climate-sensitive sectors in Central Europe, especially forestry, agriculture and water resource management.
- Integrated study of climate-sensitive regions in Europe and elsewhere, especially coastal zones, mountain regions and semi-arid areas.
- Analysis of the probability and significance of extreme climate-related events (from heavy rains to ocean current deviations resulting from climate change).
- Commencement of systematic research into the human aspects of anthropogenic climate destabilization, especially with respect to the altered utilization of natural resources (freshwater, fertile soils, biodiversity, etc.).

Climate policy research
- Analysis of implementation strategies for achieving national targets.
- Investigation of the compatibility of the Framework Convention on Climate Change with other international treaties in the field of the environment.
- Prerequisites of and conditions for a "sufficiency revolution" as a long-term, demand-side strategy for climate protection.
- Geographical and sectoral survey and analysis of technical CO_2 reduction potential existing and suspected worldwide .
- Studies relating to the verification of emission reductions.

Integrated analyses
- Scientific support for the formulation of a German (European) position for the climate protocol negotiations. In particular, the working out of an integrated (natural scientific and socioeconomic) strategy concept relating to global and national emissions reduction, taking the WBGU's Berlin Conference statement as a starting point.
- Greater research efforts into growth theory and growth policy paradigms for industrialized, newly industrializing and developing countries (sustainable pathways for environment/development).
- Ascertainment of the determinants for the adaptability of economic and social systems to strategies for greenhouse gas reduction.
- Construction of integrated global and regional models (GIMs and RIMs as simulation instruments for climate and development-related decisions).

1.6
Recommendations for Action

The results of the Climate Conference in Berlin, but also the remarks made in the previous sections, give rise to a number of tasks that have to be tackled by the scientific community and politicians in the field of climate protection, especially in the run-up to the next Conference of the Parties.

RECOMMENDATIONS FOR ACTION FOLLOWING THE BERLIN CLIMATE CONFERENCE
- There is a general need to raise awareness of the fact that international climate protection policy is a continuous process, the first steps of which were taken at UNCED in 1992 and at the Berlin Conference in 1995. If a forward-looking view is to be taken, then it is essential that the necessary targets and measures are defined and made binding on the Parties at COP-2 and COP-3. The global emission profile advocated by the Council leaves only a few years' time for the requisite measures to actually take effect, a fact that underlines how pressing the need is for resolutions to be made as rapidly as possible.
- In line with its-scenario, the Council recommends that measures be introduced within a very short implementation phase that will reduce CO_2 emissions by a constant annual rate of 1% worldwide. Only in this way is it possible to avoid having to implement a "crash" program, and the extreme demands this would place on the political sphere and the economy.
- The other greenhouse gases must be included in reduction strategies as soon as possible. Research

into global warming potentials must be strengthened and crediting mechanisms worked out in order to reduce avoidance costs while retaining the same level of ecological effectiveness.
- The instruments for making national commitments more flexible, while meeting the requirement of lower cost for equal effectiveness, must be defined and applied. The pilot phase must be used to make the benefits of Joint Implementation of CO_2 reductions a reality, by establishing criteria that evaluate the appropriateness of projects for climate protection, on the one hand, and which also create a settlement of interests between the participant states and/or enterprises. In the opinion of the Council, less than five years should suffice to successfully implement bilateral projects and to provide internationally accepted evaluation criteria.
- The Federal Government should use the pilot phase to enlarge its own stock of experience by initiating its own national projects. Public funding for private projects is necessary.
- In addition to the above, the conditions for introducing an international tradeable permit system must be investigated and identified in the protocol being worked on and in coming INC negotiations. Furthermore, the Member States of the European Union should make immediate preparations for establishing such a system throughout the EU as soon as possible.
- The Council recommends the Federal Government include the topic of environmental education in the protocol negotiations under the Framework Convention on Climate Change. The transmission of knowledge and skills is an important prerequisite for changing lifestyles, production systems and consumption patterns.
- In view of the considerable amount of finance needed to implement climate policies, the GEF must be replenished as a matter of urgency (WBGU, 1994).
- Given the fact that there is no clearly perceptible coordination between the Conventions, the Council recommends that the debate on an integrated strategy for tackling global environmental problems be intensified.

FURTHER RECOMMENDATIONS:
NATIONAL LEVEL
- Germany's self-imposed CO_2 reduction commitment announced by the Federal Chancellor at the Berlin Conference represents a toughening of the national reduction target and hence an even greater challenge. The Council therefore recommends that the "CO_2 Reduction" Interministerial Working Group adapt its catalog of measures to this

new target. Such an analysis would have to examine, in particular, the opportunities for Joint Implementation projects and the progress than can be made by reducing other greenhouse gases besides CO_2.

EU LEVEL

EU-wide climate protection measures are required, both for competitive reasons and because of the advanced degree of integration. The Member States have already devolved some of their powers to the EU level, in the field of taxes on consumption, for example, and hence do not possess unlimited legal freedom to realize their objectives at national level. For this reason, it is imperative that farther-reaching EU-wide solutions be striven for. The Council considers the following areas to be especially important:

- Joint Implementation: examination of projects, exchange on and joint execution of projects with non-EU countries.
- Emission standards: development of common standards for household appliances and for energy efficiency.
- Tradeable permits: preparation for and rapid introduction of a system for trading in CO_2 emission permits.

INTERNATIONAL LEVEL

- International organizations for the protection of the environment: consideration of the objectives of the Framework Convention on Climate Change in the respective fields of activity and the establishment of compatibility.
- Development aid: an important task in the field of bilateral and multilateral development aid is the promotion of renewable energy resources.
- Framework Convention on Climate Change: coordination between the Convention and other international treaties.
- Restructuring of international organizations: institutions are needed in interlocking areas such as climate protection and development aid, or climate protection and international trade.

2.1
Stratospheric Ozone

2.1.1
Introduction

More than 95% of the ozone on our planet is formed in the tropical stratosphere (at heights of around 20 km) by photochemical processes driven by UV radiation from the sun. The ozone is then transported by air mass transport to mid and high latitudes, where it resides at altitudes between 10 and 50 km. About 10% of the ozone penetrates from the stratosphere to the troposphere (0 to 10 km altitude) during periods of intense stratospheric-tropospheric exchange (storms) and is mostly destroyed there, so that the natural concentration of ozone in the troposphere is generally very low.

Two different global trends caused by different anthropogenic processes have been discernible for some decades now: stratospheric ozone is being depleted, whereas ozone in the troposphere is increasing (*see Section C 2.2*). The extent and the impacts of these trends show considerable regional variation, but one can speak in both cases of a global problem.

The Montreal Protocol and the subsequent amendments (London, 1990 and Copenhagen, 1992) shows possible solutions for the protection of the stratospheric ozone layer that might act as models for a future global climate protocol.

2.1.2
Implementation and Impacts of the Montreal Protocol and its Amendments

2.1.2.1
Recent Developments

– Concentrations of the main anthropogenic gases which damage the ozone layer (source gases) by causing the formation of chlorine and bromine in the stratosphere (CFCs, carbon tetrachloride, halons and methyl chloroform) have risen at a markedly slower rate. This success is attributable to the drastic reductions in emissions of some source gases as required by and implemented through the Montreal Protocol and the amendments thereto. The increase in Freon-11 in 1993, for example, was 25 to 30% less than in the 1970s and 1980s (*Fig. 21*). The maximum loading with chlorine and bromine in the troposphere probably occured in 1994, but will not occur for another 3 to 5 years in the stratosphere (IPCC, 1994a). Due to the longevity of ozone-depleting substances, however, the stratospheric ozone layer will not be able to regain its original state until the middle of the next century (*Fig. 22*).

– Global ozone depletion of approx. 3% per decade is the cumulative impact of regionally and temporally very different trends (WBGU, 1993). Over the tropics and subtropics (30° N to 30° S), i.e. in about half the earth's atmosphere, no significant ozone depletion has yet been measured. Depletion is therefore all the more severe in the other regions, with ozone depletion particularly drastic during the spring months over the Antarctic continent (the so-called "ozone hole"). However, there is a marked tendency towards depletion over mid and high latitudes in Europe in the order of 5% per decade (*Fig. 23a*). The development during the 1994/95 northern winter is shown in *Box 27*.

History of the Convention for the Protection of the Ozone Layer

– March 1985: Adoption of the *"Convention for the Protection of the Ozone Layer"* in Vienna by 21 states (including Germany). The Convention entered force on August 1, 1988. A total of 130 states had ratified the Protocol by April 1994.

– May 1985: First publication of scientific evidence for the ozone hole over the Antarctic.

– September 1987: Signing of the *"Montreal Protocol on Substances that Deplete the Ozone Layer"* supplementary to the Vienna Convention. The Protocol came into effect on January 1, 1989, and by November 1994 had been ratified by 148 states. The Montreal Protocol is subject to constant further development on the basis of reviews of the control measures in force on the basis of scientific, environmentally relevant, technical and economic data. Accelerated phasing out of the production and consumption of CFCs and other ozone-depleting substances was recommended at several Conferences of the Contracting Parties. New ozone-depleting substances were included in the reduction commitments.

– At a total of 7 Conferences of the Parties held so far (1989 in Helsinki, 1990 in London, 1991 in Nairobi, 1992 in Copenhagen, 1993 in Bangkok, 1994 in Nairobi and 1995 in Vienna), the Montreal Protocol was made progressively more stringent. Deadlines for the cessation of all production and consumption of ozone-depleting substances were defined and in some cases brought forward. A multilateral ozone fund totaling $510 million for 1995-1996 was set up to support developing countries implement the Montreal Protocol.

– The stratosphere has recovered from the radical ozone depletion caused by volcanic aerosols following the eruption of Pinatubo in June 1991, and the decrease in ozone concentrations conforms again to the long-term trend.
– The great natural variability in the circulation of the Arctic stratosphere in winter makes it still difficult to forecast future changes of the ozone level in the Arctic.
– The extent of stratospheric ozone depletion to date has led to a cooling of the troposphere, thus counteracting the greenhouse effect somewhat (*see Section C 1.2.3, Fig. 14*).
– To accelerate the recovery of the ozone layer, the latest UNEP Report (UNEP, 1994a) recommends that consumption of methyl bromide be terminated by the year 2001, and that of partially halogenated substitutes (HCFCs) by the year 2004. This proposal goes much further than the amendment of the Montreal Protocol adopted at the 4th Conference of the Parties in Copenhagen. UNEP also demands that action be taken to prevent the leakage of halons and CFCs from existing appliances, release of which might cut the regeneration rate of the ozone layer by as much as 10%.

2.1.2.2
Exceptions

After general cessation of CFC use from 1996 onwards, there will only be minor exceptions permitting the use of CFCs by the industrialized nations (result of the 6th Conference of the Parties in Nairobi, October 1994): for metered dose inhalers, critical cleaning, bonding, and surface activation procedures for the NASA Space Shuttle's solid rocket motor and analytical uses.

The signatories of the Montreal Protocol have also redefined the term "newly industrializing country". In addition to the developing countries, the following countries also receive a 10-year extension period in which to phase out CFC technology: South Korea, Kuwait, Saudi-Arabia, the United Arab Emirates, Malta and Singapore. Some eastern European countries have requested that they also will be granted in an exceptional regulation. What is extremely problematic in this connection is that, due to the extension period granted to developing countries, their consumption of CFCs between 1986 and 1991 increased by more than 50%. These countries are therefore in urgent need of technical assistance relating to the production and deployment of substitutes.

Figure 21
Near-surface
concentrations of CFC-11
from 1977 to 1993 at
different monitoring
stations and the resultant
radiative forcing. CFCs are
entirely of anthropogenic
origin and did not exist in
the atmosphere prior to
the 1950s.
Source: IPCC, 1994a

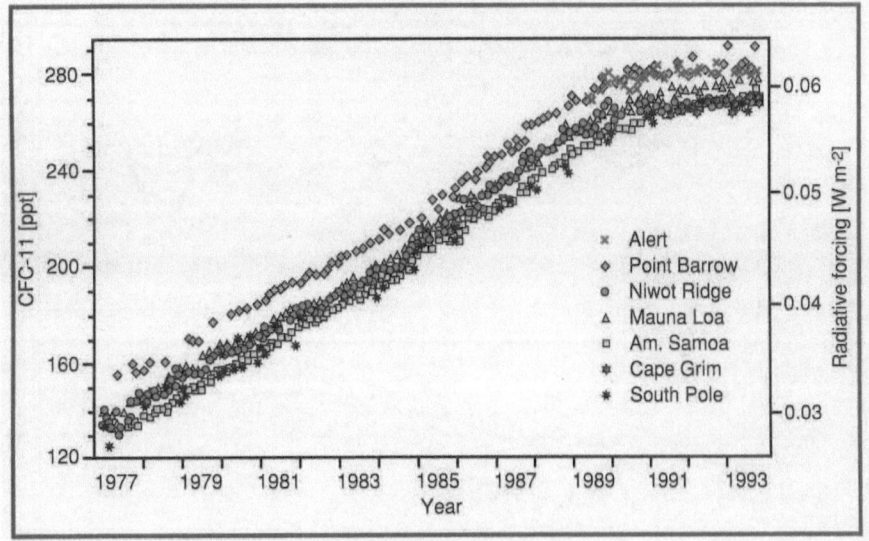

Figure 22
Trends and forecasts of
atmospheric chlorine
concentrations under
increasingly stringent
Montreal Protocol
regulations, with global
ozone depletion rates for
compliance with the
London Amendment.
Source: WMO, 1993

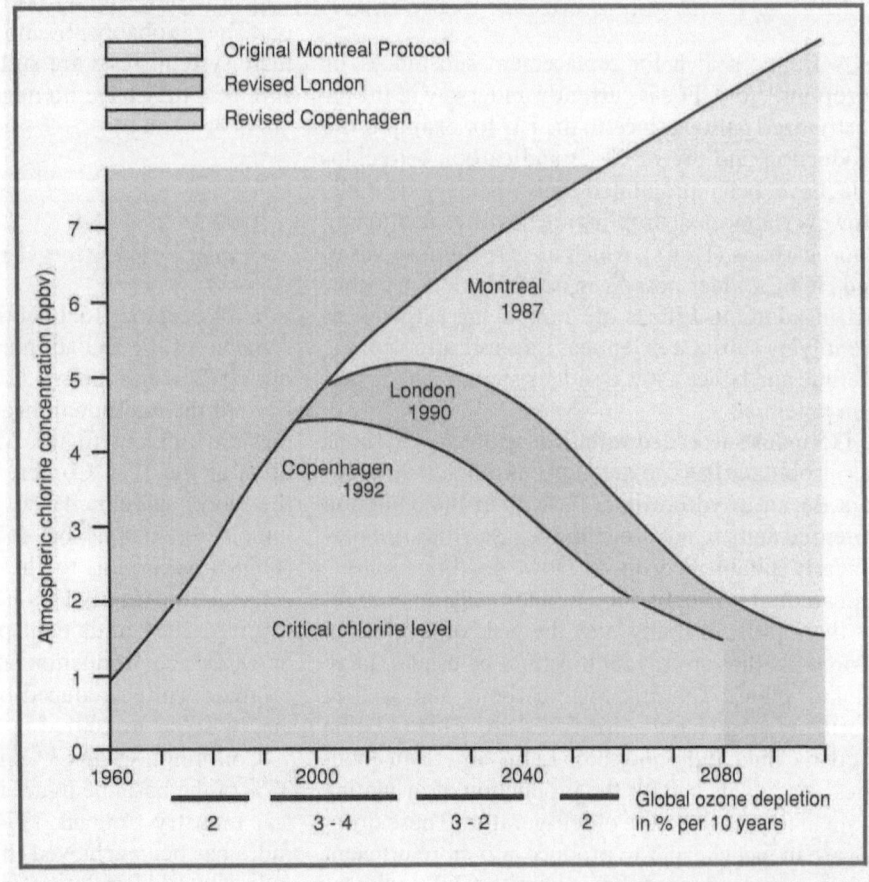

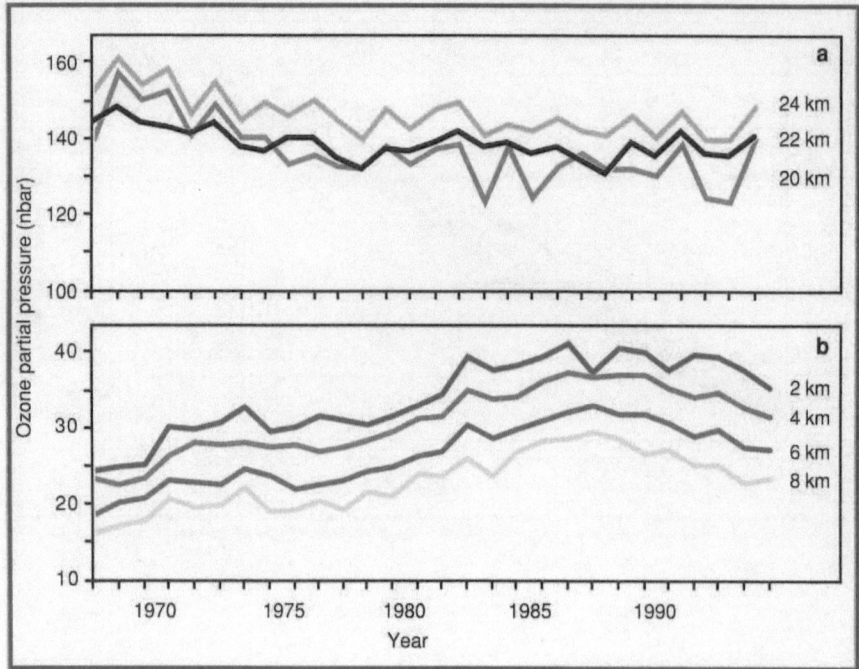

Figure 23
Annual mean values for ozone partial pressure (nbar) over Hohenpeißenberg, an observatory operated by the German Weather Service: (a) in the lower stratosphere (18 to 22 km altitude); this is the densest part of the ozone layer, where depletion is currently occurring at a rate of approx. 5% per decade; (b) in the free troposphere (2 to 8 km altitude); ozone increased here between 1968 and the mid-1980s.
Source: Claude, 1995

2.1.2.3
Substitutes

A frantic search for replacement substances, or "drop-ins", for CFCs is currently underway in the industrialized nations, since, in the EU for example, the production and use of CFCs and carbon tetrachloride have been prohibited since January 1, 1995. However, almost all drop-ins are mixtures containing fluorocarbons (HFCs), which are greenhouse gases, and HCFCs, which possess an ozone depletion potential in addition. Efforts are now being targeted at identifying substances that have a short atmospheric lifetime and hence a low or nonexistent ozone depletion potential.

Germany succeeded with the use of hydrocarbons (e.g. isobutane) as coolants for household refrigerators. Because hydrocarbons have no ozone depletion potential and are not greenhouse gases, they are particularly suitable as drop-ins. However, there is a certain amount of scepticism internationally on account of their inflammability and the risk of explosion. Moreover, these refrigerants cannot be used in large cooling plants. The first hydrocarbons deployed in Germany as substitutes for CFCs are already being used in China, India and some Latin American countries as coolants and for the production of insulating foam in the manufacture of refrigerators. These drop-ins are in fact cheaper to produce and more efficient in use compared to refrigerators with CFC coolants.

A hydrocarbon-based drop-in ("Care 30") has also been launched in Great Britain. Here, too, tests were carried out on the inflammability problem, with

no significantly higher risk being the conclusion. CF_3I appear to have similar potentials as drop-ins for halons. They probably pose no threat to ozone or the climate system. Tests are still being carried out to ensure that they have no negative effects (Solomon et al., 1994 a and b).

2.1.3
Montreal and After: The EU Initiative

On December 16, 1994 the Ministers for the Environment of the EU adopted a joint plan for phasing out HCFCs and methyl bromide. This plan goes far beyond the additional measures adopted in the second amendment to the Montreal Protocol agreed upon at the 1992 Conference in Copenhagen. Furthermore, agreement was reached on controls and import restrictions on ozone-depleting substances (ODS). According to the Council Regulation, the consumption of HCFCs is to be reduced in stages from 1.1.2004 until final prohibition from 1.1.2015 onwards. The production and consumption of methyl bromide will be reduced by 25% by the year 1995, and by 100% by 2001.

Consumption of CFCs in the EU fell in 1993 to 38% of the baseline figure in 1986 (European Chemical Industry Council, 1994); this means that much more has been achieved than the 50% reduction required by EU legislation. However, this reduction results almost exclusively from reduced consumption of aerosols, solvents and foams, which has shown a sharp drop since 1986. In contrast, more CFCs were

BOX 27

Ozone Depletion in the Arctic Stratosphere in Winter 1994/95

The ozone sonde program was continued during winter 1994/95 at the German Arctic Station (Koldewey Station) on Spitsbergen. This is a contribution to the European SESAME campaign, which is analyzing stratospheric ozone depletion. Fluctuations in ozone concentration in the Arctic stratosphere may be caused both by chemical destruction, as well as dynamic processes such as the transport and mixing of air masses with different ozone concentrations. Distinguishing between these two mechanisms is usually a difficult task, since both processes may lead to variations of the same dimensions.

During winter 1994/95, the stratospheric air over the Arctic often cooled down to what were often lower temperatures than in the previous 29 years since measurements began. One consequence of this cooling was a relatively stable circulation of the polar vortex which continued until late April 1995. These are the conditions which enable undisturbed ozone depletion within the polar vortex over a protracted period of time. Analysis of the data obtained from the Koldewey Station, which was located under the polar vortex for most of that winter, reveals a more or less constant process of ozone depletion. This depletion occurred in 1995 over a period of two months from January to March, whereas in previous years it had lasted at most five weeks. The total ozone content at the end of March 1995 was approx. 280-290 Dobson Units (D.U.), i.e. 20-30% lower compared to earlier winters. Nevertheless, this does not constitute an "ozone hole", since the total ozone content did not reach the dramatically low values of well under 200 D.U. measured over the Antarctic.

In March 1995, however, the ozone profile over Spitsbergen matched the shape of that over the Antarctic for the first time (*Fig. 24*). Measurements taken on March 20 show a pronounced minimum between 14 and 19 km altitude, indicating chemical destruction of ozone. In the forthcoming analysis of this data, the attempt will be made to separate the meteorological and the chemical causes of declining ozone levels, i.e. to distinguish between the dynamic and the chemical causes of ozone depletion (von der Gathen and Gernandt, personal communication, 1995).

Ozone distribution is closely coupled at all ti-

mes to meteorological factors, as weather conditions in the stratosphere at approx. 20 km show (*Fig. 25*): on March 20 there was a cold eddy over Spitsbergen, in which an ozone minimum (dark blue area) was measured. Another ozone minimum was found in the eddy over Western Siberia. At the same time, the stratosphere over the Eastern Siberia-Alaska-Canada-Greenland region was warm – an area where much ozone (red area) is always to be found. Also over Germany relatively high values of ozone were measured (pink) while over the Mediterranean the low values (dark blue) are found which are typical and natural for this region throughout the year.

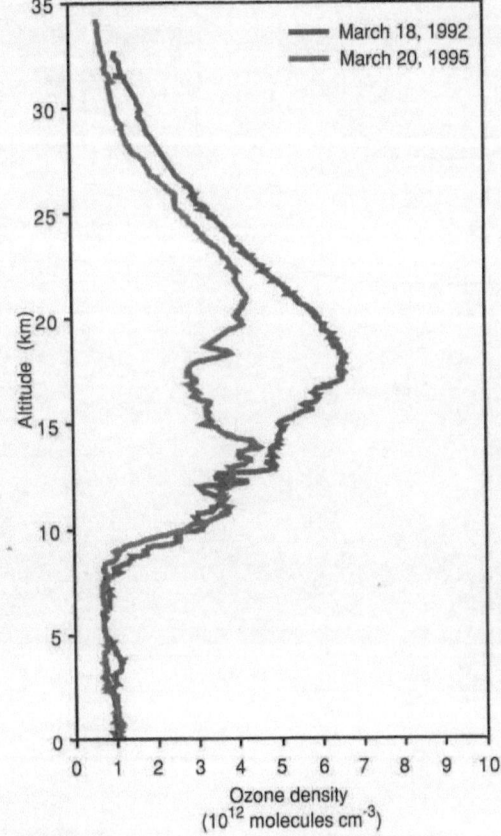

Figure 24
Vertical distribution of ozone over the Koldewey Station in Spitsbergen; red: with "ozone hole"; blue: a comparison.
Source: von der Gathen und Gernandt, AWI, personal communication, 1995

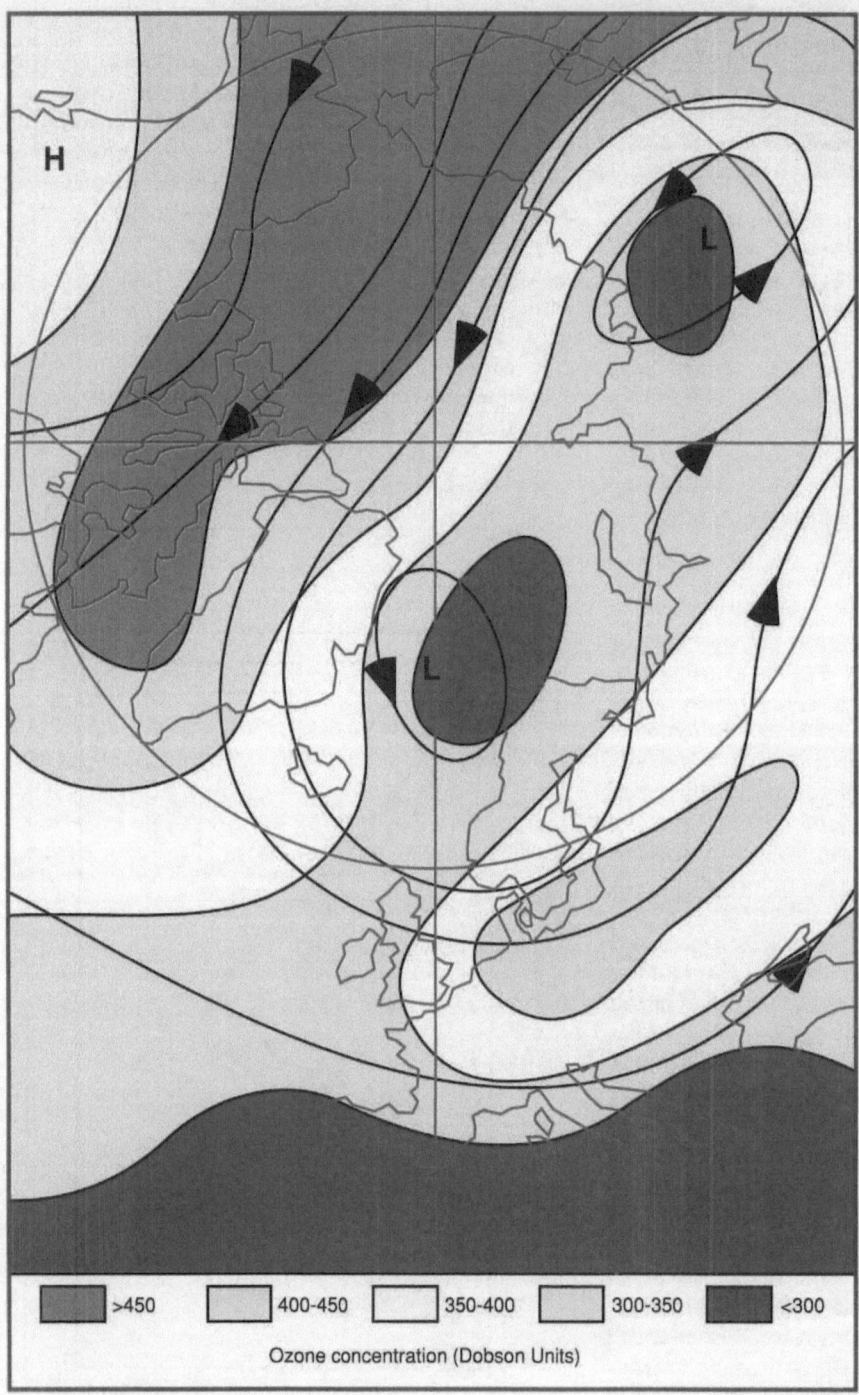

Figure 25
Ozone distribution and
weather conditions in the
stratosphere at an altitude
of approx. 20 km (50-hPa-
area) on March 20, 1995
Source: Data from Free
University Berlin and the
University of Thessaloniki

| | >450 | | 400-450 | | 350-400 | | 300-350 | | <300 |
Ozone concentration (Dobson Units)

consumed in the refrigerator and air conditioning industry in 1993 than ever before (*Fig. 26 a, b*). From
1.1.1995 onwards, i.e. a year earlier than required by
the Montreal Protocol, there is a general EU-wide
ban (except in Greece) on the production and consumption of CFCs. There are only a few, precisely defined exceptions, and the remaining consumption of
approx. 10,000 tonnes (*Fig. 26*) is to be covered with
recycled CFCs. Monitoring compliance remains a

problem, however. There is no agreement as yet on
regulations governing trade in recycled CFCs.

Germany would like to see the import of ozone-
depleting substances banned completely, in order to
stimulate the transition to and the development of
more environmentally friendly substances. In 1994,
approx. 14,500 tonnes of fully halogenated CFCs
were produced in Germany, with all production stop-
ped as from 1995 (Deutscher Bundestag, 1994).

Figure 26
CFC consumption in the
EU from 1986 to 1993.
a) by sector
b) total amounts.
Source: European
Chemical Industry
Council, 1994

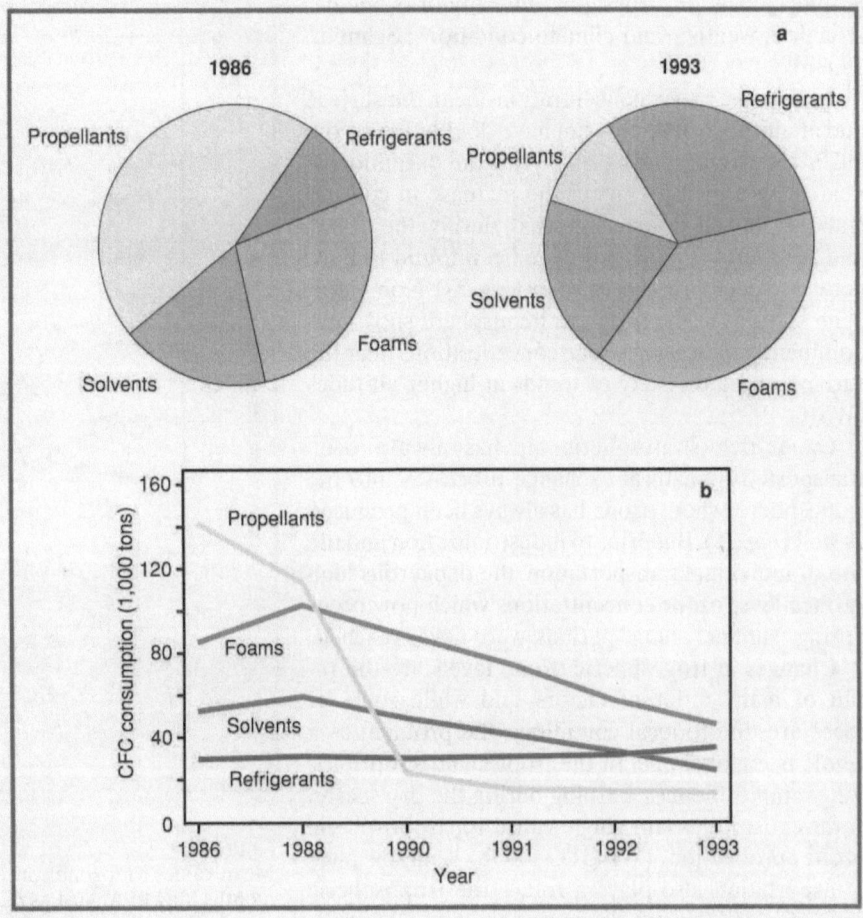

2.2
Tropospheric Zone

2.2.1
Increase in Surface Level Ozone

The unusually sunny and warm summer in Germany in 1994 raised public awareness once again for the problem of "summer smog", or tropospheric ozone. The alarmingly high concentrations of surface ozone hit the media headlines, as did various measurement campaigns and measures aimed at reducing tropospheric ozone concentrations, especially speed limits and bans on driving. Preliminary results of these campaigns indicate that the problem cannot be solved with local measures, but that action must be taken on a nationwide level instead.

Ozone is one of the strongest irritant gases on account of its oxidizing properties, and when present at surface level in the troposphere, where it comes into contact with the biosphere and hence people, it can have injurious effects, depending on concentration and exposure levels (SRU, 1994). It is now suspected to be a carcinogen. In summer months near the surface and under conditions of high insolation and substantial air pollution, additional ozone is formed, especially through the photochemical reaction of nitrogen oxides (NO_x), methane (CH_4), volatile organic compounds (VOC) and photooxidants (PAN).

NO_x emissions act as catalysts in the production of ozone under the influence of solar radiation during daytime, but at night they destroy ozone, such that the ozone concentration in "polluted" areas, e.g. in cities, can fall again at night. However, ozone accumulates in "clear air zones" through long-distance transport from industrial areas and major cities, because there is insufficient NO_x to break down ozone during the night. In clean-air regions at higher altitudes, e.g. the Black Forest or the Harz Mountains, the level of ozone-producing UV radiation is more intensive than at sea level. This explains why the highest ozone concentrations are found in mountainous areas (*Fig. 27*). Superimposed on the long-term increase in surface ozone are fluctuations, which show that photochemistry explains only part of the overall problem. The mechanisms at work are more complicated, and many more factors play a role (geographical location, orography, intensity of ozone destruction, weather and climate conditions; Schmidt, 1993b).

In Europe, ozone concentrations near the surface and at altitudes of up to 4 km have doubled since the 1950s (Staehelin et al., 1994). With the exception of near-surface measurements, the increase in concentrations slowed down somewhat during the 1980s, and at several stations has even been found to have gone into decline in recent years (*Fig. 23*). Non-European stations in the northern hemisphere show predominantly increasing ozone concentrations near the surface, and a diversity of trends at higher altitudes (WMO, 1995).

Ozone-rich stratospheric air has always been transported by natural exchange processes into the troposphere, where ozone has always been produced as well (*Fig. 28*). But prior to industrialization and the rise of individual transportation, the dangerous high surface-level ozone concentrations which now occur during "summer smog" periods, were never reached.

Changes in tropospheric ozone levels are the result of many different factors, and while some of these are due to local conditions, the problem as a whole is a global one. In the tropics and subtropics, for example, biomass burning during the dry season is the most important single cause for tropospheric ozone production (WBGU, 1993). Long-distance transportation also plays a role – the tropospheric ozone formed through dry savannah grass-clearing is carried by prevailing winds over the Atlantic to America (Andreae, 1994). While there are few series of measurements outside Europe, it is generally assumed that tropospheric ozone has roughly doubled throughout the northern hemisphere since the onset of industrialization. No conclusions of this kind can yet be drawn for the southern hemisphere (IPCC, 1994).

2.2.2
Impacts of Increased Near-surface Ozone Concentrations

Ozone in high concentrations acts as an irritant to mucous membranes, respiratory pathways and eyes, a fact that Schönbein (1799-1868), the discoverer of ozone, already realized. The *German Council on Environmental Advices* (SRU) has issued a detailed statement on the issue (SRU, 1994). The World Health Organization has also concerned itself with the problem.

In many regions of the earth, near-surface ozone is already impairing the yields of cultivated plants, hence it must be viewed as a major air pollutant (Enquete Commission, 1995b). Ozone has a direct im-

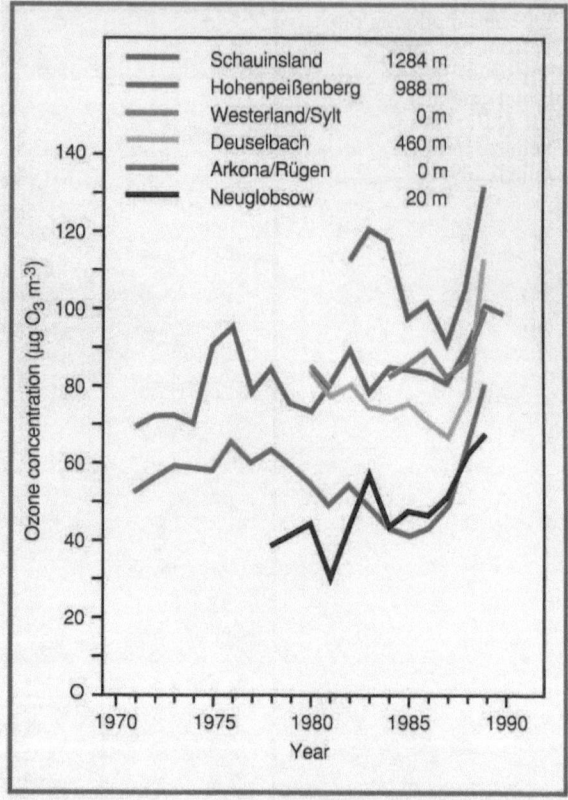

Figure 27
Mean values for ozone concentration in the summer months May to August as measured by the six German meteorological stations with the longest time series.
Source: Schmidt, 1993a

pact on leaves, causing damage to tissue and the destruction of chlorophyll. Plant species vary in their sensitivity to ozone, whereby the extent of damage is a function of ozone concentration and the duration of exposure. Such damage is often visible when the threshold value of 200 O_3 µg m^{-3} is exceeded for several hours. Ozone concentrations in excess of 100 µg O_3 m^{-3} are assumed to cause damage to sensitive plant species if they continue over periods of days to weeks (Krause, 1989). The WHO has defined concentrations of 60 µg m^{-3} (long-term value) and 80 µg m^{-3} (mean value over 10 hours) as upper limits for air quality in connection with sensitive plant species.

Regions which feature particularly intensive farming methods (North America, Europe, East Asia) also show rising NO_x emission levels, so that increased damage to harvests as a result of ozone effects during the summer months can be expected in future (Heck et al., 1988; Chameides et al., 1994). As in other fields of global change, the industrialized countries possess a higher coping capacity to change, i.e. they can compensate more easily for price increases with respect to agricultural produce. Chameides et al. (1994) have pointed out that the problem will worsen in the

Figure 28
Current and historical
measurement series for
surface ozone at various
locations in Central
Europe since 1850.
Sources: Volz and Klug,
1988; Claude and Köhler,
1994

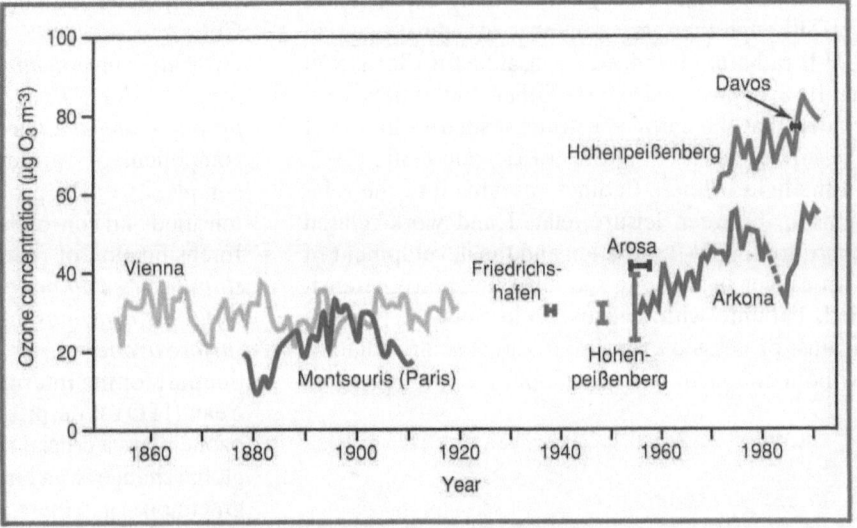

decades to come: the NO$_x$ emission levels predicted for 2025 will increase air pollution and hence the production of tropospheric ozone, and 30-75% of the world grain production could then originate from areas in which the 100-140 µg O$_3$ m^{-3} limit for tropospheric ozone is exceeded for longer periods, thus giving rise to fears that substantial crop losses may result. This situation obtains particular relevance in that the food supply situation in developing countries is expected to worsen drastically as a result of the strong population growth.

2.2.3
Changes in the Ozone Concentration in the Free Troposphere

In the free troposphere as well, i.e. at altitudes between 2 and 10 km, ozone concentrations over Central Europe have increased over the last 50 years (Schmidt, 1994; Schumann, 1994). The vertical distribution of ozone has been measured at the Hohenpeißenberg station since 1968. An increase in ozone concentration (*Fig. 23b*) at altitudes between 2 and 8 km has been clearly identified for the period up to the mid-1980s. This increase is a symptom for a global problem attributable primarily to higher levels of air traffic, the growth rates of which are in the order of 5-6% per annum. Even though tropospheric ozone is a major air pollutant and, indirectly, a greenhouse gas in the context of global change, gaps exist in our knowledge of its global distribution and variability. Any statements on trends are mainly limited to the mid-latitudes of the northern hemisphere, despite the fact that the tropics and subtropics are also affected by any increase in tropospheric ozone. During the *International Tropospheric Ozone Year* (ITOY), ad-

ditional observation stations are to be set up in the tropics and subtropics especially, where they will take regular measurements of the vertical distribution of ozone. This project forms part of IGBP/IGAC, and GEF funding has been requested by UNEP.

2.3
Dangers of UV-B Radiation

2.3.1 Effects of Increased UV-B Radiation

The UNEP Report entitled "Environmental Effects of Ozone Depletion" (UNEP, 1994b) deals in detail with the possible effects of increased UV-B radiation resulting from the depletion of stratospheric ozone. The report points out that, particularly with respect to the effect of UV-B radiation on human health, food production and natural ecosystems, there are still major uncertainties and a substantial need for research, since research in this area obtains very little support internationally.

A major cause for concern arising from the effect of increased UV-B radiation in combination with stratospheric ozone depletion involves the *increased incidence of skin cancer*. Epidemiological studies conducted worldwide provide conclusive evidence that increased epithelial skin cancer (basaliomae, epitheliomae) in fair-skinned populations correlates to exposure to the sun and that inhabitants of countries near the Equator are disproportionately affected. This increased incidence of skin cancer has been attributed primarily to changed leisure habits (sunbathing) (Schaart et al., 1993). The specific role played by UV-B radiation in the development of malignant skin tumors has not been fully explained as yet, although there are a number of indications that UV-B radiation is indeed a causal factor (Schaart et al., 1993; Setlow et al., 1993). Other studies have concluded that the entire spectrum of solar radiation at the surface causes skin cancer (e.g. Wolf et al., 1994). In the light of these findings, research into the relationship between leisure-related and work-related exposure to UV-B radiation and the development of skin cancer in the long run must be greatly intensified. Patients with sensitive skin must be warned against prolonged exposure to direct solar radiation without adequate skin protection.

2.3.2
Recent Measurements

In the first two years following the eruption of Pinatubo in June 1991, the intensity of UV-B radiation, the rays which cause sunburn, was higher than in previous years on account of the very low ozone concentrations over Germany (the minimum level of which was reached in spring 1993) (Vandersee, 1994). Measuring UV-B radiation involves a number of technical problems. Measurements obtained in the past must therefore be treated with a degree of scepticism, since in most cases they are not comparable with those obtained today. An increase in tropospheric ozone, for example, can induce a decrease in UV-B radiation at the surface, even if the total-column ozone level in the atmosphere is declining (Brühl and Crutzen, 1989). The 5% per decade increase in tropospheric ozone now observed in some regions of the northern hemisphere may have partially compensated (by 2% per decade) for the increase in UV-B radiation (Madronich, 1992). Clouds also play a critical role with respect to the radiative balance. A study carried out by Mims (1994) concludes that, as a result of the backscattering of solar radiation by cumulus clouds, the level of UV-B radiation on a cloudy day may be more than 25% greater than on a cloudless day.

2.4
Recommendations for Research

- *Continuation of the national ozone research program*, in cooperation with other international programs. In particular, support for research aimed at gaining an understanding of the natural variability of the ozone layer, without which it is impossible to predict the development of the ozone layer over the next 50 years, as well as research into the chemistry of the heterogeneous processes, essen-

tial for assessing the impact of substitutes for CFCs.
- Air traffic: *investigation of the impact of rising air traffic levels on the ozone content of the upper troposphere and the lower stratosphere* is already a component of various research foci. Given the complexity of the problem, this work must be intensified and consolidated over the long term.
- Intensification of research into the *inter-relationships between air pollution and the production and diffusion of tropospheric ozone, especially near-surface ozone.*
- Support for the International Tropospheric Ozone Year (ITOY): despite the fact that tropospheric ozone plays a crucial role in the overall context of global change as an air pollutant and as an indirect greenhouse gas, there are considerable gaps in our knowledge of its global distribution and variability. Any statements on trends are mainly limited to the mid-latitudes of the northern hemisphere, despite the fact that the tropics and subtropics are also affected by any increase in tropospheric ozone.
- *Research into the impact of air pollution and UV-B radiation on the biosphere*, i.e. on the health of humans, on plant production and on natural ecosystems. In particular, there is a lack of investigations into the long-term effect of such stresses and their socioeconomic impacts.
- *Coordination between stations measuring UV-B radiation and those measuring ozone levels.* The new UV-B radiation measurement stations should cooperate closely or be integrated at both national and international level.
- *Substitutes and the respective new technologies: further development of substitutes for CFCs and HCFCs, including new technologies* for their application, are urgently needed. The search for technologies that do not require fluorocarbons must be forced ahead in those areas not yet covered. This must also include appropriate funding for the relevant research.

2.5
Recommended Action

STRATOSPHERIC OZONE
The WBGU welcomes the Third Report by the Federal Government on Measures to Protect the Ozone Layer (Deutscher Bundestag, 1994). It acknowledges the wide-ranging support for the development of substitutes to replace CFCs and assumes that further progress can be made in this field. In the same manner, the Council welcomes the efforts made by the Federal Government to provide adequate fi-

nancial support for the Multilateral Fund of the Montreal Protocol, creating a framework for supporting selected partner countries. Measures for China and India, the two largest consumers of CFCs in this group, could prove especially effective. In the face of the continued threat posed until the middle of the next century by the damage to the stratospheric ozone layer (the so-called "ozone hole"), *it is essential that everything be done to achieve the rapid end to the production and consumption of CFCs and HCFCs in all countries*, including the developing and newly industrializing nations.

In addition, the leakage of halons and CFCs from existing appliances must be prevented at all costs. This means intensifying the recycling program, which in turn involves even greater endeavors to raise public awareness of the problem in question.

The WBGU recommends that the proposals made in the latest UNEP report on the state of the ozone layer be taken up and that appropriate measures be put into effect. The UNEP report recommends world governments phase out the consumption of partially halogenated substitutes (HCFCs) by as early as the year 2004 in order to accelerate the healing process in the stratospheric ozone layer.

TROPOSPHERIC OZONE

Changes in emission levels must be brought about through the determined implementation of measures to reduce the energy-related and nonenergy related release of precursor substances causing increased ozone production (SRU, 1994).

UV-B RADIATION

In view of the increased incidence of skin cancer, work on informing the public about the definite association between UV-B exposure and the long-term development of malignant skin tumors must be intensified substantially.

3 The Convention on the Law of the Sea – Towards the Global Protection of the Seas

3.1
Preliminary Remarks

The *United Nations Convention on the Law of the Sea* (UNCLOS), to which Germany acceded on October 14, 1994, followed by numerous other industrialized countries, entered into effect on November 16, 1994. This convention, which according to the German government represents the "most important legal instrument of the United Nations to date", was negotiated at the UN Conference on the Law of the Sea from 1973 to 1982 to provide a comprehensive regime for all types of use of the seas and, at the same time, an international "constitution" for all marine regions. November 16, 1994 thus also represents a turning point in marine environment protection, one of the focal points of global environmental policy. Even though the success of UNCLOS cannot be judged upon in this short period of time and although some of its rules are already part of international customary law or are still unclear and inadequate, a major step has been taken towards sustainable management of the seas, as has been called for in the 1987 report of the International Commission for Environment and Development.

In the following the effects of UNCLOS on the protection of marine ecosystems will be analysed and the needs for further reform discussed.

3.2
Utilization Functions of the Seas

In its 1994 annual report the Council depicted the global threat to soils based on their functions (*soil functions*) (WBGU, 1994). A distinction was made here between habitat function, regulation function, utilization function and cultural function. Such a multifunctional model (*marine functions*) could also be applied to the threat to the seas. In this section, however, the Council limits itself to an analysis of the highly complex utilization functions of the seas, which represent the main focal point of the Convention on the Law of the Sea.

Three types of utilization functions constitute the primary problem area covered by marine environment policy:
- environmental damage through the *transport function* of the seas,
- impairment of the *disposal function* of the seas through terrestrial sources of emissions and the dumping of various substances (*Fig. 29*),
- damage to the *resource function* of the seas, particularly through overexploitation of living resources.

3.3
Transport Function

3.3.1
Marine Pollution from Vessels

The classic example of marine pollution in public perception is pollution through maritime transport. Even though tanker accidents still occur and can cause environmental disasters, the greatest success up to now has been achieved in this particular area of marine environment policy (*Fig. 29*). UNCLOS requires in Art. 211 (2) the Parties to enact laws and regulations applying to ships flying their flag which must be no less effective than the existing international agreements on the protection of the environment (UNCLOS commentary).

These international agreements date back to the 50s when numerous shipping nations agreed to the 1954 London *Convention for the Prevention of Pollution of the Sea by Oil* (OILPOL). At that time this agreement was considered to be exemplary since it was the first legally binding instrument with universal regulatory application for the protection of an environmental medium against pollution. To this extent, marine environment policy could be regarded as a pacemaker for global environmental policy – and,

Figure 29
Deposition of pollutants in
the oceans.
Source: GESAMP, 1990

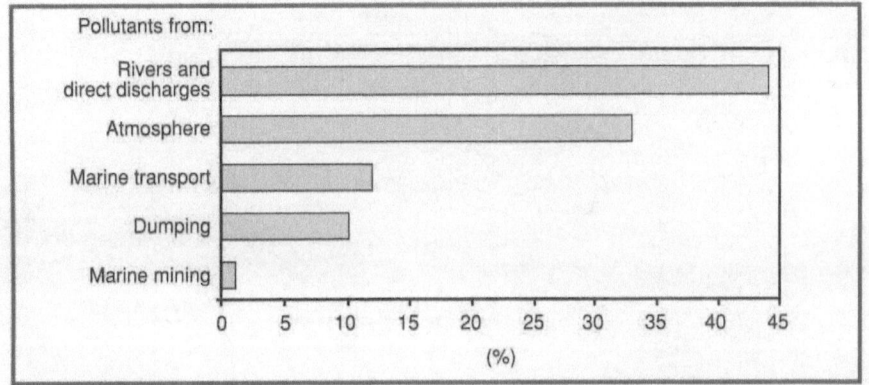

in fact, numerous principles and instruments developed for the protection of the seas can be found in later regimes of international environmental policy. The regulatory mechanisms of OILPOL turned out to be impractical, however, and the scope of application of the agreement – banning of certain oil discharges into coastal waters – proved to be too restricted.

After several revisions of the OILPOL Convention there followed a fundamental reformulation of the international regulations on shipping-related sea pollution in the 1973 *International Convention for the Prevention of Pollution from Ships* (MARPOL), London, which went into effect in October 1983 and has now been ratified by 85 countries accounting for over 92% of the world's shipping tonnage.

The MARPOL Convention in itself constitutes only a general framework with basic obligations with regard to protecting the seas against pollution by ships. Specific provisions were agreed upon in five annexes, with accession left to the Parties to a certain extent. Annex I concerning oil pollution and Annex II concerning the transport of harmful liquid substances as bulk goods became legally binding, though with a three-year transition period (Hartje, 1983). The remaining annexes concerning ship sewage (No. IV), ship waste (No. V) and concerning the transport of hazardous goods in containers and other forms of packaging (No. III) initially remained optional, but have entered into effect in the meantime, with the exception of the fourth annex.

This approach of a legally binding framework agreement with several substantial core provisions in addition to problem-specific optional annexes held the risk of "partial environmental protection", but it was the only way of establishing a universally binding regime, which is imperative due to the global nature of maritime transport. The annexes themselves, which (except for Annex IV) now apply as universally binding law in accordance with Art. 211 (2) of the Convention on the Law of the Sea, require the Parties to implement various environmental policies:

in some cases technical standards for design, equipment or crews of the ships have to be enacted, in other cases environmentally harmful behavior on the part of the crews must be made punishable by law.

Art. 211 (6) of UNCLOS grants coastal states the right to enact special regulations for particularly threatened marine areas in cooperation with and with the approval of the *International Maritime Organization* (IMO). To this extent, this article confirms the annexes of the MARPOL Convention in that several semi-enclosed marine areas were classified as "special regions". The annexes prescribe stricter regulations for these special areas; in the special areas covered by Annex I, for example, oil discharges are prohibited, except in the case of emergency. Special areas under MARPOL are: the *North Sea*, the *Baltic Sea*, the *Mediterranean Sea*, the *Black Sea*, the *Red Sea* with the Gulf of Aden, the *Persian-Arabian Gulf*, the *Southern Ocean* as well as the *Caribbean Sea*.

The following agreements contain special standards for the design, equipment, operation and crews of sea-going vessels:

– *Convention on the Protection of Human Life at Sea* (SOLAS) of 1974, whose oldest version (1914) dates back to the sinking of the Titanic,

– the *Convention on International Rules for Preventing Collisions at Sea* (COLREG) of 1972,

– the *Convention on Standards for Training, Issuing Qualifications and Watch Duty for Ship's Crews* of 1978,

– the *International Convention on Oil Pollution Preparedness Response and Cooperation*, London, which was agreed on as a reaction to the inadequate rescue actions taken after the Exxon Valdez disaster in 1990 and went into effect in May 1995.

To keep maritime traffic away from vulnerable ecosystems, states have been able to declare "particularly sensitive regions" within the framework of the IMO since 1991. The amendments to the MARPOL Convention, which went into effect in July 1993, prescribe double hulls or other structural measures having the same safety effect for all newly built tank

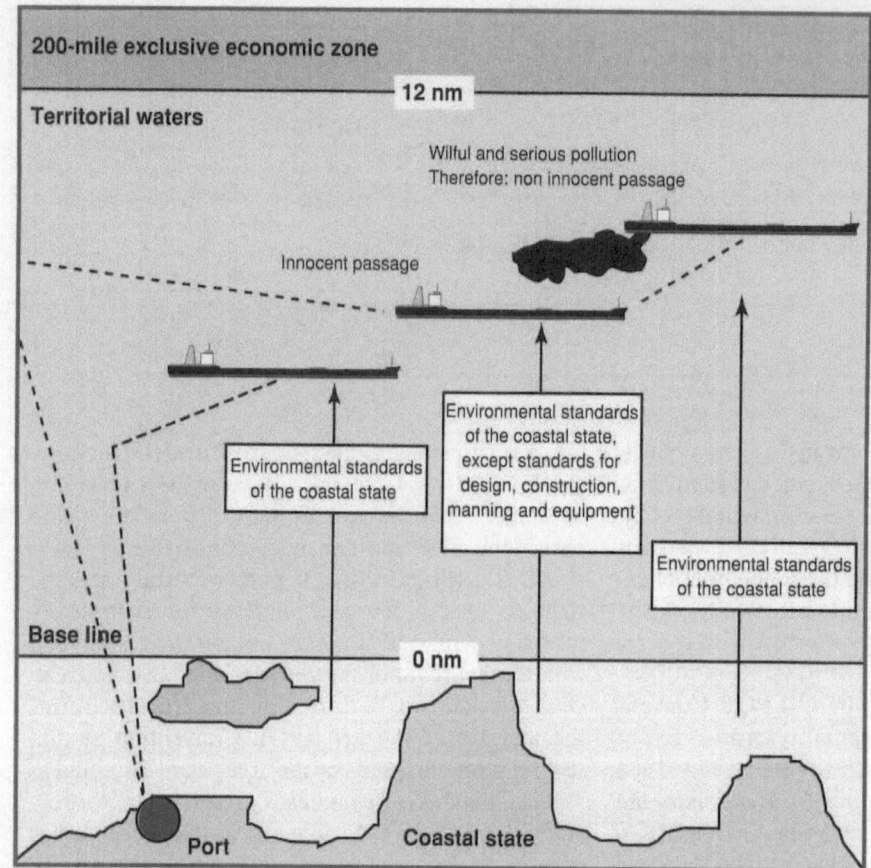

200-mile exclusive economic zone

12 nm

Territorial waters

Wilful and serious pollution
Therefore: non innocent passage

Innocent passage

Environmental standards
of the coastal state

Environmental standards
of the coastal state,
except standards for
design, construction,
manning and equipment

Environmental standards
of the coastal state

Base line

0 nm

Port   Coastal state

Figure 30
The legal regime regarding
territorial waters.
The broken lines designate
the routes of foreign ships,
the solid lines show the
extent of the coastal state's
jurisdiction in setting
standards. The legal
institute of "innocent
passage" does apply unless
inland waters of the
coastal state are entered
(left) or "non innocent"
actions are carried out,
such as "wilful and
serious" oil pollution
(right).
Source: Biermann, 1994a.

ships. Contrary to the customary procedures involved in revising the MARPOL Convention, these provisions are also to be applied to all large tankers (over 20,000 t deadweight) over 25 years old after July 6, 1995. At the same time certain older tankers are to be subjected to more stringent inspections.

In summary, a large number of international regulations for the protection of the marine environment and maritime safety have entered into effect as binding international law in recent years (Biermann, 1994b). In accordance with Art. 211 (2) of UNCLOS, most of these standards are now considered to be minimum requirements for all states involved in maritime transport (for Germany: Edom et al., 1986). Problems remain, however, with respect to enforcement of the regulations and resolution of (potential) conflicts between *flag states*, *port states* and *coastal states*.

3.3.2
Conflicts Between Flag, Port and Coastal States

Generally, implementation of environmental and safety standards is the responsibility of *flag states*. To compensate for the deficient enforcement of these duties by the flag states, the MARPOL Convention of 1973/78 granted *port states* certain enforcement rights, in particular the right to carry out certain inspections on ships. To give the crews an incentive to comply with the standards, the MARPOL annexes require the Parties to set up collection facilities in the ports for disposal of the substances concerned in the special areas. The coastal states, however, remained extensively helpless as far as taking action against passing ships without calling at their ports. They were merely granted certain emergency rights by the 1969 Intervention Agreement, which applies, however, only after an accident occurred at sea.

Whereas this allocation of legislative and executive powers has been reaffirmed by UNCLOS, it has also been adjusted to the new territorial allocation of sovereignty rights in the Convention (United Nations, 1992a).

Coastal states have almost exclusive rights up to twelve nautical miles off their coasts, though these rights are restricted to the legal institution of "innocent passage" *(Fig. 30)*. Research may take place here only at the invitation of the coastal state. The coastal states can have their own environmental standards for their *exclusive economic zone* (EEZ) of 200 nautical miles, but these standards must not exceed those

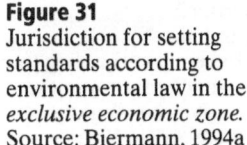

Figure 31
Jurisdiction for setting standards according to environmental law in the *exclusive economic zone.*
Source: Biermann, 1994a

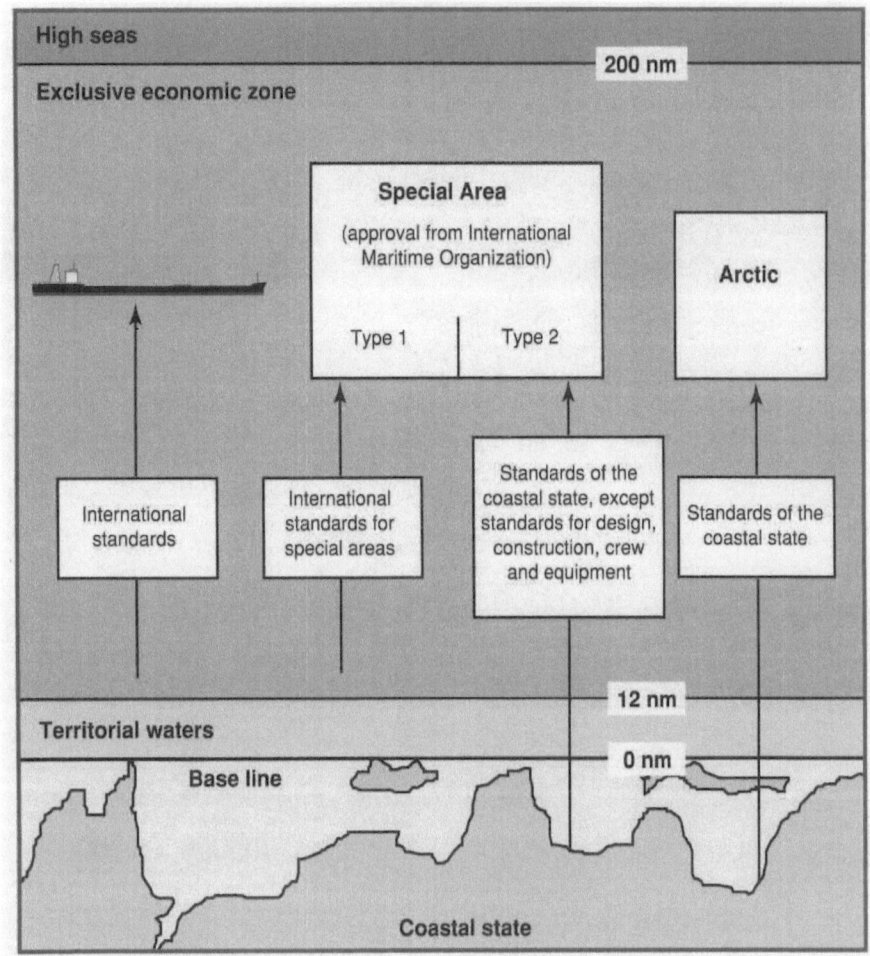

of the MARPOL regime *(Fig. 31)*. Traditional freedom of research is restricted in these economic zones, but is possible with the approval of the respective coastal states *(see Section 3.6 regarding marine research)*.

Enforcement of these standards remains the responsibility of the flag states as long as a ship does not discharge any pollutants *(Fig. 32)*. The port states can substantiate that ships entering ports have to meet international standards on the basis of the documents that have to be presented by the crews (including the oil log). Only when there is evidence of marine pollution or when there are clear indications that the documents "essentially" do not reflect the current state of the ship, may inspections of the ships concerned be carried out by the port authorities and prosecution proceedings be initiated. Prosecution on the part of the port states must be suspended, however, if the flag state takes over proceedings.

This reservation of jurisdiction in favor of the *flag states* (Art. 228) has been frequently criticized in relevant literature on UNCLOS and more recent statistics on the common practice of enforcing maritime

law standards seem to justify this criticism. For instance, of all violations of MARPOL provisions reported on between 1983 and 1991 and handed over to the jurisdiction of flag states, fines or other penalties were imposed in only 13% of the cases. In 65% of all cases the violations apparently had no legal consequences and in nearly 20% the shipowners were acquitted. By comparison, penalties were imposed in roughly 30% of the cases in which port states and not flag states conducted the investigation (IMO, 1994). However, the coastal or port states are not required to comply with the reservation clause in favor of flag states if the flag state concerned has repeatedly failed to meet its obligations (Art. 228 of the *Convention on the Law of the Sea*). This means that coastal and port states have a certain legal room for maneuver as far as prosecuting environmental violations themselves is concerned.

The current discussion regarding shipping-related marine pollution stresses several points. First, some coastal states are urging an extension of the standards prescribed in the MARPOL Convention. There are regulatory deficiencies particularly with

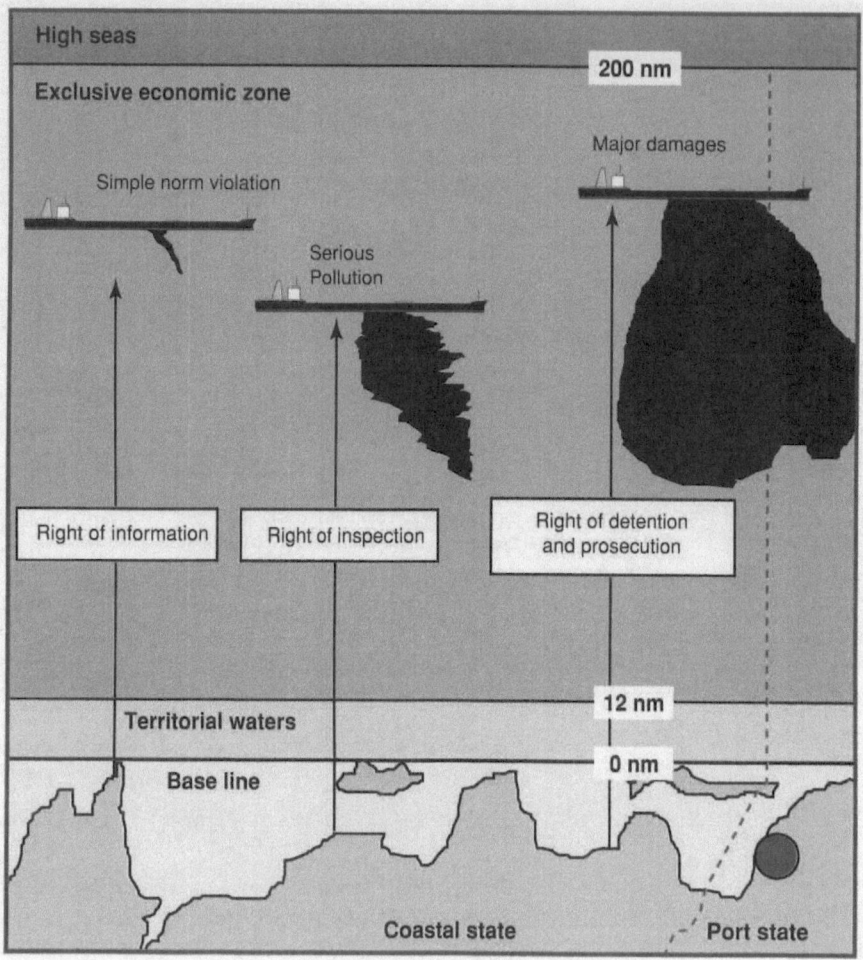

Figure 32
Jurisdiction for enforcement according to environmental law in the exclusive economic zone. If there are no clear reasons for believing that a foreign ship navigating in the EEZ has committed environmental pollution, the enforcement of environmental standards in the EEZ is the sole responsibility of the flag states or port states if the foreign ship has entered their port voluntarily. Source: Biermann, 1994a

respect to vessel-source air pollution, transport of solid hazardous goods, liability for pollution damage, dangers caused by ship fuel that leaks out after collisions, marine pollution caused by marine coating and the spread of foreign animal and plant species through the ballast water from ships with the resulting changes to regional ecosystems. Legal regulations for these environmental problems are currently debated within the framework of IMO.

Even more important than these regulatory loopholes, however, is the inadequate enforcement of environmental policy since numerous flag states do not sufficiently enforce the MARPOL provisions on their ships. A statistical assessment of shipping accidents, for example, shows that accidents are one hundred times more probable with ships of the "least safe" merchant fleets than with fleets having the highest safety standards (IMO, 1993). More recent surveys indicate that pollution caused by ships on the open sea has almost exclusively been detected by industrialized countries with a relatively effective coast guard in their waters. 25% of all violations during the 80s were detected by Germany alone; only 4% of the

violations were uncovered in waters of the southern hemisphere. It can be presumed, therefore, that environmentally harmful discharges are much more frequent in African, Asian and South American waters than in the regional waters of the North since the risk of criminal prosecution is less in the South (IMO, 1994).

The IMO is attempting to counter this enforcement shortcoming through increased inspection of ships by their respective flag states. A "Subcommittee for Flag State Implementation" was formed in 1992 for this purpose. Promotional programs for administrative agencies play a significant role in this connection, a prominent example of which is the *World Maritime University* (WMU), set up in 1983 for executive staff from the developing countries. However, the competitive pressure in worldwide maritime trade is intensifying due to the current oversupply of transport capacities, thus giving shipowners and flag states strong incentives to neglect environmental and safety standards in order to gain market advantages. In recent decades numerous shipping companies have deliberately had their ships registered in countries which have low environmental and social standards *(outflagging)* and which, as countries with "flags of convenience", are responsible for the majority of marine mishaps and probably for the majority of operational emissions as well.

Control of environmental and safety standards by the *port states* represents a fundamental alternative to or extension of the relatively ineffective enforcement on the part of the flag states. MARPOL and SOLAS as well as UNCLOS offer a limited legal framework for this. In 1982 the port authorities of the European coastal states introduced a computer-aided monitoring system that is based on spot checks of every fourth ship in the port and essentially resulted in inspection *(Paris Memorandum of Understanding)* of roughly 90% of all ships calling at European ports. These databases are now networked with the Japanese, Russian and North American port authorities. Over 130,000 inspections have been conducted since 1982, on the basis of which more than 4500 ships (3.5%) were detained because of violations of environmental and safety standards. In 1992 this figure had risen to 5.6%, in particular due to the increasing overaging of the world's merchant fleet (Plaza, 1994). The Paris Memorandum of Understanding was further tightened in July 1993 so that ships can be very frequently inspected today if their flag state has conspicuously neglected to enforce international provisions on repeated occasions.

The IMO now strives to introduce this relatively successful European system of collective monitoring of ships by potentially affected *port states* as a worldwide model. Comparable systems have already been set up for South America and for East Asia/Pacific and are being prepared for the Caribbean. A comprehensive global system of cooperation between the individual port authorities will probably be in place by the end of this century (Plaza, 1994). Supranational forms of international maritime shipping control are conceivable in the future on the basis of these worldwide databases and they may replace the structurally ineffective flag state control in the long run.

A key problem concerning shipping-related marine pollution is the situation of the poorer countries that are hardly in a position to finance extensive measures. The GEF now provides funds to finance oil disposal facilities in the ports of developing countries. In addition, the 1990 "Oil Pollution Preparedness Convention" provides for transfers to developing countries so that the latter can set up special units. However, the financial transfer effected thus far is insufficient to implement the MARPOL standards in the regional seas of the South. A new approach is therefore currently being worked up by the member states of the *Association of South East Asian Nations* (ASEAN), which demanded the establish-

ment of an international fund in a ministerial declaration in 1992 in order to support financing of the regional monitoring programs for reduction of shipping-related oil pollution and piracy. If this fund should not be set up by the international community, the ASEAN states announced introduction of special charges for passing ships to finance these measures (UN, 1992b).

All in all significant success has been achieved in combating shipping-related pollution since 1954. Tanker accidents have declined both in absolute numbers and in the volume of oil leaking out and currently cause only about 5% of the pollution of the seas through oil (GESAMP, 1993). Although quantification of the total global environmental damage due to maritime transport is difficult, there has been a great decrease in oil pollution caused by shipping operations *(Fig. 33)*. In contrast to other types of transport, such as air, road or rail transport, an environmental assessment of shipping shows a relatively positive development.

In spite of this success, transport-related marine pollution still represents an environmental problem that may have disastrous regional effects. A tanker mishap of little quantitative significance, for instance, may very well cause a local environmental disaster. In the view of the Council, therefore, there is both a need for research on the development of improved political instruments and a need for action providing for effective enforcement of the agreed regulations.

The substantial funds required of the Parties in order to implement the agreed provisions represent a special problem. The MARPOL annexes, for example, prescribe the construction of port collection facilities for oil contaminated ballast water in order to reduce the incentive for discharging illegally at sea. Thus an increase in financial aid for developing countries must be considered, either by scaling up or reallocating the funds of multilateral financing institutions, by expanding bilateral aid or by calling for the assumption of responsibility on the part of the private sector (such as the multinational oil corporations operating in the ports). Greater international cooperation in setting up and extending administrative structures in developing countries appears equally necessary (capacity building; *see Section B 3.2.2.2*).

This cooperation is currently being effected via the IMO, which should have more funds at its disposal for this purpose. This also applies to Germany, which finances 1.89% of the IMO budget (1994) at present – in contrast to the otherwise usual German UN contribution of roughly 9%. The low contribution rate can be explained by the history of the IMO, in which – as a traditional cooperative organization of the flag states – 90% of the contributions are calculated according to the tonnage carried by ships fly-

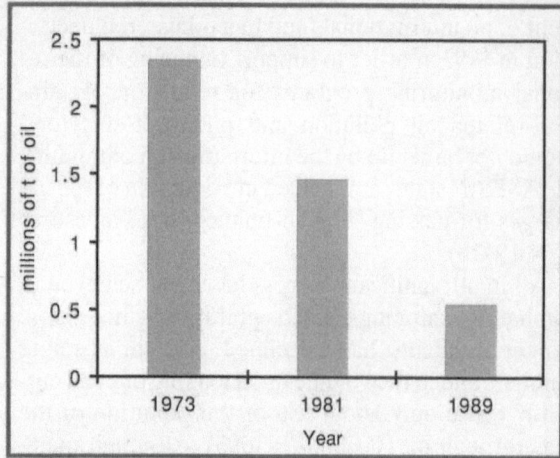

Figure 33
Oil discharge from vessels.
Source: GESAMP, 1993

ing the flag of the respective member state. As a result, smaller developing countries like Liberia, Panama and Cyprus finance the main portion of the IMO budget. In view of the new functions of the IMO regarding the environment, the previous contribution formula should be replaced in future by the general UN contribution rate.

Another point of criticism is that the previous structural standards for ships within the framework of the MARPOL and SOLAS Conventions repeatedly contained exemption clauses for ships that were built before the respective provisions went into effect. These so-called "grandfather clauses" diminish the effectiveness of the safety-relevant design standards considerably. Various measures are conceivable here: one might be, as a partial departure from the *poluter-pays principle,* an agreement that the costs of overhauling or putting overaged ships out of service have to be borne in some cases by the international community according to the *principle of burden sharing,* such as through an international compensation fund. Another conceivable alternative following market principles might take the form of *environmental charges* levied by the port authorities, depending on the age and design of the ships, in accordance with the polluter-pays principle. This system would make the operation of older ships relatively expensive and would thus result in a partial internalization of the potential threat stemming from old tankers in the price calculation of ship operators. Moreover, such environmental charges would also be possible as a unilateral measure without the consensus of the affected and polluter states. Similar environmental policy success could be achieved through reforms of the liability law, such as through introduction of a mandatory, unlimited liability insurance for

ships whose premiums could be staggered by the insurance sector according to safety standard.

Application of the environmental policy instruments outlined here would, in the eyes of the Council, additionally represent an interesting test case for market-oriented problem-solving models in other environmental fields. Since maritime shipping favors international regulations on the basis of its tradition, there are greater prospects of success in this area.

Nevertheless, besides legal enforcement by flag and port states, an extension of enforcement rights of coastal states should also be taken into consideration, since these states, as those actually affected, certainly have the greatest interest in ecologically effective regulations. UNCLOS allows the coastal states a certain and important room for manoeuvre here, which may not be taken advantage of, however, due to the technical difficulties involved in verifying pollution at sea. Reliable detection methods, such as for illegal oil discharges on the open sea, are still lacking although improvements can be expected in the near future through satellite technology and the use of special aircraft.

In addition to improvement of the detection methods for illegal discharges, the Council feels that the legal consequences of marine pollution should be made more severe since the relatively small fines usually bear no relation to the actual pollution and thus continue to represent an inadequate deterrent to illegal discharges on the part of shipowners and crews.

Furthermore, unilateral action on the part of the coastal states involved is conceivable, enabling them to enact environmental standards above and beyond the maximum provisions of the MARPOL Convention. Unilateralism has historically always been a "motor" of development in marine policy, especially since the unilateral appropriation of the North American continental shelf by the USA in 1945. Thus it was hardly surprising that the USA chose the path of unilateralism after the *Exxon Valdez disaster* in that it enacted a series of measures for its waters under the *Oil Pollution Act* (1990), going far beyond the MARPOL Convention (double hulls for tankers, for example). These measures, whose legality was at least questionable, could be enforced and have become internationally binding for all states in the MARPOL amendments of 1992. More than 50 tankers have been built according to the new U.S. standards since 1992 (Boisson, 1994). It can be assumed, therefore, that affected coastal states might take unilateral protective action in similar situations, despite all legal, economic and political doubts, even though the chances of enforcement are not always as good as those of the USA.

3.4
Disposal Function

3.4.1
Terrestrial Sources of Pollution

Sources of emissions on land are responsible for 70-80% of the total pollution of the oceans, and these emissions have already taken on threatening proportions regionally. The provisions of UNCLOS bear no relation at all to the significance of this threat to the environment: Art. 207 and 212 merely require the Parties to reduce terrestrial sources of marine pollution at their discretion and to "take account of" special international agreements in this connection. In view of the agreement that has been reached in the meantime regarding regimes for restriction of emissions with regional or global regulations applying to other problem areas – such as the *Convention on Long-Range Transboundary Air Pollution*, the *Montreal Protocol* and the *Framework Convention on Climate Change* – revision of UNCLOS already appears necessary, only just after going into effect.

Land-based marine pollution basically takes place via stationary and diffuse sources of emissions, resulting in other environmental damage so that there are overlaps with other regulatory areas of environmental policy. For example, a large portion of the pollutants is conveyed to the seas via the air and straightening of rivers leads to accelerated water runoff and sedimentation. Thus all national and international measures that contribute to a reduction of large-scale air pollution and to renaturalization of river courses have an alleviating effect on the marine environment, even if this was not the main objective. The introduction of unleaded gasoline in the USA at the end of the 70s, for example, substantially reduced the lead intake of oysters and mussels. Furthermore, waste substances are discharged into the seas via rivers. Thus protection of inland bodies of water always means protection of the marine environment, just as effective protection of regional waters can only be achieved through action to prevent pollution of the large river systems. The international *Group of Experts on Scientific Aspects of Marine Pollution* (GESAMP) regards marine pollution through nutrients (particularly nitrates and phosphates in coastal marine areas) as the most widespread marine environmental problem. In addition to environmentally harmful chemicals, the sediments carried by rivers result in damage to the marine ecosystem in coastal waters. The total amount of sediments has nearly tripled due to human activities such as deforestation and agriculture. Sediments must therefore be regarded as part of marine pollution today (GESAMP, 1990).

To a lesser extent the seas are also polluted via direct discharges, especially near ports or tourist areas. Roughly 60% of the world's population and two thirds of all big cities having a population in excess of 2.5 million are located in coastal regions. It is expected that the population of these coastal regions will rise to approx. 6 billion by the year 2020, a situation that will lead to enormous pressure and damage to coastal waters, given the inadequate environmental precautions. Besides direct marine pollution through discharge of nutrients and pollutants, urban sprawl causes tremendous damage to coastal marine ecosystems by destroying the natural biotopes along shores and banks.

Since land-based marine pollution is a consequence of regional contamination of regional waters and air and individual regional waters are polluted to a varying extent, a *regional approach* has been favored here. Separate agreements were concluded for individual regional waters, covering certain emissions and making them subject to restrictive regulations. This was done in the case of the North Sea, Baltic Sea and Mediterranean Sea, for instance, where the international regimes agreed upon in the 70s and 80s have led to improvements in water quality, at least in some categories of substances.

The *Convention on the Protection of the Marine Environment of the Baltic Sea Area* was signed by all countries bordering the Baltic Sea, including the Eastern European states, in March 1974 and, even at that time, included terrestrial marine pollution, thus making it the most comprehensive multilateral environmental treaty of that period. When the agreement went into effect in May 1980, the *Helsinki Commission for the protection of the Baltic Sea environment* (HELCOM) was set up to perform administrative and research tasks. The countries bordering the Baltic Sea agreed on reduction plans for a number of substances that are being monitored by this commission and are showing initial success. For example, pollution due to trace metals is only a regional problem now and the concentration of organic chemicals like DDT and PCB has declined. Today nutrient pollution, primarily caused by intensive agriculture and resulting in extensive eutrophication of the Baltic Sea, represents the greatest threat (OECD, 1991).

In order to "expand, strengthen and modernize" the existing regime for the Baltic Sea the Helsinki Convention was revised in April 1992 and ratified in this form by Germany in August 1994. These legal provisions do not go far enough, however. In view of the threatening state of the Baltic Sea ecosystems, a tightening of the 1992 Helsinki Convention as well as "international programs amounting to billions" were demanded at a conference of the German coastal Länder so as to limit the damage.

Comparable measures to those taken for the Baltic Sea were also agreed for the North Sea and Northeast Atlantic. The *Convention for the Prevention of Marine Pollution from Land-Based Sources* was signed in Paris in 1974 and required the countries bordering these waters to reduce discharges into rivers within the framework of a list system. The "Paris Commission" set up in 1978 was intended to monitor compliance with the treaties on the basis of national reports. The main shortcoming in the agreement was the failure to include the expansive contamination of the air. However, implementation of the 1979 Geneva Convention on Long-Range Transboundary Air Pollution provided relief for the North Sea through the substances covered in its protocols, i.e. sulfur (1985), nitrogen oxides (1988) and volatile organic compounds (1991). The overall nutrient content in the southern North Sea has now risen to double or triple the natural concentration, and even PCB pollution, which has been recorded for a long time, is still at very high levels. In 1987 a 50% reduction in emissions of substances on the "Black List" as well as of nitrates and phosphates by 1995, based on the 1985 level, was agreed on by the countries bordering the North Sea. Success in reducing nutrients and pollutants since then cannot be overlooked, but is still not enough.

The *Convention on the Protection of the Marine Environment of the Northeast Atlantic*, which was agreed on in Paris and ratified by Germany in August 1994, is supposed to combine the various treaties on protection of the North Sea and thus guarantee more effective monitoring and adaptation of the provisions. Despite success in specific cases, however, North Sea protection policy is still inadequate; roughly 40 million t of liquid and solid industrial wastes enter the North Sea every year (Freestone and IJIstra, 1990; Kruize, 1991; North Sea Task Force, 1993).

Only a few years after conclusion of the European agreements UNEP called for an application of the European approach to regional waters of the developing countries (Biermann, 1994a). The first project within the framework of the UNEP regional sea program was the Mediterranean Sea (since 1975), which is especially threatened because of its hydrological and ecological conditions (*Barcelona Convention*, 1976). Other regional sea programs followed for the Persian-Arabian Gulf (1978), the west coast of South America (1981), the coastal waters of west and central Africa (1981), the Red Sea (1982), the Caribbean (1983), the regional waters of east Africa (1985) and the South Pacific (1986). In 1985 the UNEP Governing Council accepted the *"Montreal Guidelines"*, which were drawn up by group of experts assigned to this task in 1983 and contain a detailed list of measures that represents a model for regional agreements and resembles to a great degree the UNEP regional sea agreement.

The most recent regional sea program is the action plan for the protection of the Black Sea with the *Bucharest Convention* (1992) and its three protocols. Legally nonbinding action plans also do exist for south Asian waters and the Arctic Sea and are in the process of preparation for the Sea of Japan. Nearly 140 countries are now involved in such regional sea programs. A number of recommendations for protection of the Antarctic seas have been issued within the framework of the Antarctic treaty system (Germany has been a Party to the Antarctic Treaty since 1979). The *Madrid Protocol on Environment Protection to the Antarctic Treaty* (1991), which has not entered into effect but has already been ratified by Germany, declared that Antarctica and the bordering waters were "a nature reserve dedicated to peace and science" and provided for the strictest and most extensive environmental protection regulations ever adopted for a region in an international agreement.

All UNEP regional sea programs *(Fig. 34)* more or less follow the pattern of the Mediterranean Sea pro-

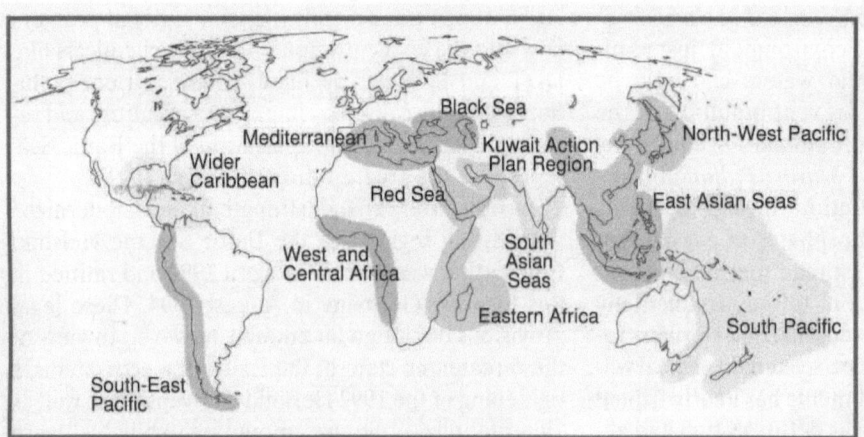

Figure 34
Regional Sea Programs of the UN Environment Programme, 1992.
Source: UNEP, 1992

tection program: according to a framework convention, the bordering states have a legally binding obligation to reduce various sources of emissions; these measures are further differentiated in protocols, accession to which is generally left up to the states. This *"framework treaty plus protocol solution"* makes it possible for the bordering states to differentiate their obligations according to capacity and environmental policy preferences, without threatening the universal binding applicability of the treaty system. Regarding institutionalization, either independent international organizations were created or the Secretariat tasks were assigned to the UNEP or other existing institutions.

The majority of the developing countries have, in the meantime, committed themselves to legally binding water pollution control measures – substantial success has not materialized to date, however. This lack of effectiveness of the UNEP regional sea programs is probably due to the considerable costs that would be incurred as a result of action to reduce emissions contributing to air and water pollution. For example, the USA and Canada alone have spent $8.85 billion on the treatment of municipal and industrial sewage as part of efforts to clean up the Great Lakes. It would be impossible for comparable sums to be provided by the developing countries, in which less than 5% of the sewage on average is currently treated at all (WRI, 1992). The World Commission for Environment and Development came to the conclusion that it "...is one thing to spend a couple of million dollars on research, and something entirely different to include the research results in development plans for the mainland and implement stringent environmental protection monitoring programs". Therefore, the Commission demanded the conclusion of "... general agreements going beyond objectives and pure research and providing for a solid schedule of investments of a genuinely effective magnitude "(Brundtland Report, 1987).

If these investments (in an ecological restructuring of economy and agriculture, integrated technology, etc.) should not materialize and the regional sea programs fail in Africa, Asia and Latin America, this failure may have global consequences. As an example, nearly 95% of the fish catch worldwide is caught in coastal waters; it is in these waters, however, that roughly 90% of all chemicals discharged from land are deposited in the sediments (UNEP, 1992). In addition, there are possible dangers to the global carbon cycle since the coastal waters contain the predominant portion of marine life and biological activity (carbon sinks). In contrast to the case of anthropogenic trace gas emissions, the industrialized countries cannot bring about sufficient relief here through unilateral measures because the most significant

quantitative production of marine biomass takes place in the regional waters of the developing countries. The Council concludes, therefore, that global marine environment policy must include the South right now and not merely deal with future emissions, as is provided for with regard to ozone and climate. Consequently, (1) international reduction agreements concerning emissions and (2) international financing mechanisms supplementing regional programs in the South are imperative.

1 With respect to *reduction of emissions,* the problems connected with land-based marine pollution are not essentially different from other damage to global commons. Therefore, the experiences and models regarding such emissions are basically also applicable to the control of marine pollution caused by terrestrial sources. In the view of the Council, treaties prescribing gradual reductions for individual substances should be agreed on. The 1979 *Geneva Convention on Long-Range Transboundary Air Pollution* (including the protocols) and the 1987 *Montreal Protocol* might act as models. The current trend of including the inland waters in the regulatory scope of the regimes should be continued since this is the only way to provide for integrated and effective management of the coastal zones. To ensure that these reduction regimes are as efficient as possible, regionally agreed emission/immission standards must be stipulated that take specific local features and preferences into consideration. This has been done quite successfully for the European regional waters up to now and should also be applied to African, Asian and South American waters in the form of integrated ecosystem management as soon as possible. Because of the relatively small number of states bordering the respective regional waters and their comparatively high degree of homogeneity, regional sea regimes would also be suitable for testing economic instruments, such as tax and tradable permit systems that have only been applied within a national framework to date.

2 Moreover, an *international financial mechanism* should be considered because market-oriented instruments alone are hardly sufficient to furnish financial resources in developing countries or to enable the necessary technological modernization, as demanded in Agenda 21. One option worth looking at is the establishment of a global financial mechanism (*Blue Fund*), through which the OECD states could support remediation of the regional seas of the South. Possible legal linkages for political measures and action might be Art. 4 (1d) and (2a) of the *Framework Convention on Climate Change*, in which industrialized states are required to protect greenhouse gas sinks, or Art. 207 and

212 of the *Convention on the Law of the Sea*, in which states are required to agree on global regimes against terrestrial marine pollution.

A separate regime concerning land-based marine pollution would have to be founded on three pillars:
– a global framework agreement *("Convention on Protection of the Seas")*,
– a global protocol on a financing mechanism,
– various regional protocols containing the specific reduction requirements and instruments to be applied.

A contractual link between the (financing) commitments of the OECD countries and the (reduction) requirements of the developing countries according to the *do-ut-des* principle would be conceivable for the framework agreement – analogous to Art. 4 (7) of the Convention on Climate Change and Art. 5 (5) of the Montreal Protocol as amended in 1990 which make contractual compliance on the part of the developing countries binding upon contractual compliance on the part of the industrialized countries with regard to finance and technology transfer. Linking a global environment fund to a dozen different regional reduction requirements or emission/immission standards, on the other hand, appears to pose a problem; offsetting quantitatively and qualitatively different forms of contractual performance by developing countries against each other and granting them payments via the environment fund on this basis would involve complicated calculation methods and dispute settlement procedures.

The Council recommends that support be given to effective regional water programs in the countries of the South. AGENDA 21 estimates the total expenditures required annually in this problem area at $13 billion by the year 2000, a sum to which the developing countries cannot contribute much and which at the same time bears no relation to the current volume of the GEF. Indeed, the term "chaos power" of the developing countries, which has been introduced into the international political discourse, may typify the future state of affairs, for a failure on the part of environmental policy in Africa, Asia and Latin America will also affect the northern hemisphere through destruction of the coastal marine ecosystems and resulting changes in global biogeochemical cycles.

3.4.2
Dumping of Waste

Dumping of toxic substances and radioactive wastes from special ships represents a specific case of land-based marine pollution that has been systematically carried out as a form of disposal for industrial wastes or sewage sludge in the past two decades. Art. 210 of UNCLOS requires its Parties to enact national regulations that are "no less effective" than global rules and standards. In accordance with the applicable revisions, this will soon lead to a worldwide ban on dumping, allowing for very few exceptions.

In accordance with a decision made at the 1972 *UN Conference on the Human Environment* in Stockholm, numerous states adopted the *London Convention on the Prevention of Marine Pollution by Dumping of Wastes and other Matter* in 1972, which stipulated a restricted ban on dumping. This marked the first time that a global system of lists of hazardous substances was specified, and these lists were also used later as an environmental policy instrument for other problems. Because of their special danger for marine organisms, the dumping of mercury, cadmium, various types of mineral oil, biological and chemical warfare agents, highly radioactive substances as well as nonbiodegradable substances has been banned since then.

The 1972 London Convention has been supplemented by a number of regional dumping agreements, some of which go beyond the former. A total ban on waste disposal now applies to Antarctica. According to the resolutions of the 16th Conference of the Parties to the *London Dumping Convention*, waste incineration at sea and the dumping of low-level radioactive substances have been banned since February 1994. Dumping of industrial waste on the high seas is to be discontinued by December 31, 1995.

In comparison to other environmental regimes, the London Dumping Convention can be regarded as a relatively successfully conceived and implemented environmental policy, even though numerous deficits have emerged since its establishment. First, the regulatory scope of the agreement was always too restricted to really achieve what is ecologically necessary. Second, economic interests were all too often given priority over ecological concerns. Wastes of low to medium radioactivity, for instance, were discharged up to 1983 and certain exemptions still apply to Russia, Great Britain and France, extending into the next century. Dumping of sewage sludge containing heavy metals is still allowed until 1998 according to an exemption clause.

3.4.3
UV-B Radiation and Climate Change

Recent studies show that marine organisms are potentially threatened by the increase in ultraviolet radiation resulting from changes in the stratospheric ozone layer (*see* Enquete Commission, 1990 and 1992; WBGU, 1993). Phytoplankton, i.e. single-cell algae, play an important role as primary producers that form the basis of the food chain and thus in the existence of primary and secondary consumers, such as zooplankton, fish, birds and mammals all the way up to human beings (Häder, 1992). Since marine algae, in contrast to terrestrial plants, possess hardly any protective mechanisms against ultraviolet radiation, they may be damaged even at low radiation doses. This means that the primary and secondary consumers in the food chain are threatened (Enquete Commission, 1990): the nutritional basis of the consumers may be reduced by the decline in phytoplankton and direct damage may occur as a result of UV radiation, posing a special danger to young fish, crustaceans and other larva. In addition, one can expect a change in biodiversity because the UV sensitivity of plankton organisms differs.

The amendments of the Montreal Protocol initiated a decisive step in discontinuing production and consumption of ozone-depleting substances (see Section C 2).

The anthropogenic greenhouse effect may also lead to radical changes in marine ecosystems (WBGU, 1993; Enquete Commission, 1992). The rise in the sea level is changing the coastal zones with their biotic environments. The shift in climate zones, including the distribution of precipitation, has an influence on the productivity of marine ecosystems. Although the primary production of biomass in the oceans is less determined by temperature than by light and nutrient conditions, the nutrient supply for the organisms depends on the presence of upwelling deep water and is thus linked closely to the general ocean circulation. It follows then that climate changes may cause significant changes in the geographic distribution and structure of marine ecosystems in the case of a shift of ocean currents. Possible synergy effects must also be taken into account here: excessive fish catches, for instance, may affect the specific adaptation processes of the marine ecosystems, and the rise in temperature of the surface water or the increasing UV-B radiation may reinforce the effects of chemical marine pollution.

It must generally be feared that the damage to marine ecosystems will reduce the carbon sink function of the seas and thus accelerate the greenhouse effect through a corresponding positive feedback effect. This interrelationship was consistently taken into consideration in the Convention on Climate Change: Art. 4(1d) requires the Parties to support the protection and improvement of sinks and stores, including biomass, forests and seas, and to cooperate in such efforts. Because the Convention on Climate Change does not contain any specific requirements for marine environment protection, however, the Council feels that this link should be established through a

protocol to be agreed upon and supplementing the Convention on Climate Change (see also the statement of the WBGU, 1995).

3.5
Resource Function

3.5.1
Marine Mining

According to Art. 208 (3) of UNCLOS, the states are required to enact environmental laws and regulations for all installations on the continental shelf that must meet the internationally agreed minimum standards. Except for Art. 5 of the 1958 *Geneva Convention on the Continental Shelf*, no international agreements have been drawn up with universal application, only a code was recommended for national implementation within the framework of the IMO in 1979. Most studies on marine pollution indicate that marine oil production is merely of local importance limited to a radius of a few kilometers around the oil rig (UNEP, 1992). A future environmental problem could occur through the disposal of such oil rigs, however (GESAMP, 1990).

Now that UNCLOS and an implementation agreement on the deep-sea mining part of the Convention have entered into effect, initiation or promotion of commercial deep-sea mining will become a potentially important field of activity, in addition to marine oil production. Such mining involves, for example, "manganese nodules", small pieces of stone the size of a fist with a high metal content forming large fields on the sea floor at depths of about 4,000 meters. The distribution question linked with exploitation of these natural resources, which have to be regarded as global commons according to the traditional law of the sea, were the primary reason for holding the 3rd UN Conference on the Law of the Sea from 1973 to 1982, whose final act was a draft of the Convention on the Law of the Sea.

The regulatory framework on deep-sea natural resources agreed on at this conference was based on a modified concept of a "common heritage of humankind" and provided for a global plan of redistribution for the revenues obtained through deep-sea mining in accordance with welfare state principles. This meant that the technologically advanced mining enterprises of the North were invited to exploit the manganese nodules while portions of the sales revenues from this common heritage of humankind were supposed to be provided to poorer countries. An *International Sea-Bed Authority* (ISA) was to be set up for implementation purposes and would advance to become a strong financial institution for international development cooperation. In spite of the amendments made during the last rounds of negotiation, this model did not meet with the approval of the industrialized nations, which – with the exception of Iceland – did not initially sign or later ratify UNCLOS after conclusion of negotiations in 1982.

Follow-up negotiations within the framework of the United Nations and of an altered international policy constellation have led to a revision of the deep-sea mining provisions of the Convention that, to a certain degree, takes into consideration the demands of the industrialized countries. In July 1994 this revision was adopted as the *1994 Agreement Relations to the Implementation of Part XI of the UN Convention on the Law of the Sea of 10 December 1982* and opened for signature. The amended regime follows a more market-oriented approach, taking into account the interests of the mining enterprises. A more detailed treatment of these somewhat complicated deep-sea mining regulations cannot be provided here, however (*see* Wolfrum, 1991). The "International Sea-Bed Authority" is being set up with a reduced scope of operations in Jamaica, the "Commission on the Delimitation of the Continental Shelf" will be established in New York and the "International Tribunal on the Law of the Sea" in Hamburg (making it the first major UN institution in Germany).

Germany played a leading role in research into the environmental impact of deep-sea mining, particularly within the framework of the *Scientific Working Group on Deep-Sea Environmental Protection* set up in 1990, and a research project in a manganese nodule area of the Peru Basin that has been financed by German funds since 1989 (Thiel and Schriever, 1994). A comprehensive environmental impact study of deep-sea mining, including all potential interactions and feedback effects, is difficult to carry out, however, and requires considerable efforts. Therefore, global environmental policy must, in the view of the Council, prepare itself in time for the confrontation with the special interests of the mining industry before commercial operations are initiated. The Federal German Ministry for Education, Science, Research and Technology is currently launching a "Deep-Sea Mining" program that will examine aspects of use, such as deep-sea mining, disposal and its environmental effects.

3.5.2
Fishing and Whaling

Exploitation of the living marine resources represents the oldest form of human use of the seas. Fishing was traditionally "free" – anyone could take as many of these seemingly unlimited marine resources. After local whale populations had been almost completely wiped out centuries ago, however, the intensification of fishing operations and the introduction of new fishing methods since the beginning of the 20th century have threatened fish stocks with overexploitation. This led to international agreements and various forms of joint management of fish stocks.

Art 1 (2) of the 1958 *Geneva Convention on Fishing and Conservation of the Living Resources of the High Seas*, for example, required the Parties to cooperate in stipulating catch quotas and preserving fish stocks. Several regional fishing regimes that specified catch quotas for individual fishing grounds or specific species complied with this requirement. To regulate whaling, the *International Whaling Commission* (IWC) was set up in 1946. A global whaling moratorium was agreed on in 1982 which has been legally binding since 1986. Norway, a Party to the whaling agreement, has lodged a reservation against this moratorium and is therefore not bound to it.

The redistribution of sovereign rights through UNCLOS that came into effect in the meantime may have far-reaching effects on the nature of fishing. Essentially it abolishes the international common ownership of fish: over 90% of all fish stocks are found within a 200-nautical-mile zone along the coasts, and UNCLOS allocates this zone to the coastal states as an "exclusive economic zone" such that the coastal state "...has sovereign rights for the purpose of exploring and exploiting, conserving and managing the natural resources, whether living or nonliving" (Art. 56). The traditional "freedom of fishing", which is restricted, however, by certain legal obligations regarding international cooperation in accordance with the Geneva Fishing Convention, continues to apply only to about a tenth of the global fish stocks living in the remaining high seas. Within the 200-mile zone the coastal state shall determine the allowable catch of fishing resources (Art. 61 (1)). In doing so, the coastal state must ensure that the stocks are utilized only to the extent of their maximum sustainable yield according to current scientific findings. On the other hand, the coastal state is required to make optimum use of the stocks, i.e. if it does not have the capacity to harvest the entire allowable catch itself, it must then give other states access to its exclusive economic zone (Art. 62 (2)).

Marine Environmental Policy of InternationalOrganizations

Approximately 200 international organizations are now involved in marine environmental policy. More than 60 such organizations exist for maritime shipping issues alone and there are about 60 in the field of fisheries. These organizations play a major role in the current development of marine environmental policy, particularly in the establishment of regional sea programs in Africa, Asia and Latin America. The Convention on the Law of the Sea that has gone into effect also attaches fundamental importance to such organizations by virtue of regular references to "responsible international organization(s)", with the help of which the states are to agree on extended regulation mechanisms (*see* Art. 197 *as well as* 200-212). For this reason the activities carried out by these organizations up to now will be described here briefly (*see also* Kwiatkowska, 1994).

Most special organizations and bodies within the system of the UN have integrated environmental policy into their sphere of activity and have thus become active in marine environmental policy; these include the FAO, UNESCO (via the *Intergovernmental Oceanographic Commission*, IOC) and the WMO. The success of marine environmental policy today can be attributed, to a certain degree, to the coordination of policy on the part of the regional organizations of the industrialized states, in particular the ECE and EU. The declaration of the Paris World Economic Summit (G7) in 1989 also contained a clear avowal to undertake greater efforts concerning marine environmental policy. However, the financial pledges of these states have remained far short of the magnitude deemed necessary by scientists and other experts.

UNEP

Marine environmental protection has been one of the main tasks of the UN Environment Programme (UNEP) since its establishment in 1972. The regional sea programs of the developing countries are frequently seen as its greatest success to date. The inadequate regulation of terrestrial marine pollution was stated to be a decisive problem for international environmental law in the UNEP program on international environmental law of 1982. However, UNEP (still) has no

rights to initiate agreements or means of shaping agreements so that it cannot directly propose environmental agreements. Nevertheless, UNEP plays an important role as a marine environmental policy actor whose importance may grow after the Convention on the Law of the Sea goes into effect. One of its current focal points is the elaboration of a strategy for the protection of coastal seas and zones. A coastal management program, jointly supported by the *Oceans and Coastal Areas Programme Activity Centre* of the UNEP as well as by the FAO and World Bank, was set up for this purpose in 1992. Among other things, training of government employees in coastal zone management is carried out within the framework of this program.

IMO

The International Maritime Organization (IMO) was established at a conference convened by the United Nations in 1948, but was called the Intergovernmental Maritime Consulting Organization (IMCO) until 1982. Now 149 states with roughly 95% of the world's trade tonnage have joined this organization. The original purpose of the IMCO/IMO was to consult the Member States and prepare agreements based on international law so as to ensure safety at sea, eliminate discrimination in the shipping sector and increase the capacity and efficiency of international maritime shipping. The organization began dealing with environmental policy problems within the framework of its objectives and competencies regarding shipping, economic maritime aspects and safety at a relatively early date. In 1973, for example, an independent *Marine Environment Protection Committee* (MEPC) was created and was recognized as a main body of the IMO in 1982. In addition, an amendment to Art. 1 of the IMCO statutes, which came into force in 1980, declared that the organization was expressly responsible for avoidance and monitoring of marine pollution by ships – the IMO is thus the only special organization of the United Nations assigned contractually stipulated environmental protection tasks. However, the IMO has been pushed aside in this sphere of activity by the UNEP regional sea program. To put it simply, the UNEP handles issues involving environmental protection of the regional seas and the IMO those concerning global marine pollution caused by maritime transport. Even though the IMO is not well-known to the public, its importance for international marine en-

▶

vironmental policy and its influence on the behavior of international actors must not be underestimated. The weight carried by the IMO is indicated by the fact that most international agreements for regulating maritime transport were signed at conferences convened by the IMO. Consequently it performs a decisive initiating function in putting environmental problems on the international agenda. The IMO was also charged with specification (for example, with regard to pollutant lists, standards, rules of conduct, etc.), implementation and monitoring of agreements.

GEF

The *Global Environmental Facility* (GEF), which is jointly managed by the World Bank, the Development Programme (UNDP) and the United Nations Environment Programme (UNEP), was established in November 1990 in order to finance international environmental protection measures. Extensive restructuring of the GEF was resolved in March 1994 (WBGU, 1994), as a result of which the decision-making mechanism was modified in favor of developing countries that had been underrepresented up to then (second phase or "GEF II"). The main purpose of the GEF is to finance projects in developing countries aimed at protection of the global environment. During the pilot phase (1991-94) 17% of the funds (of a total of $750 million) were appropriated for measures for the protection of the seas, including improved absorption capacity for ship ballast in the ports of developing countries, for projects involving pollution control of large river systems, for setting up plants and equipment for the protection of ecosystems in bays and coastal zones, for oceanographic research and monitoring projects as well as for projects for improving disaster control and prevention in connection with tanker accidents. The provisions of the "Bucharest Convention" of 1992 for the protection of the Black Sea may be-

come a precedent for a more extensive role of the GEF in marine environmental policy since they designate the GEF as the provisional administrative institution until the convention goes into effect.

NGOs

Nongovernmental agents also play a key role in the formulation of international standards and establishment of regimes relating to marine environmental policy. At the 1992 UNCED the NGOs received greater rights and means of exerting influence than had been customary in UN bodies up to that time. Among other things, several "NGO agreements" were concluded for the protection of marine ecosystems and are intended to serve as a guide for future NGO activities regarding marine environmental policy and contain a number of specific demands directed at the various governments. These agreements currently represent the most comprehensive policy declaration of NGOs with respect to marine environmental policy (Biermann, 1994a). A "statement on marine protection and demands directed at the federal government within the framework of Agenda 21" were resolved for the Federal Republic of Germany by over 100 NGOs (Project Center UNCED '92). They demand a global agreement based on international law for the reduction of emissions from terrestrial sources of marine pollution. This agreement is also supposed to regulate transfers to developing countries. It states that the South must be given financial and technological support in order to be able to protect the fish stocks within the 200-mile economic zones in accordance with sustainable use. In addition, a "toprate multilateral ocean forum" should be set up under the management of the General Assembly of the UN as part of the institutionalization of these plans and Germany should act as a forerunner in this process.

Special regulations apply to anadromous (going from a sea up a river to spawn, e.g. sturgeons, salmon) and catadromous (going down a river to the sea to spawn, such as eels) fish as well as sedentary species and marine mammals. The latter are partly internationalized according to Art. 65 of UNCLOS and fall under the competence of the "appropriate international organizations", which currently only includes the IWC. The whaling moratorium of the IWC is thus also binding for the Parties to the Convention on the Law of the Sea.

Since UNCLOS went into effect, the modern law of the Sea has, for all intents and purposes, nationalized nearly all fish stocks of the seas and revoked their status as international common property. This legal modification nota bene covers only appropriation by the respective coastal state, which has received exclusive rights vis-à-vis other states. However, the dilemma of collective assets within the framework of the national fishery industry remains, though it could be solved through regulatory measures (stipulation of quotas) or market economy instruments, such as further privatization including sale at "fishing ground exchanges", solutions which would not have been possible with traditional regimes on the use of marine resources.

Nationalization of the fish stocks in the exclusive economic zone will presumably lead to increased use of the (relatively) small stocks in the remaining high seas, including the fish species that migrate between the economic zones and the high seas. (The recent conflict between Canada and Spain/EU seems to reflect this.) These were among the developments that caused the special regime of fishing on the high seas to become an international field of conflict which has been the subject of negotiations at the UN Conference on "Straddling and Highly Migratory Fish Stocks" in New York since August 1994. In addition, the division of the seas up to the 200-mile boundary and up to the continental shelf by the coastal states poses a problem regarding distribution policy since the land-locked and "geographically disadvantaged" states come out extensively empty-handed in the allocation of property and sovereignty rights. Art. 69 and 70 grant them only certain rights in the exclusive economic zones of the neighboring coastal states.

At present the fishery industry overall is still far from attaining the maximum sustainable yield; on the contrary: most fishing grounds are acutely threatened by overfishing, i.e. management and political failure. Moreover, pollution and destruction or degradation of natural shore areas in connection with overfishing have led to damage to fish stocks, changes in marine ecosystems and declining yields. 1990 marked the first time in 13 years that worldwide fishing yields dropped. This can be explained in part by fishing bans in Europe but must also be attributed to the effects of environmental impairment (WRI, 1992). In addition, there is a threat of possible or already acute damage due to increased UV radiation *(see Section C 2.3)*.

Besides UNCLOS, which expressly includes protection of endangered species in its environmental protection provisions (Art. 194 (5)), the *Convention on Biological Diversity*, effective since December 1993, also encompasses marine organisms as part of biodiversity (Art. 2). The current activities carried out within the framework of the FAO must be welcomed in this connection, especially the *Declaration of Cancun*, which was adopted in May 1992 and demands implementation of a concept of *"responsible fishing"* in accordance with the principles of sustainable use of living resources. In this declaration the United Nations is called upon to draw up an "International Code of Behavior for Responsible Fishing" as the basis for regional fishery regimes within the framework of UNCLOS and to proclaim a "Decade of Responsible Fishing".

3.6
Marine Research

After conclusion of the negotiations on UNCLOS in 1983 the HWWA Institute for Economic Research prepared a study on behalf of the German Federal Ministry for Research and Technology in which the current and potential effects on German marine research were analyzed and measures recommended for adapting it to the new law of the sea (Borrmann and Weber, 1983). The main statements in this study have now been confirmed: the restrictions that the international community wished to be placed on the previous freedom of research were largely put into practice by the coastal states and they now exercise their rights excessively in many cases (Hempel, 1994). In other words, foreigners are not allowed to conduct research in territorial waters unless it is at the invitation of the respective coastal state. As a rule, coastal states should permit foreign research in their exclusive economic zones.

German marine research is affected by this new situation in a variety of ways. In some cases research applications are rejected, there are long waiting times for approvals and additional administrative requirements or research approvals are tied to requirements in the interest of national research or data collection. The main positive effects consist of greater cooperation with scientists and institutes, particularly from developing countries.

As a result of the modified legal situation, German marine research will presumably be directed more at the remaining areas offering room to manoeuvre (open seas, Antarctica, German Bight and Baltic Sea coast) and at cooperation within the scope of global change research. Deployment of the research ships "Polarstern", "Alexander von Humboldt" and "Victor Hensen" has proven to be particularly useful for international projects organized on a partnership basis. European cooperation and work under the auspices of the Helsinki Commission have become the pillars of German marine research; bilateral and multilateral agreements have been concluded or prepared with some newly industrializing countries. Staff and data exchanges have been increased and on several occasions development aid has been started for coastal countries.

In the view of the Council, these approaches should be used as the basis for further efforts. Particular emphasis should be placed on increasing assistance to poor coastal states in favor of an environmentally sound, sustainable use of the seas. This international cooperation should, at the same time, be based more on the principle of partnership than has been done to date and be intensified when determining damage in the case of extraordinary events. Introduction of unbureaucratic information channels in the Baltic Sea region ("red telephones") might be a model for such efforts. The Council emphasizes that the needs of liberal marine research in accordance with the new law of the sea can only be met in the long run if greater investments are made in cooperation based on partnership.

3.7
Summary and Outlook

As stated at the beginning of this section, the seas have three major utilisation functions: they serve as a transport route, via which a large proportion of international trade is carried out (*transport function*), they serve as the final store for residual products from industrial production, consumption and agriculture (*disposal function*) and they are sources of resources, whether minerals or food (*resource function*). These three types of use have caused considerable damage to the marine ecosystems in this century, damage that is now having a feedback effect on humanity. Living marine resources are severely threatened by overfishing and pollution, and the impairment of marine ecosystems will increase the greenhouse effect as the seas are affected in their role as store and sink for carbon dioxide.

On the other hand, a number of institutional provisions have been agreed upon nationally and internationally for the protection of the seas, some of which have already been implemented. In many areas marine environmental policy actors are guided by common principles, rules and decision-making procedures around which their expectations converge. International regimes whose intensity and scope of regulatory application have increased since the 70s have come into being. A networking of marine environmental policy regimes with other environmental fields has also taken place. For example, the protection of marine ecosystems was expressly incorporated into recent drafts of agreements of the UN International Law Commission regarding use of international watercourses, and the Parties to the 1992 *European Convention on the Protection and Use of Transboundary Watercourses and International Lakes* were required to cooperate with the institutions of the corresponding regional sea programs and include protection of the marine environment in their policy. In 1992 the North Sea regulations for controlling land-based pollution (*Paris Convention*) and against dumping (*Oslo Convention*) were combined and tightened in the newly formulated *Paris Convention,* an action that can be regarded as the beginning of a comprehensive, integrated ecosystem

management of regional waters, as striven for in the UNEP regional sea programs.

In the opinion of the Council, the *Convention on the Law of the Sea*, effective since November 16, 1994, offers a global legal framework for combining the individual regimes, thus providing the foundation for a functional global regime of marine protection. This represents an important step: this "constitution of the oceans" expressly designates environmental protection as the basic standard for all forms of marine use and requires the Parties to implement the relevant regulatory frameworks as minimum international standards or at least take them into consideration with respect to land-based sources of emissions. However, it still appears to be a long road to integrated environmental management of the seas (Stevenson and Oxman, 1994). Even in the endangered regional waters of the industrialized countries only partial improvements have been attainable up to now, and in the developing countries there continues to be a lack of the necessary financial and technical resources, which the industrialized countries, in turn, still do not seem willing to provide to the required extent. For this reason the Council feels that even now that UNCLOS has gone into effect, protection of the seas will require significant research and action, as described in detail in the previous sections. If international measures for the protection of the seas should not be carried out, however, far-reaching and, in some cases, irreversible damage can be expected in view of the continued rise in population in the coastal regions, the growth of industrial production and the increasing regional pollution in the catchment areas of the large rivers in the South. Even the "limitless seas" could then no longer perform or only inadequately fulfil their functions, which play such an important role for humanity.

3.8
Recommendations for Action and Research

TRANSPORT FUNCTION

Measures for identifying substandard ships and illegal oil discharges should be intensified within the framework of the Paris Memorandum in cooperation with the European Parties. The legal consequences of such violations should be made more stringent in conformity with the provisions of UNCLOS. Owner liability should be introduced since the current regulations do not have an adequate sanction effect on shipowners and flag states. The Council considers this to be a special task for the International Tribunal on the Law of the Sea to be set up in Hamburg. Interested developing countries should be offered active cooperation so as to enable them to establish com-

BOX 29

Follow-up Issues of the UNCED Conference for the Protection of the Seas

Chapter 17 of AGENDA 21 deals with protection of the seas. The following list provides an outline of the areas of activity mentioned there, their objectives and planned measures and compares them to the results achieved through the Conven-tion on the Law of the Sea in typological form. As described above, implementation of the provisions of the Convention on the Law of the Sea has not been effected in many cases.

Objectives *(AGENDA 21)*	*Measures* *(AGENDA 21)*	*Implementation* *(UNCLOS)*
Area I: Integrated management and sustainable development of coastal regions		
Working up of integrated management and development programs at national and local level	landscape and water use planning	possible for exclusive economic zones
	identification of critical regions	possible
	precautionary principle	implicitly prescribed (Art. 1 para. 4)
	emergency plans	prescribed (Art. 199)
	conservation of species diversity and productivity	demanded (Art. 194 para. 5)
	training	demanded, including special support of developing countries
	data collection and exchange	prescribed (Art. 200)
Area II: Marine environment protection		
Maintenance of life-preserving and yield-providing functions of the seas	*Terrestrial sources of emissions*:	
(pursuing preventive integrated approaches, developing economic incentives, improving standard of living of coastal inhabitants)	extension of "Montreal Guidelines"	states should "make efforts" for regulations (Art. 207, 212)
	technical and financial cooperation	prescribed (Art. 202f.)
	promote regional agreements	prescribed (Art. 197)

Objectives (AGENDA 21)	Measures (AGENDA 21)	Implementation (UNCLOS)
	avoidance and treatment of sewage	implicitly demanded (Art. 211)
	Other sources of pollution: reduction of hazardous substances	demanded
	monitoring of discharges	demanded
	Pollution through activities at sea: ratify OPRC implementation of London Dumping Convention	in force since May 1995 prescribed as minimum standard (Art. 210)
	revision of provisions for offshore platforms	demanded (Art. 208 para. 5)
	closer cooperation in monitoring illegal discharges	increased, particularly in exclusive economic zones
	protection of sensitive regions	possible for exclusive economic zones
	support of IMO	general obligation
	International Convention on Liability and Compensation Regarding Transport of Hazardous Substances	implicitly prescribed (Art. 235 para. 3)
	collection facilities for ship waste in ports	prescribed for special regions (MARPOL)
Area III: Sustainable use of living marine resources Conservation of stocks and sustainable use on high seas and in national waters	promotion of research, training and technology development for fisheries	prescribed, promoted through setting up of economic zones
	data collection and exchange	prescribed
	intergovernmental cooperation and development aid	prescribed
	ban on destructive catching methods	not explicitly but customary international law
	recognition and promotion of international institutions for the protection of marine mammals	expressly recognized (Art. 65)

Objectives (AGENDA 21)	Measures (AGENDA 21)	Implementation (UNCLOS)
	international conference on migrating stocks of fish	cooperation prescribed Art. 64)

Area IV: Management of marine environment

Data collection and preventive measures; promotion of research and international scientific exchange	intensify data collection	demanded, but possibly hindered by restriction of freedom to conduct research
	increase IOC research on carbon sinks	not yet included (conclusion of agreement 1982)
	build up GOOS	implicitly demanded
	improved scientific management	demanded
	international and regional cooperation	prescribed

Area V: Strengthening national and regional cooperation and coordination

Integration of sectoral activities and exchange of information	increased cooperation between UN organizations	demanded (Art. 197ff.)
	as well as between the latter and national institutions	demanded
	information system on regulations, exchange of data and information	demanded

Area VI: Sustainable development of island states

Support of sustainable development and adaptability of small island regions	research, planning and management development	favored by regulation concerning archipelago states (Part IV)
	protection of coastal habitats	
	develop adaptation strategies	

parable systems of port state control. Particular efforts should be made in this connection to increase the German contribution to the International Maritime Organization or replace the specific IMO contribution rate with the general UN scale in order to support the larger role played by the IMO as an international development and environmental organization.

DISPOSAL FUNCTION

Germany should commit itself to negotiating an integrated International Convention on Protection of the Seas. The latter could combine the various regional sea programs, particularly in developing countries, and contribute towards financing appropriate environmental programs of the developing countries with the help of a separate financial mechanism (*"Blue Fund"*). Such a Convention on Protection of the Seas would have to focus on land-based sources of emissions, which should include controlling discharges of pollutants via rivers and air as, well as the discharge of substances causing turbidity and environmentally harmful alteration of coastal zones. With regard to the pollution of the northern European regional waters, Germany should reduce emissions (of nutrients in particular) to a greater extent within the scope of a precautionary environmental policy and urge all states bordering the North Sea and Baltic Sea to carry out similar activities within the framework of the North Sea and Baltic Sea Conferences. The Council recommends that the German Federal Government again take the initiative and declare the North Sea to be a protected special area so as to put a fundamental stop to discharges of oil and chemicals.

RESOURCE FUNCTION

The Council recommends that efforts be made within the framework of the International Sea-Bed Authority to be set up in Jamaica in order to prevent commencement of commercial deep-sea mining operations prior to joint determination of their environmental impact. An arrangement similar to the Madrid Protocol on Environment Protection to the Antarctic Treaty is envisaged here. The Council points out that with a commencement of deep-sea mining there could be a conflict with the guiding principle of sustainable development. Due to the tremendous threat to fish stocks, Germany should commit itself to a restrictive regulation of fishing quotas and contribute to implementation of the principle of maximum sustainable yield both Europe-wide and globally. Greater efforts should be undertaken to set up appropriate institutions in developing countries within the framework of bilateral and multilateral development cooperation because a large portion of the global fish stocks is located in their economic zones.

4.1
The Genesis and Content of the Desertification Convention

In its 1994 annual report, the Council focused in detail on the problem of soil degradation. The analysis showed that soils are the vulnerable thin skin of the earth for which serious "illnesses" can be diagnosed worldwide. These "illnesses" represent a serious threat to the earth's population and biosphere that in some parts of the world is dramatic. The development of a soil-centered network of interrelationships identified twelve such illnesses (or "syndromes"), each described in a complex web of cause and effect. The analysis of the syndromes and the network of causal factors permit an improved observation of trends to be carried out and provide specific indications of political action and research planning. The results of the work continue to be of major relevance, so for a detailed description of soil-related problems reference is made here to the 1994 report.

However, one important change has occurred in the year that has since elapsed. On July 18, 1994 representatives from over 100 states met in Paris and reached agreement on the text for a "*UN Convention to Combat Desertification in Countries Experiencing Serious Drought and/or Desertification*", referred to henceforth as the "Desertification Convention". The Council stated its position on this process in provisional form in its last annual report, where also the problem of desertification was dealt with in detail both in one of the syndrome sections (the "Sahel Syndrome") as well as in the Case Study on the "Sahel Region". The final provisions of the Desert Convention are now available and should be subjected to closer analysis at this point.

4.2
Assessment

TERMINOLOGY USED

If one looks at the definitions in the Desertification Convention, one can identify two main differences compared to the 1994 annual report of the WBGU. One is that the concept of land degradation is more narrowly defined than the concept of "soil degradation" used by the Council, in that it refers exclusively to the productivity and biotic complexity, i.e. primarily on the production function of soils. Other land surface functions, such as the regulatory function, are not included. Second, the definitions also refer to the long-term loss of vegetation, a phenomenon that can also occur without the degradation of soils through water scarcity or overexploitation of plant stocks. If loss of vegetation is not accompanied by soil degradation, then it is reversible if conditions change. Because vegetation loss is usually the first step toward soil degradation, desertification can be seen as a special instance of global soil degradation (*see also* WBGU, 1994).

PROVISIONS

The Desertification Convention has created *important political conditions* by defining basic requirements for combating desertification. Among others these include increasing efficiency through bilateral and multilateral cooperation, exchange of data and mutual information between donors, public participation of the population, strengthening support through transfer of research and technology, taking local conditions into consideration and providing for active participation of affected countries.

All in all, the Convention is programmatic in character, in that binding operational and specific financial consequences could not be agreed upon, and the significance of the Convention therefore lies more in the political and psychological sphere than in any binding development programs. Nevertheless, the coming into force of the Desertification Convention means that important issues relating to bilateral and multilateral cooperation for the regions specified in the Convention will be affected. In future, therefore, the issue will be to continue the positive development of *policy formulation* prior to the adoption of the Convention in the *policy implementation* phase.

The Convention includes a number of innovations in the field of international law, in particular the notion of *"demand-oriented technology transfer"*, i.e. countries concerned define by themselves, in the context of national action programs, what the best way to combat desertification is for their specific situation, and the *bottom-up approach*, i.e. the *involvement of the local population in the development and implementation of such national action plans*. The first session of the INCD following the signing of the Convention, held in January 1995, showed, however, that both the affected country parties and the donor states have few ideas regarding actual implementation. In the following, some of the open issues are discussed.

One of the key provisions of the Desertification Convention concerns the elaboration of *national action programs* in the affected countries. These pro-grams are meant to describe the specific instruments that are appropriate and necessary for a country affected by desertification. These national action programs are the prerequisite for the involvement of the donor states, i.e. financial and technical assistance will be *"country-driven"*, as opposed to being predetermined by the developed countries.

One problem here, however, is that the structures for developing such action are often not immediately available. In view of the increasing severity of the desertification problems, the donor countries should be prepared to assist the affected countries in acquiring the capacity to develop such action programs. Germany has taken steps in that direction - the BMZ has earmarked DM 5 million for the formulation of such national action programs as part of the Urgent action Program for Africa, DM 2 million of which are earmarked for Mali, where experts will support the government in the development of its national action program over a period of three years.

The Desertification Convention is the only international convention to date that specifically requires the active participation of the population as a princi-

BOX 31

Contents of the Desertification Convention

Use of terms (Article 1)

Some terms of the Convention, a short list of which follows, deviate in their use from those used in the 1994 report on the threat to soils. For the purpose of the Conventions

- "desertification" means land degradation in arid, semi-arid and dry sub-humid areas resulting from various factors, including climatic variations and human activities (Article 1 a).
- "land" means the terrestrial bio-productive system that comprises soil, vegetation, other biota, and the ecological processes that operate within the system (Article 1 e).
- "land degradation" means reduction or loss of the biological or economic productivity and complexity of rainfed cropland, irrigated cropland, or range, pasture, forest and woodlands resulting from land uses or from a process or combination of processes, including processes arising from human activities and habitation patterns (Article 1 f).
- "arid, semi-arid and dry sub-humid areas" means areas, other than polar and sub-polar regions, in which the ratio of annual precipitation to potential evapotranspiration falls within the range from 0.05 to 0.65 (Article 1 g).

Objective of the Convention (Article 2)

- To combat desertification and mitigate the effects of drought in countries experiencing serious drought and desertification, particularly in Africa, in the framework of an integrated approach which is consistent with AGENDA 21.

Principles (Article 3)

- Decisions on the design and implementation of programs to combat desertification should be taken with the participation of populations, local communities and NGOs
- Improvement of cooperation and coordination on the part of the donor nations
- Better focus of resources where they are needed
- Full consideration should be taken of the special needs and circumstances of affected developing country Parties

Obligations (Article 4-6 - selection)

- Establish an enabling international economic environment conducive to the promotion of sustainable development (Article 4 (2) a)
- Integration of strategies for poverty eradication into efforts to combat desertification and mitigate the effects of drought (Article 4 (2) c)
- Strengthening of subregional, regional and international cooperation (Article 4 (2) e)
- Facilitate the participation of local populations, particularly women and youth, with the support of nongovernmental organizations (Article 5 d)
- Strengthening relevant existing legislation and, where they do not exist, enacting new laws and establishing long-term policies and action programs for combating desertification (Article 5 e)
- Promote and facilitate access to appropriate technology, knowledge and know-how (Article 6 e)

Relationship with other conventions (Article 8)

- Encourage the coordination with other relevant international agreements, particularly the UN Convention on Climate Change and the Convention on Biological Diversity

Regional action programs (Articles 9-15)

- Prepare, support and coordinate national, regional and subregional action programs

Scientific and technical cooperation (Articles 16-18)

- Collection, analysis and exchange of data and information (Article 16).
- Promote technical and scientific cooperation in the fields of combating desertification and mitigating the effects of drought (Article 17). To this end, research shall focus on the processes leading to and the impact of causal factors, both natural and human (Article 17 (1) a), as well as integrate valid traditional and local knowledge, know-how and practices (Article 17 (1) c)
- Transfer, acquisition, adaptation and development of environmentally sound, economically viable and socially acceptable technologies relevant to combating desertification and mitigating the effects of drought (Article 18).

Capacity building (Article 19)

- Establishment of institutions, education and the development of relevant local and national capacities, including the full participation at all levels of the affected populations (Article 18 (1) a), by strengthening training and research capacity at the national level (Article 18 (1) b), and through innovative ways of promoting alternative livelihoods, including training in new skills (Article 18 (1) h), etc.

Financial resources and mechanisms (Articles 20 and 21)

- Obligation on the part of the developed country Parties to mobilize substantial financial resources in order to support the implementation of programs to combat desertification
- Undertaking by affected developing country Parties, taking into account their capabilities, to mobilize adequate financial resources for the implementation of their national action programs.

Institutions (Article 22 - 25)

- Establishment of a Conference of the Parties (Article 22), a Permanent Secretariat (Article 23) and a Committee on Science and Technology (Article 24)

Articles 26 to 40 deal with administrative and procedural mechanisms of the Convention.

Regional implementation annexes

Annexes I - IV of the Convention are of critical importance for giving adequate consideration to respective regional needs and circumstances. These are the four Regional Implementation Annexes for Africa, Asia, Latin America and the Caribbean, and the Northern Mediterranean. Priority is attached to Africa on account of its arid regions.

The Annex for Africa includes, for example, provisions in which the affected developing countries undertake to use a decentralized approach appropriate for the respective target groups when implementing their national action programs. Subregional or regional action programs are to be set up, which can be executed more efficiently at local level; priority is attached to the treatment of transboundary phenomena (nomadic herdspeople, management of large water catchment areas, animal diseases, research, advice and support, etc.) (Annex for Africa, Articles 11 and 12). Coordination and partnership are required, and practical implementation should be facilitated by means of appropriate forms of organization. "Partnership agreements" should express, as the product of the consultation process, the political will of the participants for long-term cooperation (Annex for Africa, Articles 14 and 18).

---

ple of international law. This so-called bottom-up approach is another key provision of the Convention. The problem associated with it, however, is that most countries have little experience with such an approach and therefore face considerable difficulties in implementing it. Ideas as to the practical meaning of participation are quite divergent. Where participation is already being practiced on an active basis, it generally involves the involvement of the population in the implementation of a project designed and developed by a donor institution, perhaps through cooperation with the national agencies. The objective of the Desertification Convention, in contrast, is that the population participate in the design stage of projects. The Council welcomes this approach and advocates that a greater proportion of the funds be used in future for capacity building (*see Section B 3.2.2.2*).

Another important issue is that of land availability and land access rights.. If active participation in measures undertaken and programs for combating desertification is to be achieved, it is of great importance that the local population have a vested interest in improving the soils and the uses to which the land

is put (Toulmin, 1994). This leads to the question of land reform. The demand for land reforms is a constituent part of the Desertification Convention, so the donor nations can refer to this requirement. As already stated in the 1994 annual report, it is important to actively support land reform efforts.

Another new and important feature of the Desertification Convention concerns the agreement on *improved coordination and cooperation between donors*, one of the principal objectives. There have often been parallel projects by different countries, which duplicated and sometimes even blocked each other. In order to avoid this, and to deploy the human and financial resources more efficiently, the Convention lays down an improved coordination of developed country parties. Per country a donor state will be assigned responsibility for coordinating the relevant developing and financial cooperation activities (so-called leading agency). Such an approach makes sense, in principle, but the problem that may arise is that the concept only functions if the donor countries are willing to surrender their respective traditional positions of power, partly originating in the colonialist era.

BOX 32

Research Support to Combat Desertification – a Capacity Building Measure

Desertification is a "clinical profile" of the Earth induced by complex interrelations of causes and effects including many areas of the ecosphere and anthroposphere (*see* WBGU, 1994, the Sahel Syndrome). If desertification and its human-induced causes are to be avoided and/or combated, equivalent complex human and institutional capacities are needed. An important element here is local research, since a common reason for failure of development or environmental projects is the lack of capacity to take action. The subject of capacity building through support for research was therefore laid down in Article 19 (1) b:

"The Parties recognize the significance of capacity building – that is to say, institution building, training and development of relevant local and national capacities – in efforts to combat desertification and mitigate the effects of drought. They shall promote, as appropriate, capacity building by strengthening training and research capacity at the national level in the field of desertification and drought."

In those countries affected by desertification, the main *focus of research* has been agriculture with a concentration on the traditional sectors of livestock production and crop management, especially food crops (BMZ, 1993b). The problem of desertification is not given sufficient attention in this area, however, since there is neither a networking between the respective disciplines nor adequate consideration given to economic and sociocultural aspects.

To create an alternative for agriculture and forestry, which are threatened by overexploitation, more *technological research* is needed, a field which is underdeveloped in many countries affected by desertification. In general, surveys show that only 2-3% of research funds worldwide are deployed in developing countries.

In order to increase efficiency and improve adaptation, in the formulation of research priorities the *participation of the relevant target groups* is essential, a fact that by and large is also acknowledged by many development agencies.

National research institutions are multipliers for the implementation of internationally developed and available knowledge. *Support for na-*
tional research in affected countries – interdisciplinary and participatory – is therefore crucial, and must take the form of both financial and technical cooperation. A possible model here is provided by national agricultural research supported by a series of international agricultural research projects. The latter comprise a number of independent agricultural research facilities operating under international law and located in developing countries. Some of these centers have made important contributions to global food security through the development of high-yield varieties. Insufficient attention, however, has been paid to networking aspects and the problem of resource protection in this field (Treitz, 1990).

In Germany desertification-related research also focuses to a great extent on a single-discipline approach (i.e. soil science, geography, agricultural science, behavioral science). Too little attention is given to *interdependencies*, a crucial aspect for desertification research. The relevant public institutions still do not treat desertification as a complex of problems. For example, the Ministry responsible for the Desertification Convention, the *Ministry for Economic Cooperation* (BMZ) does not have a separate research unit in this field.

RECOMMENDATIONS

Germany's financial contribution to agricultural research in developing countries has declined in nominal terms over recent years. The increase in funds for international agricultural research is only marginal, and is less than the rate of inflation. In an international comparison, the German contribution has shown below-average growth (BMZ, 1993b). The Council therefore considers it essential that *support for research towards adapted and sustainable development* be improved significantly, alongside improvements in applied technology training, as already provided within technical cooperation projects. Support should not only be given to capacity building in developing countries in general – what is also required is the deeper knowledge and understanding in Germany, obtained through deployment of the requisite human resources, of the desertification problem and its economic and sociocultural causes.

Priority should therefore be attached to those areas of research which focus on the management of complex natural systems, in order to develop environmentally benign and sustainable concepts for land use appropriate to local diversity. The

BMBF, together with the BML and the BMZ, could develop a coordinated strategy with other research support institutions.

Another aim should be to conclude partnership agreements between research establishments in Germany and in developing countries for terms of around ten years in order to carry out special research projects in situ and to participate in the training of new generations of scientific personnel. The concept would be for working groups of up to 10 persons, who would work in close cooperation with the relevant institution in the partner country. Important in this connection, among other factors, is the improvement of the reintegration of scientists working in research and related occupations when they return to Germany. Joint projects executed by the BMBF, the BML and the BMZ would be a welcome addition in this respect.

In cooperation with Goethe Institutes in the developing countries, specialized programs could be developed for transferring know-how relating to the adaptation of environmental technologies. Cooperation with industry would be one possibility worth exploring here.

During the INCD session in January 1995 in New York, many delegates expressed worries that the Desertification Convention possesses a lower international status than the Climate Change Convention, for example, or the Biodiversity Convention. One reason for this impression is that there is little public attention to desertification in the industrialized countries. This is problematic, because one must assume that the political implementation will be all the more rapid and successful the higher the degree of public pressure that is exerted. If public awareness is to be raised, the connections between desertification, poverty and political crises must be rendered more visible, crises in which the international community must become more involved in future (e.g. Somalia, Ruanda, etc.). Combating desertification must also be understood as preventive action against potential political crises.

4.3
Recommended Action and Research

The Council welcomes the activities that are planned in the framework of the Desertification Convention as a step in the right direction. However, it considers that the Convention fails to have the necessary impact in that its focus is purely on desertification. Although it embraces part of the global problem of soil degradation, in which problems are currently at their greatest and distress is most visible. It must be pointed out that desertification is only one part of the soil problem and that similar problems for the population can result in other parts of the world as a result of soil degradation. The Council repeats the demand raised in the 1994 annual report, that more far-reaching regulations for protecting soils be created, such as a *global declaration or convention on protection of the soil*. Only such a comprehensive approach can lead to the desired solutions. The links between soil degradation and other aspects of global change must be taken into consideration in this process. The Council has already stated its position in the two annual reports of 1993 and 1994, as well as in the statement it issued on the Climate Convention in February 1995.

The Council also regrets that the wording of the Desertification Convention does not go much further than mere declarations of intent. The only new and additional source of finance mentioned therein is the GEF, and that only with considerable restrictions. The 0.7% target for development aid was not included in the Convention. In the opinion of the Council, which has repeatedly demanded that development aid be substantially replenished, there is no solid financial basis for genuinely containing desertification.

To advance the process of implementing the Desertification Convention, the Council makes the following recommendations:

- *The formal ratification of the Desertification Convention should be accomplished as quickly as possible.* The ratification periods for environmental conventions have become shorter and shorter in recent years, and a continuation of this trend would be desirable. In certain circumstances one could imagine diplomatic influence on other nations to ratify the convention as soon as possible.

- *The objective of the first Conference of the Parties to the Convention should be defined very soon, whereby extension to a global soil protection convention should be striven for.*

- In accordance with preventive crisis management, those countries should be given preferential support that are particularly threatened by the combination of poverty, desertification and political conflicts.

- Even prior to the coming into force of the Desertification Convention, *financial and especially personnel measures* are necessary among those institutions in the developing countries that are responsible for the design of national action programs.

- Research into desertification must place greater focus on the *networking of single disciplines*, both in the support of national research facilities in the respective countries, and support for international agricultural research as well as within the German research community as a whole. In view of the increasing importance of international conventions for national policymaking, thought should be given to developing a *special form of convention-oriented research*.

5.1
Current Trends in the Biosphere

Biological diversity (or "biodiversity") encompasses the number and variability of all living organisms both within a species (genetic diversity) as well as between species and ecosystems. Biodiversity has a value of its own (*see* the Preamble to the Convention on Biological Diversity; UNEP, 1992), but at the same time, through its various economic values, is an extensive resource that can be utilized by humankind (Pearce and Moran, 1994; *Box 33*). This explains why the continuing loss of biological diversity is now one of the central problems of global change (WCMC, 1992; WBGU, 1993).

Biodiversity loss has assumed a worrying scale on account of the sheer speed with which it is advancing: never in more than 65 million years has the Earth lost so many species within such a short time (Wilson, 1992). This *rapid erosion of biodiversity* is a prime example of the enormous imbalance between human demands and the capacity of the biosphere (WRI et al., 1992).

CAUSES FOR THE LOSS OF BIODIVERSITY

Loss of biological diversity is due to anthropogenic factors, usually a combination of particular synergetic factors leading to irreversible loss (McNeely et al., 1990; Ehrlich, 1992; WRI et al., 1992; WCMC, 1992; WBGU, 1993):
- The main cause is the destruction, alteration or fragmentation of habitats through conversion to rangeland or cropland or to land for settlements, transport, mining and tourism,
- industrial pollution, but also excessive application of agrochemicals by agriculture and forestry,
- overexploitation of wild flora and fauna populations for foodstuffs or raw materials; involuntary killing of marine animals caught in fishing nets,
- accidental or deliberate import of exotic species or pathogens, which can lead to the destabilization of populations or ecosystems,
- reduction of genetic diversity (genetic erosion) of

useful plants through the cultivation of high yielding varieties,
- effects of anthropogenic forcing of the greenhouse effect and depletion of the ozone layer,
- deliberate annihilation of "pests", often with unexpected impacts on ecosystems.

EXTENT AND SPEED OF BIODIVERSITY LOSS

Estimates for the number of species in the world range from 5 to over 50 million, whereby only 1.7 million species have been described to date (Wilson, 1992; WCMC, 1992). Estimates for the loss of species within the next 50 years vary between 10 and 50% of the total number of species, depending on the method used (WCMC, 1994b; *Table 8*). Thus, on the basis of these figures, the number of species lost per day is somewhere between 3 and 130, which is equivalent to an acceleration of the natural background rate by a factor between 1,000 and 10,000 (Wilson, 1992).

The *World Conservation Monitoring Centre* provides an overview of the current state of information on global biodiversity (WCMC, 1994a). However, it is not possible with the data available to assess the impact that international efforts may have had on biosphere trends since the *UN Conference in Rio de Janeiro* (UNCED) (Brian Groombridge, WCMC, personal communication, March 1995).

Biodiversity is also declining at the level of populations and genes. This trend is affecting not only wild species, but also those used in agriculture and forestry. Particularly problematic in this respect is that the original diversity of traditionally cultivated species is being replaced by a small number of high-performance varieties (*genetic erosion*). This trend harbors a threat to food security, in that the loss of genetic diversity threatens the adaptability of crops to changing conditions such as newly-occurrent diseases or climate change.

RED LISTS

In addition to determining the number of species, it is also essential to have data on the specific threat to their existence. The criteria for classifying species into categories of threat are currently being revised

Table 8
Estimated global
extinction rates of species.
Source: Reid (1992) cit. ex
WCMC (1992)

Estimate of the extinction rate	Global Loss [% pro decade]	Source
2,000 plant species per year in the tropics and subtropics	8	Raven (1987)
25% of all species between 1985 and 2015	9	Raven (1988)
At least 7% of the plant species	7	Myers (1988)
0.2 – 0.3% of all species per year	2 – 3	Wilson (1988)
5 – 15% of the species living in forests up to 2020	2 – 5	Reid and Miller (1989)
2 – 8% of the loss of all species between 1990 and 2015	1 – 5	Reid (1992)

by the IUCN (Mace and Stuart, 1994). The "Red Lists" of endangered species thus produced are intended to draw attention to the threats posed to identified species. Red Lists are particularly useful as direct indicators of the endangerment of biodiversity in various regions of the world whenever inventories of biodiversity have largely been completed. However, simply raising awareness of how species are endangered is not sufficient to provide a comprehensive protection of biodiversity. If species are to be protected with any measure of success, then action to protect habitats and especially network links between them is absolutely essential.

Habitats

By analyzing changes in, or destruction of, habitats it is possible, at least in principle, to assess directly the extent of biodiversity loss. Because habitats are so wide-ranging, precise classification of losses and changes is a very complex task, as losses can range from short-lived, minor and reversible to irreversible destruction. Despite the methodological difficulties that remain, one can still state with certainty that no success has yet been achieved in halting negative biosphere trends, let alone reversing them. Changes in land use, in particular, seriously affect both biodiversity and the abiotic components of the ecosphere. The intensification of agriculture and forestry also has negative impacts on biological diversity. While forested areas are on the increase in Europe, these are usually in the form of large-scale plantations of low species diversity and inappropriate, nonindigenous monocultures. Similar problems exist with grassland: for example, total surface area in Europe has remained constant but the transition from low nutrient supply and thus species-rich grassland to intensive cultivation through the high input of nutrients leads to loss of species. Intensification is accompanied by the destruction of wetlands, which often possess a high degree of biological diversity. Britain and Holland have each lost about 60% of their wetland areas, while the figure for California, at 90%, is even worse (Finlayson and Moser, 1991; Dahl, 1990). This trend

continues today in developing and newly-industrializing countries, although the Ramsar Convention (*see Table 9*) is helping to counteract it (Dugan, 1990). Other highly endangered habitats of global significance are marine and coastal ecosystems, such as mangrove forests (Saenger et al., 1983) and coral reefs (UNEP and IUCN, 1988), whereby the latter are particularly threatened by climate change (Wilkinson and Buddemeier, 1994).

Solutions

There are a whole series of institutions whose efforts are directed at improving our understanding of biological diversity and at turning this knowledge into action at the global level. They include the various UN bodies and the major NGOs such as the IUCN, WWF, WRI or WCMC. In 1983, the FAO responded to genetic erosion among utility plants by adopting the *International Undertaking on Plant Genetic Resources*, although this is not legally binding. In cooperation between various nongovernmental environment agencies and UNEP, the *World Conservation Strategy* (IUCN et al., 1980) and the *Global Biodiversity Strategy* (WRI et al., 1992) were developed with recommendations for action to protect, investigate and sustainably use biodiversity. Despite the paucity of data on biological diversity, it is already possible, on the basis of existing information, to arrive at global conclusions regarding the efficacy of nature conservation measures (WCMC, 1994b). Information relating to the national level is often inadequate, however, thus necessitating assistance in preparing species inventories as is also envisaged in the Biodiversity Convention.

To make up for the deficits in scientific foundations, the international scientific community is working, for example, on a *Global Biodiversity Assessment*, which is intended to provide a comprehensive source of current knowledge (*Box 34*).

One example for the practical implementation of measures to preserve habitats is UNESCO's *Man and the Biosphere* program (MAB). This interdisciplinary international program was established with

BOX 33

The Value of Biological Diversity

– *Intrinsic value*: biological diversity can be considered, as in the Preamble to the Biodiversity Convention, for example, to have intrinsic value. This value exists independently of human utilization and ensues from the demand for an intrinsic right to the very existence of biological diversity (Naess, 1986; Hampicke, 1991).

In addition, and independently of intrinsic value, a number of economic values can be assigned to biodiversity from the anthropocentric perspective:

– *Direct utilization values*: these include consumption and production values, derived from the active utilization of nature by humans.
Consumption values: directly utilized natural products that are not traded on the market are of key importance in developing countries especially. These include, fuelwood or yields from hunting and fishing for ensuring subsidence. These values have been neglected in national accounting so far. However, it is possible to assess them economically by estimating a comparable market value. Leisure and recreational value are further examples. Many evaluation studies, above all in developed countries, corroborate the *willingness to pay* among the population for these benefits (Hampicke et al., 1992). Sociocultural, scientific, aesthetic and spiritual values can also be allocated to this category.
Production values: Many natural products are not consumed directly, but serve as commodities subjected to subsequent processing. The main examples are agricultural and forestry produce. While this type of value is known from market trading, other important biodiversity outputs are overlooked in many cases – the variety of genes and biochemical products obtained from natural ecosystems is important for food security, and for medical and other industrial purposes, for example.

– *Indirect utilization values*: These values result from the functions (outputs) of ecosystems and are therefore explicitly dependent on their conservation. Examples include the protection against erosion afforded by vegetation cover, the regulation of biogeochemical cycles (e.g. the maintenance of the hydrological and carbon cycles), the fixation of solar energy by photosynthesis, the preservation of food webs and evolutive processes, the regulation of regional and local climate, but also the absorption of harmful substances (sink function).

– *Option values*: Faced with uncertainty about the possible future demand for biological diversity and the irreversibility of biodiversity loss, many people are prepared to pay for options on future utilization. An example would be the potential use derived from the permanent availability of gene inventories and natural biochemical substances.

– *Existence values*: In contrast to the above categories, existence values are formulated independently of any personal current or future utilization. They derive from the knowledge that there are particular natural systems each with their own diversity. The appreciation of the tropical forest by people in Europe is probably based above all on these fundamental values. This value features a powerful ethic dimension which originates from the sense of responsibility felt by people towards their living environment. A clear manifestation of existential values can be seen in the financial contribution to nature conservation organizations.

See WBGU (1993) on the methodological problems involved in assessing the value of biological diversity.
Sources: McNeely et al., 1990; WBGU, 1993.

the aim of developing a scientific basis for the preservation of natural biosphere resources and for environmentally sustainable use. A worldwide network of biosphere reserves were created with the active participation of local communities. This concept relates not only to particular species or populations, but also to large-scale natural and cultivated regions (WBGU, 1993; Erdmann and Nauber, 1995).

Of particular importance are the various international conventions which have been applied as instruments for preserving biological diversity over the last 50 years (*Table 9*). AGENDA 21, the political program of the UNCED process, also assigns special importance to biological diversity, devoting a separate chapter to it (*Chapter 15*). After assessing the *Convention on International Trade in Endangered Species* (CITES) in its last report (WBGU, 1994), the

Council focuses its attention in the following on the Convention on Biological Diversity, which came into force in December 1993.

5.2
Content of the Convention on Biological Diversity

The *Convention on Biological Diversity* (CBD, referred to hereinafter as the "Biodiversity Convention") is the first international agreement with binding international force that applies a trans-sectoral approach to the protection of global biodiversity (UNEP, 1992), rather than the narrow sectoral approach of previous conventions (*Table 9*), thus dealing with the subject in its entire breadth and scope. The objective is not simply that of conservation, but also "the sustainable use of biological resources and the equitable sharing of the benefits arising out of the utilization of genetic resources" (Article 1, *see Box 35*).

The Convention covers the transfer of technology and finance, as well the question of access to genetic resources. It confirms the principle of national sovereignty of the states over their natural resources, and extends this to biological diversity. In addition, however, it regulates the conditions for access to genetic resources by establishing a basic framework. Detailed lists of measures describe the contractual obligations of states for the conservation and sustainable use of biodiversity. The Contracting Parties are responsible for adopting and implementing the required measures (*Box 35*). A financial mechanism was established for providing financial assistance to developing countries so that they can achieve the objectives of the Biodiversity Convention. Contracting Parties must report on their activities as a means of monitoring implementation. The provisions for adopting protocols underscore the fact that this is a "framework convention". The Council has summarized the main points of the Convention in a previous report (WBGU, 1993; *see* Glowka et al., 1994 for detailed treatment). The following section deals specifically with issues that currently dominate the international debate. These include the problem of access to genetic resources (*Section C 5.4*), which will be handled – alongside other topics – at the second Conference of the Parties in November 1995, as well as the discussion on a legally binding instrument for the protection of forests (*Section C 6*).

BOX 34

Global Biodiversity Assessment

The scientific foundations for biodiversity are complex, and the field shows up definite gaps in knowledge. A summary of the current state of knowledge, such as that drawn up by the IPCC on climate change, has yet been produced. The preamble to the Biodiversity Convention also refers to the "general lack of information and knowledge regarding biological diversity" and the "urgent need to develop scientific, technical and institutional capacities to provide the basic understanding upon which to plan and implement appropriate measures".

With the support of the GEF (approx. US$ 3 million), the UNEP started the Global Biodiversity Assessment (GBA), but this operates without any formal link to the Biodiversity Convention. The objective is to carry out a peer-reviewed independent scientific analysis of current topics and explanatory approaches on the global aspects of biodiversity, with the cooperation of more than 1,000 scientists worldwide. On the basis of this analysis, gaps in knowledge and controversial themes are to be identified, but also areas specified where there is consensus within the scientific community. Current issues regarding the protection and the utilization of biodiversity in forestry and agriculture, fisheries and tourism, as well as the interaction with climate change and with social and economic dimensions are included. The questions as to the scope of biological diversity, where it is located and what status is accorded to its protection are not a component of the GBA; these have already been published in a detailed report (WCMC, 1992).

The GBA aims at examining the following fundamental issues:
– How can the biodiversity of genes, species or ecosystems be measured and compared?
– What functionality does biodiversity possess for ecosystems?
– How can extinction rates be predicted?
– How can biodiversity be protected?
– What economic value does biodiversity possess?
– How much biodiversity does humanity need for survival?

The results of this initiative are expected in the form of a major report by the end of 1995.

Table 9
Legally binding, global conventions of relevance for the biosphere: a selection.
Sources: Ojwang, 1994; IUCN Environmental Law Centre

Convention	Where concluded	Date	Entry into force in Germany
Convention relative to the Preservation of Fauna and Flora in their Natural State	London	Nov. 8, 1933	not signed
International Convention for the Regulation of Whaling	Washington, DC	Dec. 2, 1946	July 2, 1982
International Convention for the Protection of Birds	Paris	Oct. 18, 1950	not signed
International Convention on Fishing and the Conservation of the Living Resources of the High Sea	Geneva	April 29, 1958	not signed
Convention on the High Seas	Geneva	April 29, 1958	Aug. 25, 1973
Ramsar Convention on Wetlands of International Importance Especially as Waterfowl Habitat	Ramsar (Iran)	Feb. 2, 1971	June 26, 1976
Convention for the Conservation of Antarctic Seals	London	June 1, 1972	Oct. 30, 1987
Convention Concerning the Protection of the World Cultural and Natural Heritage	Paris	Nov. 23, 1972	Nov. 23, 1976
Convention on International Trade in Endangered Species of Wild Fauna and Flora (CITES)	Washington, DC	March 3, 1973	June 20, 1976
Convention on the Conservation of Migratory Species of Wild Animals (CMS)	Bonn	June 23, 1979	Oct. 11, 1984
United Nations Convention on the Law of the Sea (UNCLOS)	Montego Bay	Dec. 10, 1982	Nov. 16, 1994
International Tropic Timber Agreement (ITTA)	Geneva	Nov. 18, 1983	March 21, 1986
Convention on Biological Diversity (CBD)	Nairobi	May 22, 1992	Dec. 29, 1993

5.3
Development and Current Status of the Convention Process

There have been growing demands since the 1970s for a broad-based, cross-sectoral convention for the protection of biological diversity as a key component of international environmental policymaking. Between 1984 and 1989, the *World Conservation Union* (IUCN) discussed draft articles for such a convention. Under the leadership of UNEP, this process was taken a stage further through the establishment of a special working group in 1987. In 1991, this UNEP working group was reconstituted as a formal *Intergovernmental Negotiating Committee* (INC). At five INC sessions, after difficult negotiations, a convention text was brought forward. Under the time pressure of the forthcoming UNCED conference in Rio de Janeiro in 1992, consensus was achieved; the resultant text of the Convention was signed by 157 countries and the EU during UNCED. After ratification by more than 30 states, the Convention entered into force 18 months later, in December 1993. Today it has been ratified by more than 110 states – including Germany and the EU, but not the USA.

After only two sessions of the *Preparatory Committee* (ICCBD), the first Conference of the Parties to the Convention on Biological Diversity took place in late 1994 in Nassau, Bahamas. The main achievement of the Conference was to set up, staff and finance the various organs of the Convention (Secretariat, Subsidiary Body on Scientific, Technical and Technological Advice), i.e. to give them the capacity to take action. The second Conference of the Parties shall take the final decision regarding the location of the Secretariat. The financing mechanism for the Biodiversity Convention shall remain the GEF on a provisional basis; guidelines for the allocation of funds were drawn up. The Conference also started work on the establishment of a *Clearing-House Mechanism* for technical and scientific cooperation (Article 18). However, work on the issue of "safe transfer, handling and use of any living modified organism resulting from biotechnology" (biosafety) has been slow to get started; the drawing up of a protocol will be examined by a working group. The first Conference of the Parties signaled to the CSD that the Convention could contribute to the conservation and the sustainable use of forests. However, this subject is not mentioned in the medium-term programme of work agreed upon for the Convention. This program of work is a guideline for operations over the next three years. In addition to various other points, the access to genetic resources, the links to the FAO's *"International Undertaking on Plant Genetic Resources"* (*see Section C 5.4*) and marine biodiversity are on the program for 1995 and are being prepared at a session of the *Subsidiary Body on Scientific, Technical and Technological Advice* (SBSTTA). The second Conference of the Parties will be held in Indonesia in November 1995.

The first Conference of the Parties succeeded in creating the basis for continued work. Of particular

---

BOX 35

The Convention on Biological Diversity

Objectives (Article 1):

The objectives of this Convention, to be pursued in accordance with its relevant provisions, are

- conservation of biological diversity,
- sustainable use of its components,
- fair and equitable sharing of the benefits arising out of the utilization of genetic resources, including by
  - appropriate access to genetic resources,
  - appropriate transfer of relevant technologies, taking into account all rights over those resources and to technologies,
  - appropriate funding.

Principle (Article 3)

States have

- the right to exploit their own biological resources ,
- the responsibility to ensure that they do not cause damage to the environment of other states.

National commitments (Articles 5–14)

The Contracting Parties shall

- cooperate with other Contracting Parties in respect of areas beyond national jurisdiction and on other matters of mutual interest for the conservation and sustainable use of biological diversity,
- develop national strategies for the conservation and sustainable use of biological diversity,
- integrate the conservation and sustainable use

of biological diversity into relevant sectoral or cross-sectoral plans, programs and policies,

- identify and monitor components of biological diversity,
- adopt measures in support of in-situ conservation (e.g. system of protected areas),
- adopt measures for ex-situ conservation (e.g. gene banks, botanical gardens),
- adopt regulations for the sustainable use of biological diversity,
- adopt measures that act as incentives for the conservation and sustainable use of biological diversity,
- support research and training,
- encourage and propagate understanding and public awareness of the importance of conserving biological diversity,
- introduce procedures requiring environmental impact assessments.

International regulations (Articles 15–22)

The Convention contains international regulations on the following aspects:

- access to genetic resources,
- access to and transfer of technology,
- exchange of information,
- technical and scientific cooperation,
- handling of biotechnology,
- financial resources,
- relationship with other international conventions.

Articles 2–5 contain definitions, general principles and international measures. Articles 23–42 regulate the administrative and procedural mechanisms of the Convention.

---

importance for the implementation in the signatory states are the production of national reports on the status of biodiversity and the development of strategies for the integration of the Convention's objectives into national policymaking. In this connection, the Council calls on the Federal Government to present the national report and the national strategy for Germany as quickly as possible. It is too early as yet to assess the success of the Biodiversity Convention, because no results can yet be expected on the current status of the convention process. Future reporting by the Contracting Parties will facilitate evaluation. One positive aspect is that the financing mechanism is already being applied to projects aimed at achieving the objectives of the Biodiversity Convention, and

that the Conference of the Parties decided on eligibility criteria. The Council considers it important for the future negotiation process that a protocol on biosafety be formulated and adopted without delay and that an instrument be developed for protecting forests (*see Section C 6*). Parallel to this, the FAO's "*International Undertaking on Plant Genetic Resources*" is being adapted to the Biodiversity Convention. Discussions are being held on integrating this undertaking, which is especially aimed at mitigating genetic erosion of useful plants, as a protocol to the Convention.

The impacts of the Biodiversity Convention should be measured not only in terms of the negotiation process. The discussions relating to the Convention has sharpened awareness in society for the seriousness of species and biotope loss. This is all the more important, in that attaining the Convention's objectives cannot be left entirely to national authorities, but also requires the active support of environmental lobbies and the public at large.

5.4
Key Focus: Access to Genetic Resources

BIODIVERSITY AS A VALUABLE NATURAL RESOURCE

During the colonial era and in the post-World War II period, the export of biological material from developing countries to the industrialized nations was not perceived as a topic of much relevance. Plant material was deemed a free resource to which anyone had access, or was the property of colonial administrations. The consequence was that "biodiversity" was accorded no value as a resource. This led, among other things, to a situation in which neither the indigenous peoples and local communities, who in many cases had contributed their traditional knowledge about the use of plants or had developed local varieties of useful plants, nor the governments of the countries of origin had any share in the profits made from external utilization. This free access to genetic resources contrasted to the otherwise usual payment for the use of other natural resources such as coal, oil or ore.

The picture today is much different. Demand by industrialized countries, especially from agribusiness and the pharmaceutical industry, for the genetic resources which exist in the world has grown enormously. The developing countries possess the greatest genetic variety – more than 95% of worldwide production of the 20 most important agricultural plant species stem from the genetic material from developing countries (Kloppenburg and Kleinman, 1987). For several decades access to genetic resources has become an increasingly problematic subject of discussion within North-South relations. Concepts such as "developmental justice", "conservation", "sustainable development", the "equitable sharing of benefits" and the "rights of indigenous peoples" have meanwhile made access to genetic resources a burning international issue.

In agriculture and forestry, genetic diversity of utility plants and animals is of cardinal importance. To ensure the necessary increase in yields for the maintenance of food security and the adaptability of species used to changing conditions (e.g. newly occurring diseases, resistant pests or climate change), new genetic material must constantly be introduced into the species used. These genes have their origin in the rapidly declining diversity of traditional local species and varieties and in the wild relatives of used species. Single genes can be of great value as raw material for agribusiness. The demand for genes of wild species will further increase as a result of new methods of genetic engineering, because today genetic material can be freely exchanged between different species, e.g. between bacteria and plants, whereas classical breeding methods were restricted to the crossing of genes from closely related species.

Medicine, too, owes many of its most successful drugs to the active substances in wild species. Traditional Chinese medicine makes use of approx. 5,000 plant species for therapeutic purposes. About 60% of the world's population depend on traditional herbal medicine. Naturally occurring active substances play a major role in new product development by the modern pharmaceutical industry – in the United States, some 25% of prescriptions are filled with drugs whose active ingredients are extracted or derived from plants (amounting to sales of US\$ 15.5 billion) (Reid et al., 1993).

Recent pharmaceutical research, in its systematic search for new agents, has turned its attention once again to wild fauna and flora on a major scale (Eisner, 1989; Gettkant, 1995). Enormous progress has been achieved with regard to the technology for screening large numbers of samples for medicinal properties, through automation and new biotechnological methods. Another factor is the growing realization that action is urgently needed, because biodiversity resources are declining rapidly, as is the valuable traditional knowledge of indigenous peoples about the properties of wild species. Finally, the synthesis of agents in the laboratory is limited to relatively simple molecular structures, so the pharmaceutical industry is now dedicating greater attention to the complex agents found in wild flora and fauna, where evolution has been "experimenting" for millions of years with biochemical substances.

BIODIVERSITY PROSPECTING

Numerous governmental institutions and private enterprises are engaged in the task of collecting, identifying and analyzing genetic resources. This research into and collection of biological material and its processing for potential industrial use is called "biodiversity prospecting" (Reid et al., 1993).

CONVENTION REGULATIONS

The Biodiversity Convention reflects this international discussion, establishing a new framework for regulating access to genetic resources among the Contracting Parties (Article 15). States are recognized, for the first time in international law, as having sovereign rights over their genetic resources and the authority to determine access to genetic resources. The Contracting Parties are not permitted, however, to impose restrictions on access to genetic resources for environmentally sound uses by other Contracting Parties. Access to genetic resources "shall be subject to *prior informed consent*" of the Contracting Party providing such resources and "on *mutually agreed terms*". These conditions are intended to protect the country of origin against granting access to genetic resources too readily, without being informed about the scope and impact of biodiversity prospecting (Glowka et al., 1994).

Benefits arising out of the utilization of genetic resources should be shared in a fair and equitable way. The Convention provides various mechanisms for sharing such benefits (Articles 15, 16 and 19):
- sharing in a fair and equitable way the results of research and development,
- fair and equitable sharing of the benefits arising from the commercial and other utilization of genetic resources,
- scientific research with the full participation of Contracting Parties providing genetic resources, where possible within the territory of such Contracting Parties,
- access to and transfer of technology to providers, including technology protected by intellectual property rights.

Conceivable forms of financial compensation include:
- advance payments,
- payment for each sample supplied,
- profit-sharing in the royalties for products which are later marketed, depending on the affinity of the final product to the initial substance and the contribution made by the provider (Laird, 1993).

This sharing in benefits can also include nonmonetary forms of compensation, particularly social welfare benefits (such as health care provision) and capacity building measures.

The Convention recognizes indigenous peoples and local communities as the holders of knowledge relevant for the conservation of biological diversity (see Articles 8j, 10c, 17 and 18). Article 8j contains rules providing for their involvement:
- The knowledge, innovations and practices of indigenous and local communities embodying traditional lifestyles relevant for the conservation and sustainable use of biological diversity shall be respected, preserved and maintained.
- Wider application of traditional knowledge shall be promoted with the approval and involvement of the holders of such knowledge.
- The equitable sharing of the benefits arising from the utilization of such knowledge, innovations and practices shall be encouraged.

The framework provisions of the Convention regarding the access to genetic resources must be translated into national legislation. This process, especially that of legal enforcement, will have a critical influence on the effectiveness of the regulations (Hendrickx et al., 1994).

Access to genetic material collected prior to the Convention coming into force and which is stored in ex-situ collections (botanical and zoological gardens, gene banks), principally in the industrialized nations, is not covered by the new provisions, as the Convention is not applied retrospectively. In future, however, new regulations for the utilization of genetic material will have to be agreed upon, especially with regard to the freely accessible gene banks managed by the FAO. In this connection, the adaptation of the FAO's *Global system for the Conservation and Utilization of Plant Genetic Resources* to the provisions of the Biodiversity Convention are already in the preparatory phase.

BILATERAL CONTRACTS FOR BIODIVERSITY PROSPECTING

Access to genetic resources is increasingly becoming the subject of bilateral contracts for biodiversity prospecting. Such contracts define financial and technical compensation for the provision of access to genetic resources – i.e. the permission to collect or supply samples of biological material. Negotiated between equal partners, they generally represent major progress towards achieving the Convention objectives, compared to the previous practice regarding the export and utilization of genetic material without any form of compensation.

An institution from an industrialized country (collector country) will typically negotiate a contract for the supply of genetic resources with an institution in a developing country, usually from the tropics (supplier country or country of origin) (Laird, 1993). Those involved in biodiversity prospecting may be governmental or nongovernmental organizations. The parties to such a contract, on the collector and/or supplier side, might be any of the following:
- states, represented by authorities or a public-law institute,
- private, profit-oriented enterprises as users, "brokers" or collectors,
- nonprofitmaking institutions or groups, e.g. research institutes or (on the supplier side) indigenous and/or local groups.

More than two hundred enterprises and research establishments are currently examining plant and, in some cases, animal species from natural ecosystems for their potential uses for medical purposes (Reid et al., 1993; ICCBD, 1994). This work is governed by contracts – sometimes concluded before the Biodiversity Convention came into force – which regulate access to genetic resources (*Box 36*).

These examples show that biodiversity prospecting can, under certain conditions, represent one way of achieving the objectives of the Convention. Criticism is sometimes leveled at particular aspects of cooperation, such as the confidentiality of agreements, the lack of involvement of the indigenous population in the case of INBio and NCI, or the inadequate consideration given by Shaman to national sovereign rights (Zerner and Kennedy, 1994). Nevertheless, these pilot projects are pioneering in their efforts to combine both the utilization and the conservation of biodiversity. By analyzing different approaches to biodiversity prospecting, it is possible to identify the opportunities and risks implicit in such projects and to elaborate the criteria that should be taken into account when formulating contracts for biodiversity prospecting.

OPPORTUNITIES CREATED THROUGH BIODIVERSITY PROSPECTING

There is no market mechanism in operation that focuses on the long-term nonmonetary value of biodiversity (WBGU, 1993). This lack of consideration given to the value of biological diversity puts sustainable use at a disadvantage compared to the destructive utilization of natural ecosystems to satisfy short-term interests. In contrast, contracts for biodiversity prospecting provide an opportunity to turn the conservation of nature into an economically attractive alternative by attaching a value to the resource "biodiversity".

Biodiversity prospecting should always be seen, however, as but one component in a comprehensive strategy for the conservation and sustainable utilization of biological diversity (Reid et al., 1993). By establishing this link between conservation and sustainable use, as is achieved by INBio, for example, funds can be channeled directly from contractual agreements into the conservation of biodiversity and the drawing up of an inventory of such resources. Indirect positive effects are also produced because prospecting agreements may channel funds to the respective local level by means of appropriate clauses providing for the involvement of indigenous or local communities. This can create alternatives to the conversion and/or nonsustainable use of natural ecosystems, and hence provide local incentives for preserving biodiversity (WRI et al., 1992).

Developing countries can advance in this way from a position as mere recipients of transferred technology and knowledge in this field to "donors" of genetic resources and indigenous knowledge (*see Section B 2*; Lesser and Krattiger, 1994). By paying for genetic resources, the traditional knowledge of in-

digenous communities obtains an economic value, thus enabling its preservation. Precisely how such equitable compensation should be arranged is a controversial issue, however. Protecting the intellectual property of indigenous groups is an area in which major deficits exist (Shiva, 1994; Posey, 1994).

Provided that biodiversity prospecting contracts also involve the transfer of technology and know-how, some countries could succeed in establishing favorable conditions for the independent utilization of their genetic resources (Reid et al., 1993; Gettkant, 1995). The long-term effects that could be attained are certainly greater than those for a country that confines itself purely to the supply of genetic material. Interest and cooperation on the part of pharmaceutical and biotechnology companies increase the opportunities for "donor countries" to build their own research capacities and thus to open up new, sustainable pathways of development. Such a process could be realized in a series of steps:
- capacity building for collection and information processing of biological diversity and the management of natural resource protection,
- improvements in information management (data on the abundance and distribution of the respective species, or traditional knowledge about its effects),
- development of screening methods aimed at offering added-value products (screening of samples in the country of origin to isolate active compounds),
- establishment of biotechnological research and development, up to and including the independent production of cosmetic or pharmaceutical products.

RISKS OF BIODIVERSITY PROSPECTING

Biodiversity prospecting can also harbor substantial risks for populations, species and indeed the stability of ecosystems (INF, 1994). These threats generally increase in proportion to the amount of biological material that is extracted:
- The collection of small samples to screen for active compounds is the initial phase of biodiversity prospecting projects, mostly with little impact on the ecosystem.
- If such a sample is found to contain an active substance, more material of the same plant must be available for detailed analysis. If the species in question is rare or confined to a particular locality, the amount required may already involve serious impairment of the ecosystem. Furthermore, valuable genetic diversity may be lost through the harvesting of local populations on a larger scale.
- If a marketable product is derived from a particular substance, wildland material is often used as the source of supply. Due to the sheer volume re-

Contracts for Biodiversity Prospecting – Three Examples

The INBio – Merck agreement

This pilot project represents a new approach in the field of biodiversity prospecting and the first attempt to organize financing for the conservation of biodiversity and biodiversity inventories in exchange for rights of use (Sittenfeld and Lovejoy, 1994)

The partners:

Merck & Co., Inc. (New Jersey, USA): a subsidiary of Merck, an internationally operating pharmaceutical firm.

Instituto Nacional de Biodiversidad (INBio, Costa Rica): a private, nonprofit organization in Costa Rica, with tight links to the relevant ministry (*see* Gámez et al., 1993; Sittenfeld and Gámez, 1993). INBio generates income by commercially exploiting biodiversity resources while at the same time giving consideration to conservation of nature. Commercial contracts for biodiversity prospecting concluded with INBio must meet the following criteria:

- monetary and nonmonetary payments for the provision of samples and extracts,
- payment of a direct contribution to the cost of maintaining the National System of Conservation Areas in Costa Rica,
- royalties paid on net sales to industry of products derived from Costa Rica's biodiversity resources.

All profits generated by INBio are reinvested in nature conservation.

The agreement:

The agreement was signed in September 1991 (INBio, 1994) and renewed in July 1994 with a similar content (Reid et al., 1993; Sittenfeld and Lovejoy, 1994). In the context of the global biodiversity debate, the agreement provides both parties with considerable public-relations potential. Costa Rica profits from the improved administration of the National Park program.

Services performed by INBio:

INBio supplies Merck with approx. 10,000 plant and insect samples in extracted form obtained from the nature conservation areas of Costa Rica. These samples are collected and identified by INBio staff and supplied to Merck for its exclusive use for the term of the contract. Merck obtains the right to patent any products derived

from the material. INBio retains the right to conclude agreements for joint research with other companies.

Obligations on the part of Merck:

- training of four Costa Rican scientists in Merck's research center,
- US$ 1 million payment to INBio for the two-year contract term, including a US$ 100,000 contribution to the National Parks Fund,
- US$ 135,000 for laboratory equipment and the establishment of a modern analytical laboratory at the University of Costa Rica,
- payment of around 2-6% of the net proceeds from sales of products developed through cooperation with INBio.

This contract is the most discussed and analyzed biodiversity prospecting project, and a pilot project of major importance. It cannot be transferred to other countries without adaptation, in that Costa Rica is predestined in many respects for such cooperation. Its rich biodiversity heritage, political stability, relatively good environmental laws and resources (25% of the land surface is designated as national parks and nature reserves), as well as its comparatively high standards of education make it an exceptionally suitable partner.

Cooperation between Shaman, Inc. and indigenous communities

The partners:

Shaman Pharmaceuticals, Inc.: a Californian pharmaceuticals company. Product development is based on ethnobotanical principles, thus generating interest in cooperation with indigenous groups. Shaman follows a policy of mutual benefit sharing with indigenous partners and of conforming with the principles for conservation and sustainable use of biodiversity. Shaman's approach is complex, in that they operate several projects with different partners.

Local indigenous communities: these are various small groups, one example being a Quicha community in Ecuador, or Yanomani Indians in the Amazon region. These groups are interested in direct monetary rewards, but also in other types of benefits, for example improved health care or capacity building.

The contracts:

Services performed by the indigenous groups:

Cooperation with and support for Shaman in ethnobotanical biodiversity prospecting. Shaman interviews indigenous specialists and collects samples. In some cases, sustained and long-term supply of natural products to Shaman.

Counter-performance on the part of Shaman:

Shaman pays direct compensation for services rendered in accordance with the needs of the partners. The company pays part of the profits it generates into a fund administered by the *Healing Forest Conservancy* (HFC). This nonprofit organization was founded by Shaman when the company itself was established, and supports, with the advice of an independent international committee and representatives of the relevant regions, all local communities and countries in which it is active. In this way, indigenous partners receive direct and immediate compensation without having to wait for potential marketing of a newly-developed product after five or ten years. The funds are invested in projects aimed at the conservation of biological and biocultural diversity. The basic principle and aim is that nonproduct-related support for all communities provides an incentive for the broad-based protection of biodiversity resources. Shaman is currently developing guidelines for pharmaceutical companies and other industries that work directly with the inhabitants of the tropical rainforest (King, 1994). The company pays for the supply of biological material, providing technical assistance and knowledge in the process. They promote the growth of local trade and commerce to secure a long-term supply of natural products from sustainable use and for the extraction of active substances.

Agreements concluded by the National Cancer Institute with developing country organizations

The partners:

National Cancer Institute (NCI): a government body in the United States operating a drug research program that has been screening natural substances since 1960 in the search for a cure for cancer and AIDS.

Government organizations in developing countries: agreements are concluded, through intermediaries (e.g. botanical gardens in the USA), with government agencies in countries of origin.

The contracts:

Services performed by the countries of origin: supply of biological material to the NCI, granting of rights to patent registration for products developed under the contract.

Obligations on the part of the National Cancer Institute:

- royalties: the licensee of any active substance developed under the terms of the contract must negotiate with the country of origin over the payment of royalties. The licensee must pay a certain amount towards the conservation of biodiversity, even if the substance is produced synthetically.
- capacity building: scientists from the country of origin can work for a period as guest researchers at NCI. The objective is to build capacities for an indigenous drug research program in the country of origin.
- contributions for the local indigenous population: if the institution in the country of origin utilizes the ethnobotanical knowledge of the local population, then the latter is involved in planning the collection strategy. The NCI maintains secrecy on all relevant knowledge until both partners agree to publicize. Before doing so, however, permission must be obtained from the traditional healer or local community concerned.

quired and its associated market value, there is a major risk of overharvesting the natural stocks.

These threats and risks are exemplified by the fate of the Kenyan shrub called *Maytenus buchanani*, source of an important anticancer compound, which was almost entirely depleted as a result of overharvesting (Oldfield, 1984). The international response in 1994 was to include four species of medicinal plants in the Annexes to the *Convention on International Trade in Endangered Species of Wild Fauna and Flora* (CITES).

The more intense the utilization, the more important it is to have information on the species itself and the ecosystem providing its habitat in order to assess the limits of sustainability. If requirements can no longer be met sustainably from the natural habitat, the answer might be to cultivate the species instead. In many cases, however, large-scale investments are needed to develop the appropriate cultivation methods. If such reproduction is carried out in the countries of origin on small farms, this can create and alternative source of income for local people. The last step would be to synthesize the active compound in the laboratory, thereby decoupling the marketing of a biological resource from the ecosystem in which it originated. Arrangements must therefore be made to ensure that a share of the royalties are used for the conservation of nature in the country of origin (*see* INBio-Merck and NCI, *Box 36*).

CONCLUSIONS
Biodiversity prospecting contracts

The analysis shows that it is neither possible nor meaningful to look for universally applicable regulations due to the diversity of

- contracting parties (national governments, indigenous groups, private companies, research establishments),
- conditions (stable country with a clearly defined legislative framework, or a large state apparatus beyond any form of precise control),
- ultimate uses (drugs, agricultural resources, cosmetic products, etc.).

Applying a predefined generalized contractual model would not be appropriate for the specific needs of communities and states (Laird, 1993; Zerner and Kennedy, 1994).

Nevertheless, criteria can be derived that should be observed when drafting contracts for biodiversity prospecting in order to attain consistency with the objectives and provisions of the Biodiversity Convention. However – as required by the Convention – there must be national laws governing all biodiversity prospecting contracts, because private contracting parties are not specifically covered by the provisions of the Convention. For that reason, it is crucially important that the Convention be translated into national law and enforced through legal practice (Hendrickx et al., 1994).

Special consideration must be devoted to the following points (*see also* INF, 1994):

- observance of the sovereignty rights of states over their genetic resources (prior informed consent, mutually agreed terms),
- guarantees for the short-term and long-term share of benefits for the countries of origin,
- linkage between biodiversity prospecting and measures to conserve and inventory biological diversity,
- compliance with the principle of sustainability, particularly when large amounts of biological material are harvested,
- information and involvement of indigenous and local people in the formulation of contracts and participation in the profits generated by biodiversity prospecting,
- linkage with technology transfer, development of research infrastructure and capacity building. Support for the independent management and exploitation of genetic resources by the country of origin.

If the points listed above are complied with, then contracts for biodiversity prospecting may provide a

valuable contribution towards achieving the objectives of the Biodiversity Convention. Hopes should not be exaggerated, however. The preconditions for such contracts and the size of the anticipated profits mean that not every developing country will have the opportunity to deploy biodiversity prospecting contracts as an instrument for the conservation and sustainable utilization of biodiversity. Combined with other forms of sustainable use, however, such as "agroforesty" or "tourism", they can point in the direction of environmentally sound management of natural resources.

UNRESOLVED ISSUES

One issue that is currently generating considerable debate is whether an *international standardization of access regulations* is desirable or not as a means of preventing dumping practices and assuring adequate sharing in benefits on the part of the suppliers of genetic resources, as well as compliance with the other guidelines of the Convention. At the international level, the proposal has been made that an independent institution carry out the organization, harmonization, simplification and monitoring of biodiversity prospecting contracts (Lesser and Krattiger, 1994). Whether such an institution is needed, and whether its function can be performed by one of the bodies created under the Biodiversity Convention (SEI, 1994), or perhaps better through an international organization, requires more detailed investigation.

It is not always possible to specify precisely the *country of origin* of a sample, in that the distribution of species is independent of national frontiers. Were INBio to supply genetic resources from which Merck then derived a successful product on the market, they would have considerable difficulty proving that the origin of the basic biological material was Costa Rica rather than some neighboring country.

There are also a number of open issues with respect to *cooperation with indigenous groups*, such as how the "responsible" indigenous group is defined in terms of social or topographic criteria, or whether an indigenous organization and/or a person responsible for the group should be selected as a contact instead. This is closely linked to the question as to who are the actual "holders" of ethnobotanical knowledge. Are they the individual shamans or medicine women who possess the rights to dispose of such resources, or are we dealing with the commonly possessed knowledge of a community? How should one proceed when the same traditional knowledge is spread out over an entire region that has a wide variety of different indigenous peoples? Are concepts for intellectual property rights even applicable across cultural boundaries in the first place? These questions are debated at present on the international plane, but at the same time it is an area where considerable research is needed.

5.5
Recommendations for Research

The research recommendations relating to biodiversity that were put forward in the 1993 annual report have lost none of their topicality and relevance (WBGU, 1993). Of particular relevance are those concerning the relations between biodiversity and ecosystem function, the adaptation of species and ecosystems to climate change and the question of economic values of biodiversity. The remainder of this section is dedicated to recommendations that are derived from the central topic of this annual report.

BIODIVERSITY
- Support for global comparative biogeography to obtain a better understanding of how biomes and/or ecozones are distributed on the earth.
- Development of concepts for recognizing structural changes to landscapes that impair the diversity of species and habitats.
- Development of early warning systems for particularly endangered species and/or ecosystems, with the aim of rapid implementation of protective measures.
- Investigation of the impacts of legal, economic and (agro-)political conditions (e.g. global level, EU) on biological diversity.
- Research into how a broad societal discourse on the conservation of global biodiversity can be set in motion, thus contributing to a clarification of the aims and instruments of the Biodiversity Convention.

ACCESS TO GENETIC RESOURCES
- Determination and assessment of the extent of genetic erosion as well as its medium- and long-term effects on humanity and environment.
- Acquisition of more profound knowledge on the basic taxonomic principles (also with the help of molecular-genetic methods) for conducting research on the gene pools of useful plants.
- Further development of methods for testing the genetic stability of samples (e.g. in gene banks) during regeneration.
- Making use of further potential benefits of plants, especially with regard to the development of environmentally sound products, as well as extending the cultivation spectrum of cultures as suppliers of renewable raw materials.
- Application of economic assessment methods to genetic resources.

- Effect of patent and species protection systems on the conservation of biodiversity and genetic resources.

FORMULATION OF BIOPROSPECTING CONTRACTS
- Studies on the type and scope of the mutual information of the Parties regarding the objectives of projects and method of collecting material (prior informed consent) as well as on the participating social groups.
- Examination of the basic legal conditions in agreements in order to protect the rights of indigenous and local communities, their knowledge, their customs and practices.
- Development of ideas on monitoring measures to prevent unauthorized access to biodiversity.
- Working up proposals on legislation and incentives, both in countries where there is demand and in the supplier countries, so as to guarantee compliance with the guidelines of the Convention concerning bioprospecting.
- Conducting studies as to whether it is meaningful to set up an independent international institution as an information and intermediary agency (facilitator) for mediating, coordinating and monitoring bioprospecting agreements.

5.6
Recommendations for Action

NATIONAL IMPLEMENTATION OF THE
BIODIVERSITY CONVENTION
The Council views the worldwide loss of biodiversity as one of the most pressing problems involved in global change (WBGU, 1993). Accordingly, the Council attaches great importance to measures for practical implementation of the Biodiversity Convention. A recommended guideline for development of a national strategy aimed at implementation of the Convention is to place special emphasis on the cross-sectional nature of the Convention (*see also* WRI et al., 1992). Many fields of governmental action, from research, planning, agriculture and forestry to problems of world trade and development cooperation, are affected by the Convention. In this context the Council urges that Germany's national report and national strategy be presented as soon as possible. Implementation at all levels requires, however, an awareness and knowledge of the problems and possible solutions. Considerable deficits exist in this respect according to the Council, which recommends that discussion over objectives and contents of the Convention be taken beyond specialized reports and bodies to all parties concerned, even in state and local governments. Offensive public relations and me-

dia work would contribute to raising awareness of the life-threatening loss of biodiversity, as has been done in the case of climate problems.

Setting up a focal point in Germany for facilitating the exchange of information and technology transfer – for bioprospecting agreements, too – is recommended to support the Clearing-House Mechanism of the Convention.

EXPANSION OF NATURE RESERVES

The national implementation policy should be oriented to the "Lübeck Principles of Conservation" (LANA, 1993). This includes appropriate handling of biological resources and a transition to sustainable land use in all sectors. Expansion of protected areas within the framework of the European concept of "Natura 2000", involving the networking of nature reserves, as well as through the setting up and maintenance of biosphere reserves represent action for effectively counteracting the loss of biodiversity in Germany. Revision of the Federal Conservation Act and consistent implementation and application of the European conservation guidelines (particularly the fauna-flora-habitat guideline) are necessary and long overdue.

AGRICULTURAL SUBSIDIES

Subsidies at the wrong place in the agricultural or forestry sector torpedo efforts to protect biodiversity (so-called *perverse incentives*) through market distortion. The present system of agricultural subsidies within the EU must, therefore, be urgently coordinated with the objectives and contents of the Biodiversity Convention (Art. 10). In addition, implementation of the principle of environmentally sound use in European agriculture and forestry would act as a model for the other Parties to the Convention.

DEVELOPMENT COOPERATION

Direct financing via the GEF is only one means of promoting the aims of the Biodiversity Convention internationally. The other instruments of technical or financial aid, especially with regard to development cooperation, should also focus more on the Convention.

ENVIRONMENTALLY SOUND WORLD TRADE

Measures providing for environmentally sound world trade could make a significant contribution to saving biodiversity. In particular, regulations within GATT/WTO should be coordinated with the objectives of the Biodiversity Convention (*see Section C 7*).

ACCESS TO GENETIC RESOURCES

Developing countries should be assisted in setting up a scientific infrastructure, inventorying their biodiversity, conservation management and capacity building for independent utilization of national genetic resources (extraction methods, biotechnology, etc.).

Aid should also be provided to create focal points in the countries of origin aimed at improving the effectiveness and control of access to genetic resources as well as to build up capacity for negotiating, implementing and monitoring bioprospecting agreements.

6.1
Current Trends

Forests are the dominant form of vegetation in the biosphere, covering roughly one fifth of the Earth's land area, and they represent a valuable ecological and economic resource (WBGU, 1993). Human use has always changed forest ecosystems, but the middle of this century marked the beginning of a radical increase in the intensity and extent of utilization, primarily in the form of destruction through clearing of woodland for cultivation (Sharma, 1992; WBGU, 1994).

Only about half of the tropical forests are still in a natural state. If destruction through clearing for cultivation purposes, improper use, settlement, etc. continues unabated, the area covered by tropical forests will decline from the current figure of approx. 1.8 billion ha to 600 million ha by the middle of the next century. Only in regions that are difficult to access will forests be able to maintain themselves. The problem of forest destruction and degradation is likely to grow in the near future since anthropogenic environmental changes impair the further development of forests. This may lead to large-scale collapse of forest ecosystems (Enquete Commission, 1995).

Although the main emphasis of public interest in recent years has been on the clearing of tropical rainforests for cultivation, the problem goes far beyond that: the felling of the sensitive boreal and mountain forests (e.g. in North America, Siberia and in the Himalayas) also holds considerable ecological risks. In this case it is the outright clearance of forests and the use of heavy-duty machines that not only cause harm to biodiversity, but also have serious effects on the soil (WBGU, 1994). Irreversible losses for biodiversity take place especially when primeval forests are cleared on a large scale, as is being done in tropical countries as well as in rainforests along the Pacific coast (Canada, Chile). The secondary forests that grow after the clearing, frequently reafforested with monocultures, offer a habitat to only a fraction of the species originally living there.

The forest areas in the industrialized countries, by contrast, have extensively stabilized or even expanded. The main concern here is the "new type of damage to the forests" (*Waldsterben*), essentially caused by the deposition of pollutants from the air. In Germany only a third of the forest area has not suffered damage (BML, 1993), in Poland and Great Britain only 5-10% of the area is considered to be unimpaired (UN-ECE and CEE, 1992).

A reversal of these global trends is not foreseeable at the present time. This makes the lack of binding instruments of global environmental policy for the protection of the forests based on international law all the more aggravating. After the failure to draw up a corresponding document at the UNCED in Rio de Janeiro in 1992, where only a nonbinding "Forest Declaration" was adopted, this issue continues to be of the utmost relevance.

6.2
Status of International Discussion

The international community has been striving to come to a global arrangement about forests for a long time now. A binding document failed at the 1992 UNCED due to, among other things, the resistance of many developing countries that claim national sovereignty for their forests and resources. The result of the Rio Summit was merely a "*Nonlegally binding authoritative statement of principles for a global consensus on the management, conservation and sustainable development of all types of forests*" (Forest Declaration, *Box 37*). The debate over an international instrument has been intensified recently. If the necessity for a global arrangement is affirmed, then the question of aims and instrumental structure must be clarified. On the one hand, the issue of forests could be dealt with in a separate convention (*Forest Convention*); on the other hand, it would be possible to regulate the use of forests in a protocol to the Convention on Biological Diversity (*Forest Protocol*).

The positions of the various states on the question of what concept would be appropriate for regulation

BOX 37

Contents of 1992 "Forest Declaration"

Objectives:
- contribute to the management, conservation and sustainable development of the forests,
- provide for the multiple and complementary functions and uses of the forests,
- examination of forestry issues and opportunities taking into consideration sustainable forest management.

Principles:
States have
- the sovereign right to utilize, manage and develop their own resources pursuant to their own environmental policies,
- the responsibility to ensure that no damage is caused to the environment of other States.

Elements:
- development and strengthening of institutions and programmes for the management, conservation and sustainable development of forests and forest lands,
- recognition of the vital role of forests in maintaining the ecological balance, as storehouses of biological resources and sources of genetic material,
- promotion of a supportive international economic climate conducive to sustained and environmentally sound development of forests,
- taking action towards reforestation, afforestation and forest conservation,
- support by the international community to strengthen the efforts of developing countries to develop their forest resources, taking into account the importance of redressing external indebtedness,
- provision of new and additional financial resources to the developing countries,
- promotion and facilitation of access to environmentally sound technologies and corresponding know-how,
- support of international cooperation,
- support of trade in forest products based on nondiscriminatory and multilaterally agreed rules and procedures as well as encouragement of reduction of tariff barriers and impediments to trade,
- avoidance of unilateral restrictions to international trade in timber or other forest products,
- control of pollutants, particularly airborne pollutants.

---

of forest use are not clearly defined, with the exception of the countries strongly oriented to the export of timber. Finland and Canada, for instance, demand a Forest Convention. Most other countries are cautious about making concrete statements on the issue of forests. The German federal government has made it clear that it is committed to protection of the forests, but has not yet indicated what form it should take.

The NGOs, too, have yet to agree upon a uniform position internationally. Some fundamentally reject an international arrangement since the effectiveness of such global agreements is generally called into question. Others explicitly favor an arrangement within the framework of a Forest Protocol (Greenpeace, 1994). Even the organizations representing indigenous ethnic groups have not formed a definitive opinion in this question.

In 1994 the first Conference of the Parties to the Convention on Biological Diversity declared to the CSD that it would be willing to make a contribution to the protection of the forests (UNEP, 1994). However, the issue is not explicitly mentioned in the Convention's first medium-term work program (1995–97).

Parallel to these institutional developments, there are a number of intergovernmental initiatives concerning forests. The focus of these initiatives is on the definition and development of criteria and indicators in connection with "sustainable management". The CSD treated the issue of forests in connection with Chapters 11 and 15 of AGENDA 21 at its third meeting in April 1995. An *Intergovernmental Panel on Forests* (IPF) was set up with the aim of promoting multidisciplinary action in conformity with the Forest Declaration at the international level. In addition to assessing the existing institutions and instruments, the IPF will also examine the question as to the necessity of new, binding agreements on protection of the forests according to international law.

6.3
Forest Protocol Within the Convention on Biological Diversity

The definition of biological diversity is very comprehensive and includes the diversity of ecosystems (Art. 2). The various ecotypes of forests are, of course, also covered by this definition, from boreal forests and those in temperate zones to tropical and evergreen mountain rainforests. Thus the convention is basically applicable to forests, which means that the Parties have an obligation according to international law to apply the standards of the Convention to the forests on their territory.

All types of forests play a major role in the conservation of biodiversity (Myers, 1992; Dudley, 1992; WRI, 1994) and are thus key regulatory elements of the Convention. In order to implement the Convention appropriately, the forests must be included.

A Forest Protocol would have to take into consideration the objectives of the Biodiversity Convention which encompass both the conservation and sustainable use of biodiversity. First of all, the negotiations regarding a corresponding protocol would give the concept of "use" a comprehensive meaning that would not only consist of commercial use of timber but also aspects such as land use and land rights issues, nontimber products and subsistence use. Second, "conservation" would be included on an equal basis. The Convention places great emphasis on the link between conservation and sustainable use. Therefore, a one-sided narrowing of the issue to include only one of the two aspects would not be possible within the framework of a Forest Protocol. Regulation of both aspects within one framework could contribute to a resolution or avoidance of conflicts between conservationists and forest users seeking to obtain appropriate revenues for their products. Conversely, a regulation favoring a certain type of use, as must be feared with a new Forest Convention, would undermine the objectives of the Biodiversity Convention. Separation of the two aspects into different legal instruments additionally holds the danger of conflicts over objectives. At present there is no forum that is required to provide for a binding treatment and resolution of conflicts between two conventions.

If the Biodiversity Convention is compared to Chapter 11 of AGENDA 21, one can see that the four program sections in Chapter 11 that contain a detailed description of conservation and sustainable use of forests, including use of timber, are contractually specified in Art. 6 to 10 of the Biodiversity Convention. This means that the political implementation of AGENDA 21 demanded by the CSD has already

been stipulated on a legally binding basis by the Biodiversity Convention.

The Convention – like the Convention on Climate Change – is a framework convention. It expressly provides for protocols as implementation and specification mechanisms: in accordance with Art. 28(1) the Parties have to cooperate in drawing up protocols. This does not mean, however, that Parties can be forced into recognition of a binding protocol against their will. Art. 32(1) stipulates that only Parties to the Convention can become members of a protocol (Glowka et al., 1994). The mechanism of a framework agreement with protocols is intended to make a gradual regulation possible, taking into account the respective needs of the different states.

The role of indigenous peoples and local communities and their traditional knowledge is recognized in AGENDA 21 and a strengthening of this role is required. The Biodiversity Convention gives consideration to this aspect in the Preamble as well as in Art. n 8j, 10c, 7(2) and 18(4). The Convention fundamentally recognizes the dependence of traditional ways of life on biodiversity. Profits attained through indigenous knowledge must be distributed justly. States have to allow indigenous and local communities to take part in the decisions affecting them (Art. 8j) and preserve their knowledge and practices (Art. 10c). The knowledge of indigenous and local communities is to be taken into account in the exchange of technology and scientific cooperation. These provisions thus furnish a basis for a Forest Protocol, consisting of a detailed regulation protecting, securing and stipulating the rights of indigenous and local communities.

Some developing countries (particularly Malaysia and India) complain that international efforts are based on a "global interest" in forests. This was put forward during the negotiations on the Forest Declaration in Rio de Janeiro in 1992 (Johnson and Cabarle, 1993). These countries fear that a global arrangement would impair their national sovereignty concerning use of natural resources. The doubts on the part of these (and other) developing countries are directed at any type of regulation, no matter whether a Forest Protocol or a separate Forest Convention. Within the Biodiversity Convention, however, the fears regarding loss of sovereignty are not well-founded as it particularly recognizes the sovereignty of states to make use of their own biological resources according to their national environmental policy in Art. 3 while Art. 15(1) explicitly affirms the authority of states to determine access to genetic resources based on their sovereign rights. National sovereignty is not impaired by the Convention on Biological Diversity but acknowledged. As any contractual arrangement, the Biodiversity Convention, of course, also means a qualification of this sovereignty, binding it to the provisions of an agreement based on international law through its ratification. Since admission to a protocol is voluntary (Art. 32), however, the sovereignty of the states is not impaired by this in principle.

UNEP as the administrative UN body for the Biodiversity Convention would have responsibility for a Forest Protocol, whereas a Forest Convention would probably be administered by the FAO. Consideration of ecological aspects is undoubtedly more likely under UNEP auspices than under the primarily use-oriented FAO. A protocol would strengthen the role of the Biodiversity Convention and thus that of UNEP, leading in turn to greater consideration of ecological aspects internationally. The same applies to the distribution of responsibilities at the national level (ministries in charge).

A financial mechanism already exists for the Biodiversity Convention, the *Global Environmental Facility* (GEF). On the one hand, a Forest Protocol would be able to utilize the funds of the GEF and, on the other hand, the Parties would be able to develop new or additional financing mechanisms for the protocol.

6.4
Convention for the Protection of the Forests

The current initiatives on forests as well as the interest groups that demand an arrangement separate from the Biodiversity Convention primarily advocate regulation of timber use. The following, therefore, is not based on a protection-oriented convention but on one whose contents are still undefined.

All arguments in favor of a new convention are founded on the hope of improving the current situation of the forests in this way. The different interest groups naturally have different conceptions about what "improve" means. The following arguments are rather speculative regarding the possible regulations since what can be achieved in negotiating a new instrument depends on the governments . With the exception of greater openness for possible Parties, these arguments are not of a legal nature but primarily of a political and economic nature.

A waiving of national sovereignty is not expected in a new Forest Convention so that the fundamental principle of sovereignty over natural resources that has applied since the 1972 conference in Stockholm will not be changed in all likelihood.

Furthermore, there is hope that the legal position of indigenous and local groups will be strengthened through a Forest Convention: land use rights could be justly granted and a right to codetermination on the part of indigenous groups might be recognized on the basis of international law. This depends on the political will of the negotiating parties, which does not seem to be pointing in this direction at present, however. In addition, a corresponding result can be equally achieved within the framework of the Biodiversity Convention, which has already laid the foundation for improving the position of these groups (Preamble, Art. 8 j and 10 c).

A special advantage of a newly negotiated Forest Convention would be that all states could become members without any requirements, while according to Art. 32(1) ratification of a Forest Protocol would be tied to adoption of the Convention. It must be noted here that, given the very broad participation in the Convention (more than 170 signatory states, more than 110 Parties), this circumstance would not represent a major obstacle in the negotiations, especially since the countries having great interest in a Forest Convention (timber producers such as Malaysia, Finland, India, Canada and Sweden in addition to consumers such as Japan, Germany and the USA) have signed and (except for the USA) also ratified the Biodiversity Convention. Moreover, an equal right to be consulted on the part of the countries that are not Parties to the Convention and negotiation of an "attractive" Forest Protocol might be an incentive for the still hesitant states to sign the Biodiversity Convention.

Another argument put forward is that new negotiations for a Forest Convention may bring more interest groups to the negotiating table than through the Biodiversity Convention. This does not make sense, however, since in accordance with Art. 23(5) any interested party can take part in negotiations within the framework of the Biodiversity Convention anyway. Furthermore, every country is entitled to include interest groups in their delegation, as is customary practice.

6.5
Conclusions and Recommendations for Action

FOREST PROTOCOL / FOREST CONVENTION
The dramatic pace of forest destruction requires rapid, immediately effective action. Since both the objectives and basic conditions of the Biodiversity Convention have been contractually stipulated and the major institutions for implementation have been set up, a Forest Protocol would presumably take less time for negotiation than drawing up a completely new Forest Convention (Lyke and Fletcher, 1992). If the obligation to conserve the forests were not incorporated into a new Forest Convention, its environmental policy significance would be questionable. A regulation of forest use separate from the Biodiver-

sity Convention may lead to its decisive weakening and marginalization.

It is argued that the protection of biodiversity would not have to be regulated within a Forest Convention since the Parties are required to comply with the provisions of the Biodiversity Convention anyway – even with regard to other international agreements (Art. 22). However, based on the way national and international cooperation functions today, one cannot expect right from the beginning that the fundamentally binding norms of the Biodiversity Convention regarding conservation and sustainable use of biodiversity will be observed after negotiation and implementation of a Forest Convention. The Biodiversity Convention is a framework convention whose norms are broad and open to many possible forms of regulation of forest use. The argument of required compliance with the Biodiversity Convention does not appear to hold water in view of this background. Moreover, it is not very probable that conservation aspects will be established in a Forest Convention, given the great interest of many timber-exporting countries in a use-oriented Forest Convention. This means it is also unlikely that the multitudinal value of forests (forms of vegetation with great biodiversity, CO_2 sink, significance for the provision of fresh water, etc.) will be given appropriate consideration alongside timber production in a Forest Convention.

Conversely, the timber-exporting nations (e.g. Canada) argue that a new convention would prevent the various aspects of forests (climate protection function, biodiversity, use for development and the economy) from being "split up" among different protocols and instruments. The counterargument to this is that key aspects of dealing with forests, i.e. sustainable use and conservation of biodiversity, are mentioned as equal goals in the Biodiversity Convention.

FORUMS FOR INTERNATIONAL COOPERATION

The Vienna Convention for the Protection of the Ozone Layer and its protocols, the Framework Convention on Climate Change and its implementation, the recently adopted Convention to Combat Desertification and the Convention on Biological Diversity already require a great commitment on the part of the Ministries of the Environment and Development as well as departments and institutions. A Forest Convention would result in further fragmentation of the capacities available through international cooperation since more extensive cooperation than in the past cannot be expected.

It must be kept in mind that the human resources of the developing countries, in particular, are limited. Many countries are only able to send a single representative to negotiations. A Forest Convention separate from the Biodiversity Convention would increase

personnel needs even more. A Forest Protocol would also mean greater labor requirements, but they would be less in comparison to a new convention because the negotiation and administration of the protocol could be handled through the national structures established by the Biodiversity Convention.

The latter permits all interested nongovernmental organizations to take part in the Conferences of the Parties as observers (Art. 23/5). The participation of the NGOs is a result of the Rio process, which is reflected in Chapter 27 of *AGENDA 21*. Comparable participation in a new Forest Convention would first have to be negotiated. Art. 23(5) of the Biodiversity Convention guarantees the NGOs a formal status that would also be applied in a Forest Protocol.

FORESTS AS AN INTEGRAL PART OF BIODIVERSITY

The Biodiversity Convention with its broad approach and its objectives covering both conservation and sustainable use most certainly offers a suitable framework for a binding instrument of forest protection based on international law. Nevertheless, a Forest Convention might also represent a meaningful approach, but only if its objectives place equal importance on protection or conservation of forests as on use, in addition to the other major points. Thus far appropriate objectives and contents of a new Forest Convention supported by a consensus have not emerged and the duration of this process is not foreseeable. This does not necessarily mean that a Forest Protocol will be much easier to reach since in both cases the same representatives of the various interests would be involved in the negotiations. In any case, however, a Forest Protocol would be bound to the objectives and contents of the Biodiversity Convention, which has already met with great approval, to the surprise of many. In addition to the advantages of a Forest Protocol already mentioned, one should not underestimate the fact that the stability and significance of the Biodiversity Convention would be reinforced by successful conclusion of such a protocol. However, it is uncertain whether future Conferences of the Parties to the Convention will decide on a regulation of forests based on the current political situation. It must be pointed out that the issue of "forests" is an integral part of the issue of "biodiversity" and that an arrangement within the framework of the Biodiversity Convention, whose principles have to be complied with in political and administrative measures, too, appears meaningful. *For these reasons the Council recommends that the German federal government commit itself to a Forest Protocol within the framework of the Convention on Biological Diversity.*

7.1
Globalization of Economic Activity

The economies of the world are becoming increasingly integrated. This is evidenced in particular by
– globalization of markets,
– increased interdependence of trade,
– greater mobility of labor, capital and knowledge,
– internationalization of production and increasingly complex ownership patterns,
– the predominance of multinational corporations,
– the formation of large economic blocs (above all the EU, the Nafta, the Mercosur and, in future, the *Association of Pacific Economies* (APEC)), each featuring a major degree of internal coordination of economic policy.

The consequences of these developments, simply put, are
– increasing international exchange of goods and commodities, resulting in growth of transport activities,
– increasing global growth rates, mainly in eastern Asia (equilibrium effect). The term equilibrium effect generally refers to the impact on economic growth rates. This is compounded by structural effects involving changes in sectoral, regional and company size structures.

Intensification of competition with respect to internationally traded goods leads to
– growing international division of labor, with increasing sectoral specialization of countries participating in trade (structural effect),
– increasingly rapid geographical diffusion of knowledge and technologies and, in certain fields,
– a mutual convergence of preferences on the part of the population – relating to lifestyles, for example, or environmental awareness.

These develements reflect substantial global changes, and they are connected to various regional and global environmental effects (*see* WBGU, 1993 for a definition of global environmental effects). Therefore, the issue of 'Trade and Environment'

has become increasingly prevalent in public discussions.

Negative environmental impacts (environmental stresses) can be expected if, for example
– emissions and/or waste increases due to a higher level of transport operations (growth in the number of vehicle kilometers per year) and/or the equilibrium effect (boost in economic growth),
– resource consumption rises or
– industries consuming scarce environmental resources shift their operations to countries with lower environmental standards.

Potentially positive impacts (mitigation of environmental stress) can arise if
– growth effects in countries severely affected by poverty create scope for more environmental protection,
– exchange of goods leads to faster diffusion of innovative, low-emission technologies or
– a higher level of environmental awareness is generated via the transfer of knowledge associated with the exchange of goods and factors.

These phenomena are closely linked to general economic developments, such as the declining significance of transport costs as a percentage of the amount of added value, but are also an expression of political decisions, such as the integration policies of the various economic blocs. Another dimension here is that of global institutional frameworks, one of which is the General Agreement on Tariffs and Trade (GATT). This agreement
– has had a crucial influence on world trade since 1947,
– established, following adoption of the international trade agreements of the so-called Uruguay Round in Marrakech on April 15, 1995, a new phase of world trade liberalization in which environmental protection shall play a greater role in future, and which
– has attracted frequent criticism recently in connection with environmental issues.

Implementation of the new GATT agreement got underway in early 1995 following ratification by the U.S. Congress. In consequence, the Council now includes the so-called GATT regime (Helm, 1995) among the topics to be dealt with. As a first step, the policy of GATT will be subjected to critical appraisal, whereby the principal focus concerns the interests and positions of the respective contracting parties. This will be followed by a description of the Marrakech decisions, and after that GATT will be assessed in terms of broader environmental issues. (The term GATT is used in the following sections, despite the fact that this was formally superseded by the WTO on January 1, 1995.) A transition can then be made to selected recommendations for action and research.

7.2
The GATT Regime

7.2.1
Brief Outline

GATT is a multilateral trade agreement, formally speaking, although de facto it functions as an international organization (Senti, 1994). Its supreme organ is the Session of Contracting Parties consisting of representatives of each GATT signatory, which may pass resolutions by majority vote. Until the end of 1994, the GATT Secretariat was an administrative body without political powers, located in Geneva, that monitored compliance with the GATT provisions. Changes to the rules may be made at periodic bi- or multilateral "rounds" of negotiations between the representatives of the member states. There is also a special procedure for resolving disputes arising from trade restrictions by member states and/or for dealing with issues involving interpretation of the agreement's provisions, in which ad hoc committees of independent experts (so-called GATT panels) issue recommendations that become binding on member states once they are accepted by the Session of Contracting Parties. All of these agreements, institutions and procedures are commonly referred to as the GATT regime. The so-called Uruguay Round (the 8th and final multilateral round of negotiations) was opened in 1986 under this GATT regime, and finally came to an end in Marrakech in April 1994 (date of adoption) after years of tough bargaining. Following ratification by the signatories, it entered force on January 1, 1995, thus opening a new phase of GATT policymaking that will be examined more closely in a later section.

To this day, GATT rests on the belief that free trade, and the international division of labor and/or specialization that ensues, serves to increase the welfare and prosperity of populations worldwide. The

principles for achieving this liberalization of trade are those laid down in the Agreement:
- that each signatory apply its trade regulations in the same manner to all other signatories ("*most-favored nation*" *requirement*),
- all "like" products must be granted the same treatment as domestic products (*nondiscrimination or* "*national treatment*" *requirement*) and
- reciprocal application of all GATT rules among the contracting parties (*reciprocity*).

In laws and treaties it is not just the key articles that are important, but especially the various exception clauses. This also applies to GATT. In customs unions and free trade zones, as well as towards developing countries, exceptions may be made to the principle of most favored nation status. Article XI provides for exceptions to the ban on quota restrictions, and Article XX opens up the option for trade restrictions in order to achieve political goals of higher priority. Under Article XI, export bans may also be imposed on important goods and thus also on certain environmental assets or resources (such as animals). In addition, Article XX allows for trade restrictions "necessary to protect human, animal or plant life or health" or relating to "the conservation of exhaustible natural resources", i.e. in pursuit of environmental aims. The general stipulation of GATT is that these restrictions may not give rise to arbitrary discrimination between countries or to hidden trade barriers.

When evaluating GATT from the environmental perspective, one has to take into consideration that the agreement was originally intended – in 1947, after plans to establish an *International Trade Organization* had failed – as a provisional international treaty for liberalizing world trade. It was aimed at making the free trade paradigm a central element of world trade and at the same time at counteracting segmentation of the world economy, as had occurred to the detriment of all during the interwar period. No consideration at all was given to environmental protection, since neither at national nor international level was this a political issue. The fact that the term environment does not appear in GATT is therefore a reflection of the particular concerns of that time, namely reconstruction of an efficient world economy and the desire to achieve this by trade liberalization.

In general, however, the GATT agreement was and continues to be open for further amendments and additions. There has always been scope for supplementary agreements or new interpretations of particular Articles (e.g. by the GATT panels), as well as the permissibility of trade measures in certain exceptional cases. What is also interesting in this connection is that the GATT Secretariat presented a study as early as 1971 on "Industrial Pollution Con-

1947 formal adoption of GATT (General Agreement on Tariffs and Trade), October 30, 1947

1971 Trade and Environment Working Committee founded

1971 "Industrial Pollution Control and International Trade" study

1990 GATT panel decision on Thailand's import barriers on foreign cigarettes

1991 First GATT panel decision on the tuna-dolphin case (*Box 39*)

1991 Reactivation of the GATT Group on Environmental Measures and International Trade. The tuna-dolphin conflict marks the beginning of public debate on environmental acceptability of trade.

1992 GATT states its basic principles on the subject of trade and the environment in its "Trade and Environment Report", arguing that trade and environment are essentially compatible, in that free trade enables the growth rates necessary to finance environmental protection.

1993 Completion of the Uruguay Round. En-

vironmentally relevant changes are primarily those in the Agreement on Agriculture, according to which payments made in the context of environmental programs are explicitly permitted, as are environmentally specific exceptions in the TRIPS agreement (*Trade Related Intellectual Property Rights*), which stipulates that inventions can be excluded from patenting if serious damage to the environment might be the result.

1994 Second GATT panel decision on the tuna-dolphin case (*Box 39*)

4/1994 Ministerial Decision on Trade and Environment adopted in Marrakesh: environmental aspects of trade are anchored in the new WTO, especially through the definition in the Preamble of sustainable development as an objective, through the establishment of the Committee on Trade and Environment and the admission of environmental experts in conflict resolution procedures.

6/1994 First public hearing of NGOs at GATT headquarters, further consultations between the WTO Environment and Trade Committee and NGOs are planned.

---

trol and International Trade" (*see Box 8 and 39 on environmental activities under the GATT regime*). The paucity of environmental references in GATT was due primarily to the specific structure of interests on the part of the member states.

7.2.2
Interest Structures Within GATT

INTEREST TYPES
Before the text of the General Agreement, the panel rulings to date, or indeed the institutional structures within GATT can be properly understood, it is essential to consider the different interests of the member states. While trade liberalization probably improves a country's overall welfare, it is also true that, with sectoral and regional variations, there are not only winners but also losers in this process – with the latter demanding protective measures from their respective governments. In the past, individual states made repeated efforts to gain competitive advantage by means of protective measures (trade restrictions), or to coerce other states to react in a certain way by

applying trade policies. In other words, despite the free trade paradigm, GATT has had to deal with protectionist actions throughout its history.

To what extent trade restrictions can be enforced depends largely on the international positions of the individual states concerned. States with a high volume of foreign trade have displayed the greatest capacity for imposing such restrictions, with the result that important producer groups in these countries have repeatedly demanded new protectionist measures from their governments. In states with little foreign trade, interest in protectionist measures on the part of producers has usually been less, since one could assume from the outset that protectionist regulations would be difficult to enforce or would trigger countermeasures by important trade partners.

A basic and simplified distinction can be made between the following groups of states as regards these positions of power and interest:

– the *world gravitation centers*, i.e. those states or combinations of states with relatively high and highly diversified foreign trade volumes,
– those states with a major interest in foreign trade on account of initial natural or economic condi-

tions (e.g. the members of the so-called *"Cairns" Group* during the Uruguay Round) (this group, which derives its name from the Australian city where the first session of the member states was held, was formed during the Uruguay Round, included the principal exporters of agricultural produce and opposed protectionist measures aimed at influencing these commodity flows. Its members are Argentina, Australia, Brazil, Canada, Chile, Columbia, Hungary, Indonesia, Malaysia, New Zealand, the Philippines, Thailand and Uruguay),

- states with a powerful interest, on account of their high economic growth potential, in the intensified integration of world trade (especially the *growth economies in Asia*),
- states which function primarily as *suppliers of basic commodities* (above all the OPEC nations and the states in the former Soviet Union),
- states with a priority interest in inexpensive imports, on account of their low economic growth potential in the short term (e.g. *Black Africa*).

This is a simplified classification to the extent that allocation of a particular country to either of the two categories *"protectionism"* or *"free trade"* will depend on the sector or sectors concerned, and need not apply across all sectors. The USA, as a hegemonial power, has promoted and secured a liberal world trade, but this has not prevented it from pursuing protectionist policies in agriculture and textiles, for example.

However, this rough classification shows the diversity of interests brought together at the GATT negotiations and the respective "clout" of their representatives. The dominating position of the world gravitation centers within the world trade system renders it virtually impossible to make decisions against their interests, since they can threaten to exclude others from their markets. For representatives of other groups of states, the realization of such threats would always mean economic drawbacks that could not be offset by any unilateral measures. This explains why mainly the world gravitation centers were able to impose and enforce numerous barriers to free world trade (Grossmann et al., 1994). Their intervention has resulted in the exclusion of entire sectors from the GATT regime (agriculture, textiles, aircraft construction), and the imprecise operationalization of the antidumping and antisubsidy measures offered scope for actions fired by protectionist ambitions (Mavroidis, 1993). This also explains the numerous exceptions for regional associations and hence the continued formation of economic trading blocs (Borrmann and Koopmann, 1994). The latter took advantage of available options, above all for preferential agreements with individual states on the lifting of trade restrictions, for example the preferential treatment afforded by the EU to the so-called ACP states, i.e. the previous colonies of the EU member states in Africa, the Caribbean and the Pacific (as regards preferential treatment regulations, *see* DIW, 1994).

The concentration of GATT efforts on the elimination of tariffs also led to a rise in nontariff trade barriers (nontariff trade barriers are measures that cannot be construed as customs duties, e.g. quality standards for products and processes, import quotas or restricted access to official approval procedures for the commencement of production and marketing activities. Whereas in the case of customs duties the barrier to trade is effected through a direct increase in the cost of the goods, nontarifficated restrictions do not exert a direct influence on price). Typical of these are "voluntary" self-restrictions by exporters, by means of which the world gravitation centers especially have been able to exert influence on other states.

This corresponded with the way in which the implementation of decisions within GATT itself was organized. A panel comprised of independent trade experts adjudicated on violations of GATT rules. However, the recommendations of such panels did not obtain binding force unless all contracting parties gave their unanimous approval, thus permitting the powerful trading nations to block any trade restrictions not in their favor. Because the possible sanctioning of violations must be carried out in the respective sector by the states bringing the action, disciplinary effects can only be brought about by the world gravitation centers, since a high market potential is required for trade restrictions to have any noticeable impact on the volume of foreign trade of the states against which action is taken. Despite the free trade postulate, the GATT regime has seen a fragmentation of trade flows and numerous regulations that bolster the dominant position of the world gravitation centers – developments that GATT was originally supposed to prevent.

In contrast, the countries with low per capita income exploit their exemption from the principle of reciprocity by not reducing barriers to trade in their trade laws. This lack of willingness to make concessions, on the one hand, and the strong negotiating position of the world gravitation centers, on the other, has led in the past to multilateral negotiations achieving little genuine progress beyond the reduction of certain tariffs.

IMPLICATIONS FOR ENVIRONMENTAL POLICY

The interests outlined above reveal clearly why environmental concerns have played a minor role in GATT so far. As already shown, protection of the environment may not have been a socially or politically relevant topic when the Agreement was originally drawn up (adoption was on October 30, 1947) but the Agreement itself was always open for relevant amendments, additions and revised interpretations. Furthermore, the development of GATT to date has been characterized by tough bargaining for further liberalization of world trade alongside permanent defensive measures against new forms of protectionism. A virtual blockade against the liberalization of world trade was engendered above all through the formation of economic blocs, which were able to negotiate special arrangements for specific areas of the economy or selective agreements on account of the power they wielded. Reference was sometimes made to Articles XX(b) or XX(g) in this connection, which permit import restrictions "necessary to protect human, animal or plant life or health" or "relating to the conservation of exhaustible natural resources". As actual practice shows, these exceptions can indeed be applied to environmental protection objectives. However, the vaguely defined scope they offer must be utilized and well-founded. That there have been few initiatives in this connection is not attributable to GATT alone, but above all to the reticence of the major economic blocs in the field of environmental policymaking.

Another aspect is that GATT has treated these few environment-related initiatives in a very restrictive fashion until now, and that serious interpretational problems have arisen in this context. The restrictive attitude of GATT was partially an expression of the risk, which cannot be excluded, that these exceptions might be misused in order to pursue a covert form of protectionism. One example concerned the import ban on cigarettes imposed by Thailand (*Box 39*). In the latter case, problems arose with the interpretation of the requirement of Article XX(b) – measures which are "necessary to protect human, animal or plant life or health" are permitted – and Article XX(g), which stipulates that measures must relate to "the conservation of exhaustible natural resources" (Helm, 1995). The "necessary" and "relating to" conditions cause substantial interpretational difficulties. This is compounded by the fact that important principles in the field of environmental policy, in particular the "polluter-pays principle", have not yet been accepted by GATT as valid arguments.

Another prevalent example of restrictive GATT decisions was the "Tuna-Dolphin Case" (*Box 39*), in which the import ban imposed by the USA on Mexican tuna fish was rejected by a GATT panel. The arbitration verdict of the panel has not yet been accepted by the contracting parties and therefore has no value as a GATT precedent (as at April 1995). The

---

BOX 39

Important GATT Panel Rulings of Relevance to the Environment

1990 – USA vs. Thailand: import ban on American cigarettes

With the exception of a small number of licensed brands, Thailand imposed import restrictions on foreign cigarettes until 1990, justifying these barriers with Article XX(b) GATT, according to which measures "necessary to protect human (...) life or health" are permitted. Foreign cigarettes, the argument ran, contained more harmful additives. However, a GATT panel ruled that the ban was not "necessary" as required by Article XX, in that there were less GATT-inconsistent means of achieving the stated goal of protecting the health (OTA, 1992), and was therefore inconsistent with GATT.

1991/94 – Mexico vs. USA: the tuna-dolphin case

One of the most prevalent conflicts in the trade/environment field is the famous tuna-dolphin case. In February 1991, a GATT panel was called to decide on a U.S. trade embargo on Mexican tuna fish. The USA justified its embargo by claiming that Mexican tuna fishermen were using methods that caused excessive dolphin kill ratios as defined by the *U.S. Marine Mammal Protection Act* (MMPA). This law prohibits the import of fish or fish products from countries that do not apply regulations similar to that in the USA for protecting marine mammals. The panel found that the import ban operated by the USA was an unacceptable violation of GATT principles. In a second panel decision on the same issue in 1994, the panel again found the embargo to be inconsistent with GATT, but based this on an important qualitative change to the effect that it is possible under GATT to strive for the protection of the environment and natural resources extraterritorially by means of trade measures.

case is interesting above all because the issue in question does not involve the assessment of products or their quality (including environmental impacts), but production methods (or fishing methods in this case), thus raising the fundamental question as to what extent certain nations may "impose" production processes on others via import bans. Considerable reservations have been expressed within GATT regarding the use of trade policy as an instrument for extraterritorial environmental or social policy.

The situation, to summarize, is that restrictions and open issues continue to exist in the trade/environment field. In essence, the only environmental concerns which have been respected so far have been those relating to the international exchange of goods or which have a bearing on global environmental assets. Another issue now raised is the extent to which multilateral environmental agreements (MEAs) should imply obligations for the GATT regime, or to which important environmental principles – the "polluter-pays principle", for example – should apply within GATT as well. The definition of dumping, normally interpreted to mean non-cost-adequate price differentials between domestic and foreign products aimed at dominating foreign markets, also needs reviewing. Clarification is needed as to whether the term is applicable to cases in which certain countries deliberately opt for low environmental standards in order to acquire international competitiveness (eco-dumping). A central issue here is the extent to which environmentally harmful production processes can be made the object of protective measures.

7.3
The Marrakech Decisions

The decisions taken at Marrakech in 1994 are designed to bring about important structural changes to the GATT regime. In brief, the following changes were made:
- Protection and preservation of the environment and the principle of sustainable development are explicitly included in the Preamble as important objectives.
- The newly established World Trade Organization (WTO) represents an important organization for implementing and monitoring the GATT rules.
- Further reductions in tariffs (e.g. by about a third for industrial products) were agreed upon, and improvements were made to existing rules for free world trade.
- The GATT rules shall also apply in future to the service industry.
- Intellectual property rights have been extended to cover trade with copyright, patents and further industrial property rights.
- Direct investments abroad shall be facilitated.
- Product-related subsidies in the agricultural sector shall be cut by one fifth, market access to agricultural markets will be facilitated, and subsidies for agricultural exports are to be cut.
- Existing import restrictions in the textile field are to be dismantled.
- Liberalization of government tendering practice will be extended.
- Dispute settlement procedures have been reformed.

The inclusion of environmental protection as an explicit goal in the Preamble to the Agreement came at a late stage and was a contentious issue. The USA, in particular, demanded the inclusion of social and environmental standards shortly before final adoption in Marrakech, causing concern among developing countries especially where experience with NAFTA (so-called eco-protectionism) was still fresh in their minds. In the Council's opinion the integration of environmental aspects into the GATT regime has yet to be accomplished. Consideration must be given to the fact that trade policies geared to environmental protection will only lead to significant change if major flows of goods are affected.

The task of the World Trade Organization since the beginning of 1995 has been to monitor compliance with GATT and, in place of the previous negotiation rounds, to organize its further development. This enhancement of the WTO's powers vis-à-vis the GATT Secretariat therefore makes the WTO the third pillar in the international trade system alongside the *International Monetary Fund* (IMF) and the World Bank. It must also dedicate its efforts to the objective of environmental protection as laid down in the Preamble. A Committee on Trade and Environment makes recommendations to the WTO on how to achieve these objectives. Permissible trade barriers may not include measures that involve "arbitrary or unjustifiable discrimination" of individual states. There are no criteria as yet for defining the terms "arbitrary" or "unjustifiable", or at least the views of member states diverge considerably on this point.

The world gravitation centers are increasing their efforts to standardize environmental norms with respect to goods and services, as well as to production processes. Arguments in support of such efforts are the preservation of competitiveness (prevention of "eco-dumping") or the application of globally necessary protective regulations to all states (Young, 1994). Countries with low per capita incomes, on the other hand, refer to their rights to protect their locational advantages and stress the primary responsibility of the world gravitation centers for global environmental problems (Mehta, 1994). In order to arrive at effective agreements, coordination is needed between the member states on the criteria for determining the utilization of global environmental functions, as well as agreement on the instruments that are then applied. In view of the complicated justification for antidumping measures and the scope for interpretation that results as a consequence, it is unlikely that all GATT member states will accept the need to take action in the environmental policy field. The consequences are either to dispense with permitting trade barriers in favor of environmental protection, or to interpret permissibility options in a protectionist manner (especially on the part of the world gravitation centers).

The procedures for arbitration or decision making may be of even greater importance for environmental policy. Amendments to the decision-making process are aimed at curtailing the dominant influence of the world gravitation centers (Großmann et al., 1994). In particular, this involves an explicit prioritization of countries with low per capita incomes within the dispute settlement procedure and hence a departure from the principle of unanimity. If violations against GATT principles are objected to, negotiations between the parties in the conflict, possibly with the involvement of a mediator, shall be preferred to other instruments. A settlement procedure before a court of arbitration cannot then be rejected by affected states, and decisions made can only be changed by a unanimous decision of the GATT contracting parties or by a judicial appeal body called in to review the legal correctness of the decision. If claims to compensation are established in such a procedure and the state against which the action has been brought is not prepared to settle such claims, then measures and/or sanctions can be taken by WTO members across all sectors, also in connection with other treaties. Experience shows, however, that sanctions do not really take effect until at least one world gravitation center implements such measures. A reduction in the influence of individual members also serves the objective achieving of greater transparency of decision-making processes through consultations with independent experts, especially environmental experts and representatives of nongovernmental organizations (whereby no statements have been made as yet on the legitimation of particular groups or individuals).

Another important aspect for environmental policy is the declared will to include previously excluded areas (especially agriculture and textiles) in the general reduction of tariffs. However, the levels of tariff reductions and special regulations that exist for particular states put limitations on the positive impacts, especially for countries with low per capita incomes.

There is even the possibility that net importers of agricultural produce among the developing countries will suffer a worsening of welfare as a consequence of reduced agricultural subsidies and agricultural surpluses (Nguyen et al., 1993). However, since the general level of tariffs is also declining in the industrial field, there is a chance of more diversified integration of developing countries into the international division of labor. This may lead in such countries, on account of the structural effects, to long-term changes in the utilization of environmental resources – changes, however, which are difficult to forecast.

The continued existence of exceptions for regional associations (customs union, free trade areas) might lead, however, to any further measures to dismantle trade barriers on the part of world gravitation centers vis-à-vis states with low per capita income being made dependent on compliance with political or social criteria (e.g. under the EU's General System of Preferences). Because individual groups of states can pursue normative-political goals in relations with particular states, the issue is raised in the global context as to the legitimation for such a transboundary policy. This raises, in turn, the question regarding the extent to which one country's political will can be "forced" upon other countries through trade policies.

In addition to the inclusion of services (with provisions for many exemptions), the new rules governing the protection of intellectual property create greater incentives for international undertakings to transfer their know-how. The same applies to the facilitation of direct investments. This will likely benefit companies in the world gravitation centers (Oppermann and Baumann, 1993), whereas in countries with low per capita income problems with access to essential commodities (e.g. seed) can be expected, obstructing an efficient approach to the use of environmental resources.

In conclusion, the agreements adopted at Marrakech have opened up opportunities for change in the direction of more environmental protection. Global environmental assets had been excluded from the GATT regime before, and globally relevant externalities could only be internalized on condition that they met the restrictive requirements of the relevant clauses governing exceptions or protective measures. Change may therefore come about in this area. However, to what extent the new dispute settlement procedures will actually operate in favor of more environmental protection remains to be seen. The objective of creating free trade relations, the classical GATT goal, was a priority concern in Marrakech as well. A boost to growth is expected that will generate production increases worldwide in the order of US$ 750 billion annually (GATT Secretariat, 1993; for estimates on the consequences for welfare *see* Goldin et al., 1993). This growth will also have environmental impacts. However, there is some contention as to the regional distribution of these productivity and environmental impacts, one factor being the differences in protectionistic opportunities of different states. To that extent, the institutional changes achieved through WTO disclose the possibility for the integration of other issues, and hence that of environmental protection, into the world trade regime. Much will depend, however, on the actual behavior of the major economic trading blocs and industrial nations, especially the structural changes in production and consumption brought about by trade liberalization.

7.4
Free Trade and the Environment

7.4.1
Free Trade Classical Arguments

A summary analysis of GATT with particular reference to environmental policy raises two fundamental issues:
- Is free trade, the prime objective of GATT, compatible with the requirements of environmental protection?
- To what extent are there options for improved integration of environmental protection within GATT?

The following remarks focus on the first question, the second being dealt with in a later section.

As far as the first issue is concerned, the positions range from the assumption of a harmony of goals between free markets and improved environmental protection (supply of environmental functions) (GATT, 1992) to the assumption of irreconcilable conflict (Daly and Goodland, 1994). Following the classical argument, trade activities – from the extraction of commodities through production, processing, transport, consumption to waste disposal – are desirable for society because they lead to macroeconomical efficiency if their benefits exceed the total costs, including positive and negative impacts on the environment. The main problem here is how to operationalize the evaluation criteria (Klemmer et al., 1993). Especially in a global context it is impossible to fully identify, quantify and monetarize all socially relevant costs and use components of trade activities – only rough approximations are feasible. Nor is it possible to determine the desirable volume, extent and the structure of world trade. For this reason, it is essential to investigate the extent to which the relevant participants can be induced to change their be-

havior in the direction of greater efficiency by establishing certain institutional conditions that also take environmental factors into account. The following *criteria* could be of particular significance for such an institutional framework:
- There is an extensive and equal integration and processing of available information on the value of traded goods, rights and services, as well as the environmental functions involved (although one must realize that these preferences can only be formed and articulated at the immediate individual level).
- Incentives are provided to the trading parties to develop improved processes for raw material extraction, production, processing, transport, consumption, waste disposal and in the performance of trade activities in order to achieve greater efficiency in the utilization of resources.
- The transaction costs (for introducing institutional procedures, identifying the relevant participants and required information, ensuring coordination between participants as well as follow-up checks and possible sanctions in response to noncompliance with agreements) must not exceed the benefits of the procedure in question (including welfare effects due to intensified international division of labor, reduction of environmental stresses resulting from poverty and/or prosperity, more powerful incentives for developing and transferring know-how).

The expansion of international trade would then lead to the possibility of internalizing in prices additional information about the scarcity of resources, and hence also the environment in other states. This would result, as a consequence of the different conditions, in divergent potentials for producing goods or services, which in turn would be expressed in different cost structures worldwide. The adjustment of the supply of goods produced in individual countries with different cost structures to existing global demand through world markets would enable the structure of supply to be based on the comparative cost advantages of the respective economies within the international division of labor. This would also mean certain states having locational advantages, as a result of their natural conditions, for those economic activities which involve especially intensive use of environmental functions. These locational advantages would then be reflected in lower prices for products from such states on the world markets.

This is the essential conclusion drawn by the advocates of international free trade and thus the basic goal of GATT: free trade is supposed to lead, via improvements in the exchange of goods and division of labor (specialization), to a global increase in welfare which, as a consequence of the intensified competition thus triggered, accelerates the generation and geographical distribution of technical know-how (growth and development effect). The crucial question as regards the impact of trade on global environmental assets is whether or not all relevant information is really expressed in the prices of the world market, and whether competition is assured. The next section therefore addresses these issues in more detail.

7.4.2
Global Public Goods and Externalities

Transaction costs can block the development of markets and prevent the price mechanism from performing its coordination function. This applies above all, within the global change context, to the global commons and the occurrence of severe external effects (Wießner, 1991; Altmann, 1992; Anderson, 1992). Defining and allocating rights of action and disposal in relation to public goods generally involves prohibitively high transaction costs. When allocating the right to take action relating to environmental functions, it is not possible, within certain geographical limitations, to exclude individuals from utilizing environmental functions (*free rider dilemma*). In order to prevent individuals or states from consuming environmental functions without giving due consideration to the scarcity of such functions (the problem of overexploitation), it is conceivable and makes sense to allocate action and disposal rights to collective bodies. The latter may be national or international collectives. In the case of environmental functions that, as a result of *global* ecosystemic integration, noone in the world can be excluded from utilizing (such as climate stabilization through the atmospheric carbon cycle), it is therefore necessary that the allocation of utilization rights be coordinated within a global collective. The Framework Convention on Climate Change and the decisions taken at the Conference of the Parties are such collective decisions. Without such coordination, the scarcity of global commons will not be accounted for in the distribution of goods, rights and services via markets – prices would therefore remain too low.

In addition to the costs for defining appropriate action and disposal rights, costs for negotiations between all relevant participants must be taken into account when adapting the world trade system to aspects of global change. Trade transactions may trigger off the impairment of environmental functions, the consequences of which might involve material, immaterial or health injury to many individuals in different countries of the world – for example, transporting a consignment of fruit by air from the south-

ern to the northern hemisphere, with concomitant emissions of pollutants. (If one attempted – pursuing the example a little further – a negotiation between the injured parties with those participating in the fruit trade by including such impacts in the price for the fruit, then the costs for identifying the parties, analyzing the specific cause-effect relations, quantifying and monetarizing the relevant impacts, as well as negotiating, settling and enforcing agreements would exceed the respective benefit for the individuals concerned, to the extent that the transaction would not be made in the first place.)

The Council takes the view, on the basis of this logic, that there is a need for institutional action aimed at internalizing transboundary or global externalities, whereby the costs for taking the required action must also be taken into account. If external effects are fully or partially internalized, markets could perform their coordination function on a global scale as well. In this case, one could continue to assume that free trade improves welfare.

7.4.3
Competition Failure

If the function of markets is the fullest possible processing of relevant information, organizing the coordination of this process through competition enables suppliers of products to be selected on the basis of efficiency. Only those suppliers who are able to meet the needs of consumers at the lowest cost will prevail and make profit. Anticipation of profits induces permanent efforts on the part of suppliers to improve the efficiency of their products. In the global context, incentives to develop and apply more efficient processes are generated through the transfer of commodities, investments and knowledge, also in countries which do not have the research and development capacities to solve existing problems. To evaluate the functional capacity of competition within global trade, it is therefore necessary to investigate whether all relevant determinants of efficiency are taken into account in the selection process.

There are growing claims in industrialized countries that competition is being distorted by the different extents to which impacts on environmental functions are included in individual states (U.S. Office of Technology Assessment, 1992). With respect to the issue of so-called eco-dumping , however, a distinction must be made between

- impacts on those environmental conditions and resources that have to be taken account of within global trade and
- locational advantages enjoyed by individual states as a result of different environmental conditions

and heterogeneous preferences regarding the handling of environmental resources within the states.

Reference has already been made to the necessity of regulating the *global commons* and the *internalization of global externalities*. If this need for global institutional action is not complied with and the scarcity of resources is not fully reflected in prices, competitive advantages ensue for those suppliers whose utilization of global environmental functions is not compensated for through the price mechanism. This leads to distorted competition or even the failure of competition.

Where environmental impacts are not global or transboundary in nature, however, there is not necessarily a need for an international standardization. If, in such cases, individual countries attach low preference to the protection of their own environmental assets because the economic situation demands that greater priority be given to supporting economic growth, for example, then from the viewpoint of competition policy there are essentially no objections to lower national environmental standards, since the costs this involves in the form of a lower supply of environmental functions are borne by the country that makes use of its locational advantages. Global harmonization of standards with respect to these nontransboundary externalities would imply a disregard for the individual preferences of countries with low per capita income.

The rationale for global interventions aimed at ensuring competitiveness is often based on the growing importance of globally operating companies (*see* UNCTAD, 1994 on the dependence of approx. 150 million workers worldwide on multinational employers). The market power of individual companies arising from inadequate global coordination of competition law can indeed lead to a situation where those companies prevail whose success is attributable less to an efficient structure than to opportunities they have for restricting freedom of trade for other market participants – e.g. through contractual obligations, strategies for combating potential and true competitors – (Fikentscher and Heinemann, 1994). The result for global environmental effects is, on the one hand, that the selection of suppliers does not depend on how efficiently they handle resources; on the other hand, there are fewer opportunities for the further development of processes that improve the supply of global environmental functions, as a consequence of less variety on the markets and the difficulties in establishing innovations. However, implementation of a global competitive framework in which environmental factors are also taken into account is faced with enormous problems due to the different conditions in the individual states.

7.4.4
Normative Interventions

In addition to the factual conditions already mentioned, which led to market and competition processes not fulfilling their functions with respect to the inclusion of global change aspects within international trade, institutional interventions can be grounded on the fact that coordinating claims to global environmental functions on the basis of efficiency violates certain normative goals in respect of which a basic consensus, based on common values, exists worldwide. This consensus relates in particular to distribution and safety issues.

A "fair" or equitable share of world markets for all participants is a demand that is often advanced in connection with distribution. Reference is made to the prices for raw materials in trade between nations with high and nations with low per capita incomes, where the latter as chief producers of such commodities tend to suffer from unfavorable terms of trade (*see* Simmons, 1993 on the problem of a global definition of equity). However, the demand for a "fair" or equitable share is opposed by the fact that there is no common set of values worldwide on which to base the definition of what constitutes "fairness" in this context, hence there can be no normative justification for global price structure interventions aimed at more equitable distribution.

As far as safety is concerned, the issue arises as to whether world markets give adequate consideration to risks against which individuals, by international consensus, must be protected. One example for protection against health risks would be the Convention on the Prevention of Major Industrial Accidents adopted by the *International Labor Organization* (ILO), which contains globally applicable minimum standards for worker protection. Reference is often made, in the context of global change, to the necessity of preventing irreversible environmental damage as a minimum requirement for sustainable economic activity (Dahl, 1994). The question to be clarified here, namely how to reach a globally accepted definition of the need to take intergenerational risks into consideration – especially against the background of divergent environmental and economical development potentials – means that one cannot expect the topic of safety to become a globally accepted issue, at least not at present. (Klemmer, 1994).

To summarize: there is a need for interventions in the world trade system to make sure that global environmental functions are taken into account, especially with respect to global commons and in the definition of externalities of global or transboundary importance. Such interventions are an appropriate method of ensuring the operation of both underlying market and competition functions. Other normative problems at the global level involve, most importantly, the compliance with minimum standards for employee and health protection. Institutional interventions to this end only achieve their purpose if the costs do not exceed the benefits. The costs for introducing the relevant regulations can be high in the global context because of the substantial divergence in national preferences regarding the value of environmental functions. For this reason, therefore, the extent to which global environmental functions are to be protected as global commons, the value of globally relevant externalities and the necessity for regulations governing competition will all be assessed differently from one state to the next. The mere identification of deficiencies in the allocation mechanism as regulated through competition and markets does not automatically lead to the conclusion, however, that institutional interventions should be generally preferred; what is needed instead is an analysis of the extent to which the GATT regime can be improved such that environmental factors are taken adequately into account.

7.5
Conclusions and Recommendations for Action

Our analysis of the interconnections between trade and environment has shown that the free trade paradigm in GATT can be accepted in principle, but that there is a need for action aimed at improving international coordination on matters of global environmental importance. The key issue is whether a stronger focus on efficiency on the part of the institutional framework can and must be achieved through further amendments within the GATT regime or in some other manner. A welcome step forward would be if the GATT regime could be reformed to take greater account of global environmental factors (*Box 40*). Most of the following recommendations are therefore focused in this direction.

Within the GATT regime, the disparate capacities of member states to enforce their particular interests has prevented a situation arising in which all member states possess the same options. It is therefore an open question as to whether the structural changes brought about by the establishment of the WTO, and the efforts to define what constitutes "dumping", subsidies or countervailing measures by a particular state, are leading to a reduction in the protectionist endeavors of states or regions. If the attempt is made to reach a globally accepted definition of when interventions in the world trade system to protect the environment are admissible or desirable, then the diffi-

culties mean that the final outcome is unlikely to be based on efficiency considerations; instead, there is a risk of additional potentials being created for the application of protectionist measures and for the misuse of environmental arguments in support of same.

The Council believes that there is an urgent need for agreement on the meaning of the term *eco-dumping* and on the efficient organization of the *dispute settlement procedure*. There is an overlap here between the research needed and the concrete action that is required. To that extent, the German Federal Government should provide an impetus to research in this area, in order to arrive at relevant solutions as quickly as possible.

Another fundamental problem of the current GATT regime lies in the fact that there is insufficient interaction with international environmental agreements and that some of the GATT rules are in opposition to the commitments of those agreements. This will likely cause problems not only for the Convention on Biological Diversity, but possibly also with regard to the Climate Convention. Options for sanctions should therefore be allowed against states that do not meet their commitments – especially CO_2 reduction commitments. This would act as an effective lever for enforcing decisions that have rather, up to now, the characteristics of voluntary agreements.

In order to give international environmental agreements priority over GATT rules, the contracting parties already possess the option, in certain exceptional cases, of granting waivers under Article XXV:5, of exempting individual members from certain GATT obligations, i.e. to deem trade measures enacted under selected international environmental agreements or conventions as GATT-consistent (Helm, 1995). Such a waiver requires a two-thirds majority and a quorum of least fifty percent of the contracting parties.

The Federal Government should therefore work towards a commitment on the part of all GATT signatories to examine ways of including international environmental agreements under the waiver provisions, and possibly even to reduce the consensus threshold required for such a move. In the view of the Council, the Montreal Protocol, the CITES Convention and the Basle Agreement could fall under such a waiver. Because the waiver solution must remain limited to exceptional cases, however, thought should also be given to extending Article XX(h) to embrace international environmental agreements.

In the short term, steps must also be taken to conclude agreements on global interventions to protect environmental functions. Given the difficulty in reaching agreement on the necessity of interventions, it may be advantageous to focus on specific problems for which the existence of a global dimension would

BOX 40

Elements of an Ecological Reform of the GATT/WTO Regime

Whereas the 1947 GATT agreement contained no explicit reference to the natural environment, the Preamble of the Agreement of the World Trade Organization (WTO) explicitly states that the optimal use of the world's resources should be in accordance with the objective of sustainable development and the protection of the environment. However, this general statement of aims has yet to be realized.

This report lays stress on the active use of the possibilities given via the GATT Articles III and XX which permit *product-related measures* for protecting human health and the natural environment, excluding, however, trade-restricting measures aimed at influencing environmentally problematic *production standards*. Action is needed here, and this can only be accomplished through a reform of the GATT agreement.

GATT Article XXV:5 permits the granting of a waiver, which offers the possibility of assigning priority to international environmental agreements over GATT. A more comprehensive solution should be sought for here in view of the urgent protection of the environment. NAFTA (North American Free Trade Agreement) could serve as an example here, since it determines that in case of a conflict between the NAFTA and (certain) environmental agreements the latter shall have priority.

Such measures can be justified not only in terms of environmental policy, but also within international law: the General Assembly of the United Nations, for example, declared in 1988 that the climate is a common concern of mankind – and the limitations to the traditional notion of national sovereignty were confirmed in 1992 in the Rio Declaration on Environment and Development, which says that "states have ... the responsibility to ensure that activities within their jurisdiction or control do not cause damage to the environment of other states or of areas beyond the limits of national jurisdiction" (Principle 2).

To ensure that such measures applied to international environmental problems do not result in abuse, well-defined criteria are needed, however, in order to have the capacity to examine the eligibility of environmental measures. Such criteria can only be defined in a normative sense. The following criteria should be brought into the debate on the ecological reform of the GATT/WTO regime:

1 *There has to be a justified environmental interest against those states against which trade measures are to be applied.*
   This would apply to all international environmental problems, such as ozone, climate, water and soil problems; the decisive factor would be the existence of transboundary impacts.
2 *When measures are instituted, the party concerned must itself make an appropriate contribution to solving the environmental problem in question.*
   This would avoid a situation in which states shape measures in such a way that the costs are fully or primarily borne by other states.
3 *Trade measures must relate to specific environmental goals, be reasonable in scope and non-discriminating.*
   These three basic principles are already anchored in the GATT regime and should also be applied to measures for protecting the environment.
4 *States must show willingness to accept cooperative solutions.*
   This would take into consideration the fact that cooperative solutions must be given preference over unilateral measures in the interest of a comprehensive international protection of the natural environment.
5 *Income from import charges imposed to prevent eco-dumping should be earmarked for solving those environmental problem to which the charges relate.*
   Earmarking resources in this way would reduce the incentive to misuse the system, firstly, and, secondly, additional funds for solving environmental problems would be made available.

These – or similarly worded – criteria could serve as a basis for discussion of an international ecological reform of the GATT/WTO regime, which at the same time meets the objectives of effective global environmental protection and the goals of having a world trade system that is as open as possible.

have to be examined. Such an examination should be based strictly on the principle of fiscal equivalence, i.e. aim for geographical identity of users and cost-bearers, which leads to the range of problems actually requiring global treatment being somewhat confinable. Regional associations of states should be made greater use of in coordinating the response to international environmental problems (one example being the NAFTA rules).

Only when there is agreement on the cases in which global intervention must occur at all does the need arise to translate the established necessity for intervention into concrete measures. In view of the diversity of national conditions, recourse may be made during implementation in many cases to national regulations if globally determined values for the quality of environmental functions have been laid down for individual states. The approach to implementation should also be adequate to the problem, in that certain environmental media or causal agents require the application of specific instruments (*see* Klemmer et al., 1993, on this issue). In the case of CO_2 emissions, for example, which are globally distributed throughout the atmosphere, emission permits could be traded between emitters – power stations, for example. Soil, in contrast, is an environmental medium with a wide variety of functions for humans and nature, which would therefore require a more complex instrumental structure (Werbeck and Wink, 1994). All in all, therefore, it makes less sense to take an overall view of the "international trade – environment" network of interrelationships, which would only produce generalized statements of little relevance for institutional implementation, than to carry out a more differentiated analysis of the different phases of a trade activity – extraction of raw materials, production, processing, transport, consumption, waste disposal – with their specific implications for the supply of functions of individual environmental media. By regulating specific, discrete problems, further-reaching and more comprehensive solutions may then be striven for.

In view of GATT's low potential as a means of regulating conflicts between trade and global environmental assets, the Council recommends that the Federal Government also consider the establishment of an independent organization to which the surveillance of existing environmental agreements, but also in the long term additional responsibilities for enforcing and further developing international environmental agreements, should be assigned (*see* Esty, 1994 for the basic features of such an organization). Most importantly, this institution should cooperate closely with the WTO. However, if one considers the conflicting interests among individual states regarding the assessment of action required to globally reg-

ulate the protection of the environment, it is at least questionable whether individual states would be willing to surrender part of their sovereignty to a global institution.

7.6
Recommendations for Research

- Development of criteria for ensuring the environmental conformity of GATT and the GATT-conformity of global environmental measures.
- Definition of what is meant by so-called "green protectionism" and the derivation of criteria for uncovering such practices.
- Definition of "eco-dumping" and the derivation of criteria for justified import restrictions. The core issue concerns whether immission or emission standards should be taken as the basis for defining eco-dumping criteria.
- Evaluation of the new dispute settlement procedure (game theory analysis) for global trade and environmental conflicts.
- Identification of globally relevant commons and possibilities for including them in the GATT regime.
- Scope and/or options for deploying informational instruments, such as national or international environmental auditing (analogy to Made in Germany).
- Possibilities for applying and improving "waivers" and/or GATT Article XX(h).
- Possibilities for introducing the "polluter-pays principle" and the principle of liability into the GATT regime.
- Development of a catalog of criteria for assessing the geographical structure of globally relevant environmental impacts as a result of increasing trade volume.

General Conclusions and Recommendations

This report deals with "ways towards global environmental solutions". First of all, the adoption, improvement and national implementation of international conventions as one such approach should be further pursued. Initial efforts in this direction were made by the Berlin Climate Conference in 1995. This conference demonstrated the indispensability of international agreements – the ultimate outcome must be an effective Climate Protocol – but at the same time revealed their limitations: effective international accord will only be achieved if certain preconditions already exist or are subsequently created in the signatory states.

Therefore, the Council considers the creation of the requisite "societal conditions for solving global environmental problems" as the second crucial issue. Within this far-ranging but poorly researched field, the following topics need particular emphasis:

1 Global environmental policy will only be successful if the political decision makers in the individual states are supported by a population whose environmental awareness and willingness to act in an environmentally sound manner move it to demand that respective solutions be implemented. The Council therefore recommends that environmental education be improved both at home and worldwide. This involves a stronger orientation of public and private education and training systems towards global environmental problems, and includes strengthening institutions that support environmental education worldwide, e.g. UNESCO. The Council suggests that the Federal Government take adequate action so that environmental education becomes an established agenda item at international environmental conferences. It should also be put on the agenda at the next Conference of the Parties to the Climate Convention.

2 The scope available for mitigating global environmental problems depends strongly on population trends. The slight decrease of world population growth should be seen as justifying even greater efforts, since they seem to be effective. Of central importance in this connection are the eradication of poverty, a guaranteed provision of care to the aged, the improvement of the social and societal position of women, as well as greater efforts to improve vocational training and health care provision.

3 Migration pressure is increasing in many regions of the world as a result of population growth, poverty and environmental degradation. Migration flows continue to be directed at neighboring regions that are often afflicted by the same problems. In future Europe might be affected to an increasing degree. The Council suggests – not least against the background of immigration and asylum policies in Germany – that the causes of migration be combated more vigorously than before in the countries of origin, above all in the most severely affected developing countries. German development aid therefore must not be allowed to shrink any further, but instead must be increased significantly in the long run.

The major part of this report is dedicated to international agreements as a promising way to solving global environmental problems. The various sections include many special recommendations for action and research from which several general conclusions can be derived:

1 Faster progress in global environmental policy can be achieved when states or groups of states take over a vanguard role with respect to certain solutions. In 1994 the Council had already proposed, and repeats that proposal here, that preparations be made for a CO_2 tradeable permit system within the European Union. The pilot phase for Joint Implementation of the Climate Convention should be started without delay and actively shaped by Germany.

2 The system of international agreements must be expanded and systematized without unduly increasing the number of conventions. Conventions already exist with respect to climate, biodiversity, desertification and the protection of the sea; these conventions must now be implemented. With regard to the ecosphere, effective regulations governing forests and soils must be formulated. As far as the protection of forests is concerned, the Council recommends regulation by a protocol under the Biodiversity Convention. The Desertification Convention should be made part of a wider convention on the protection of soils.

The GATT/WTO regime will need to be restructured if the WTO fails to give adequate consideration to environmental protection. Conventions are probably unsuitable, however, for dealing with the population problem or with environmental education. But the stated objectives and measures in the various conventions and international agreements should be linked to each other more closely than is currently the case. The Council has demonstrated this need in relation to the Climate Convention. Coordination of financial support aimed at reducing global environmental stress must also be improved.

3 The instrument of international agreements should be developed further, since it is the precondition for successful global environmental policy. This does not mean that further development needs to be based solely on formalized conven-

tions with specialized institutions and multilateral funding. The Desertification Convention, for example, would not have been signed without the bilateral funding option.

Positive results so far are primarily a greater awareness concerning possible threats and agreements on formal procedures. These must be followed now by the practical implementation of conventions because the major trends in global change that the Council analyzed in its 1993 annual report, namely

- population growth,
- long-term changes of the atmosphere,
- loss of biological diversity,
- degradation of soils and
- scarcity and contamination of water,

show no signs of amelioration. On the contrary, the pressure to act is further increasing.

Annex: Notes on the Inverse Scenario in Section C 1.3

1.1
Tolerable Stress for Nature and Society

The potential consequences of marked climate change – to the extent that they can already be identified and evaluated in the first place – take a multitude of different forms: the spectrum extends from the threat to human life through floods or heat waves to the loss of aesthetic values through extinction of species (*see Section C 1.4.2*). If the different qualities of stress were properly taken into account, then the possible climate change impacts would have to be presented within a high-dimensional action space. Due to the gaps in available knowledge, however, such a presentation could only pretend to have scientific precision.

If fundamental values are instead taken as a basis, approximate but well-substantiated conclusions can be obtained directly. The Council constructs its "inverse scenario" solely on the basis of the following maxims:
– preservation of Creation,
– prevention of excessive costs.

The first principle can be specified in relation to the climate problem as follows: human-induced disturbances of the atmosphere are considered no longer tolerable if they cause the mean global temperature T to leave the Quarternary range of fluctuation. This would mean that global climatic conditions would be deviating markedly from those that have shaped the coevolution of humanity and the ecosphere and which have thus produced the environment as we know it today.

From this principle we derive the explicit requirement for temperature T (°C)

$$T_{min} \leq T \leq T_{max}$$

where T_{min} = T_{min} (glacial) – 0.5 =   9.9   ,
T_{max} = T_{max} (interglacial) + 0.5 =   16.6   .

T_{min} (glacial) = 10.4 °C is the (smoothed) minimum temperature during the last glacial period (Wuerm), whereas T_{max} (interglacial) = 16.1 °C is the respective maximum temperature during the preceding interglacial period (Eem). The tolerable temperature window is obtained by extending the natural range between these extremities by 0.5 °C at either end, i.e. the range of acceptable temperature is demarcated rather generously. For comparison, the mean global temperature T_0 today is 15.3 °C.

The second principle should make immediate sense; the problem here is how to monetarize the cascade of effects that result from climate change (*see also Annex 2.2*), and how to interpret the word "excessive". The decline S in percent of *Gross Global Product* (GGP) caused by climate change shall serve as a one-dimensional cost indicator in a first approximation. It is not our intention to discuss here the advantages and disadvantages of GGP as an economic indicator.

Economists assume as an initial approximation that resultant costs of climate change in the order of 3-5% of annual GGP over a period of several decades would cause severe disruption of social relations with far-reaching sociopolitical consequences. By way of illustration, one should note that the current annual volume of transfer payments from west to east Germany is about 5% of GNP. Furthermore, one should realize that a moderate value for S would also involve temporary but extreme stresses for many regions in the world on account of the considerable geographic inhomogeneity of climate impacts. The second limiting condition is therefore

$$S \leq S_{max} ,$$
where $S_{max} = 5$.

The resultant range of acceptance $\mathcal{A}$ in the 2-dimensional stress-temperature space is shown in *Figure 35*.

$\mathcal{A}$ is not over-confined in any sense: coping with the stress potential it implies represents an enormous challenge for a growing and largely underdeveloped global population. The definition of a fundamental tolerance window is an operationalization of the objective in Article 2 of the Climate Change Convention and forms the starting point for all further analytical steps.

1.2
Admissible Climate Change

Since even the best available coupled ocean-atmosphere general circulation models (GCMs) can still not provide any reliable predictions of future precipitation patterns, wind fields, etc., climate dynamics are characterized here purely as changes in temperature T(t), where t is the time variable.

We also make the (plausible) assumption that the "climate cost function" S depends solely on the global mean temperature T and its temporal derivative

$$\dot{T} = \frac{d}{dt} T(t) , i.e., that$$

$$S = S (T, \dot{T}).$$

This reflects the idea that the speed of a given climate change is a major determinant of the adaptation costs.

T and $\dot{T}$ demarcate the so-called phase space of climate dynamics. The quantity of all climate changes T(t) that conform to the range of acceptance $\mathcal{A}$ fills a

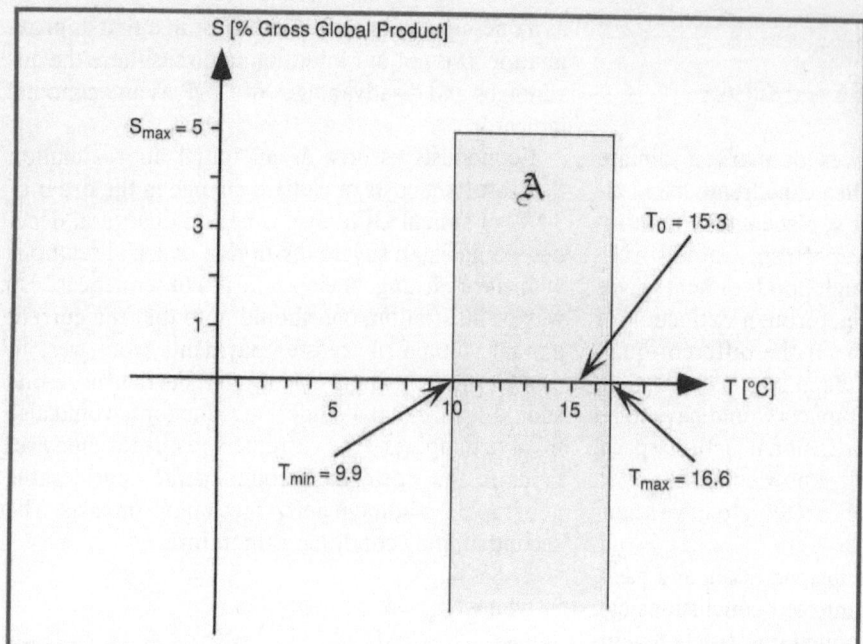

Figure 35
Tolerable range of
acceptance $\mathcal{A}$ for a
"temperature window". T_0
is the current mean global
temperature.
Source: WBGU

sub-area $\mathcal{B}$ of the phase space; this domain $\mathcal{B}$ must be determined in the next step.

The following definitions are made in order to simplify the task:

Zero time t = 0 is the year 1995. New variables are introduced according to

$$X = T - 0.5\,(T_{max} + T_{min}) = T - 13.25$$
$$Y = \dot{T} = \dot{X}$$

It follows from this that:

$$X_{max} = T_{max} - 13.25 = 3.35 \ ,$$
$$X_{min} = T_{min} - 13.25 = -3.35 \ ,$$
$$X_0 = T_0 - 13.25 = 2.05 \ .$$

The unit for Y was chosen as °C per decade; the current value Y_0 is approx. 0.07. The current state of climate dynamics is thus represented by point $P_0 = (X_0, Y_0) = (2.05, 0.07)$ in the X-Y plane.

Due to the transformation of variables carried out above, we must consider S (X, Y) rather than S (T, T.) in the following analysis. This cost function for $Y \geq 0$ is chosen as follows:

$$S(X,Y)= \begin{cases} S_{max}\left(\dfrac{Y}{Y_{max}}\right)^2 (X - X_{min}), & X_{min} \leq X \leq X_{min}+1 \\[3mm] S_{max}\left(\dfrac{Y}{Y_{max}}\right)^2, & X_{min} \leq X \leq X_{max}-1 \\[3mm] S_{max}\left(\dfrac{Y}{Y_{max}}\right)^2 (X_{max}-X)^{-1}, & X_{max}-1 \leq X \leq X_{max} \end{cases}$$

and

$$S\,(X, Y) = \infty, \text{ if X is outside the range } [X_{min}, X_{max}].$$

INTERPRETATION OF THE APPROACH FOR THE COST FUNCTION

Within the admissible temperature interval $[X_{min}, X_{max}]$, S should reflect only the costs of adaptation to climate changes. These costs are most likely to depend nonlinearly on the speed of change Y; the simplest assumption is quadratic growth.

In the center of the temperature window, i.e. in the range $[X_{min} + 1, X_{max} - 1]$, the adaptability for a given value of Y should show little or no variation. The situation changes, however, as we approach the boundaries of the window: a temperature increase would probably be more and more difficult to adapt to as the right-hand (hot) boundary is approached, but progressively easier as the left-hand (cold) boundary is approached. To simplify the model, these relationships are expressed by multiplication factors that are directly or inversely proportional to the distance from the temperature threshold values. The damage to the human organism caused by rising body temperature would be described in a similar way for states of undercooling or high fever.

If these simplest possible yet nontrivial assumptions are applied, the shape of S (X, Y) is fully determined but for a "calibration constant" Y_{max}. The latter is obtained as follows:

– Most integrated assessments of the annual and global direct costs of doubling CO_2 emissions by the middle of the next century are of the order of 1-2% of GGP (see, for example, Nakicenovic et al. [1994] and the references therein). This rise in CO_2 over such a period corresponds in the climate model described above to a mean temperature increase of somewhat less than 0.2 °C per decade.

- The annual cost incurred through general environmental degradation in Germany is considered by experts to be 2-3% of GNP.
- Estimations to date of the resultant costs of climate change do not take systematic account of either extreme events (droughts, floods, tropical storms, fluctuations in ocean currents, etc.) or possible synergies between the various trends of global change (e.g. interaction between anthropogenic greenhouse effect and soil degradation).

If all these factors are included in the analysis, then a figure of

$$Y_{max} = 0.2$$

is not overpessimistic: inside the admissible temperature window, a temperature gradient of 0.2 °C per decade would actually cause adaptation costs of up to 5% of GGP annually.

Since adaptation to global cooling should lead to problems of similar in severity to adaptation to global warming in a first approximation, $S(X, Y)$ can be projected using the symmetry requirement

$$S(X, -Y) = S(-X, Y)$$

to the range $Y < 0$.

We are therefore looking for the domain $\mathbb{D}$ in the X-Y space of climate dynamics that does not lead to any departure from the range of acceptance $\mathbb{A}$ in the T-S stress zone. On the basis of the assumptions being made, the boundary

$$Y = \Gamma(X)$$

of $\mathbb{D}$ results directly from the equation

$$S(X, \Gamma(X)) = S_{max} .$$

For the foreseeable future of the climate, the first quadrant $(X, Y \geq 0)$ of the climate phase space is of particular importance. In that area

$$\Gamma(X) = \begin{cases} Y_{max} & , 0 \leq X \leq X_{max} - 1 \\ Y_{max}(X_{max} - X)^{0.5} & , X_{max} - 1 \leq X \leq X_{max} \end{cases} .$$

The admissible domain for climate dynamics is shown by the shaded area in *Figure 36*.

Various assumptions about the climate damage function $S(X, Y)$ are mere *educated guesses*. This is particularly the case with the quantitative relation between Y_{max} and S_{max}. The limitation of the tolerable climate domain $\mathbb{D}$ still makes sense, however, even when macroeconomic factors are left out of the reckoning. Defining the acceptable temperature range $[X_{min}, X_{max}]$ reflects to a certain extent the long-range objective of preserving life on earth, while the limitation of Y relates to short-term preservation aspects. Ecosystems would probably have major difficulties adapting to temperature gradients of more than 0.1 °C per decade (Enquete Commission, 1995a). If the elasticity of natural systems in the center of the window is taken to be double the value normally assumed, and if the special conditions that arise when

the temperature boundaries are approached are taken into consideration, then one obtains the tolerable climate domain $\mathbb{D}$. This domain would certainly be narrowed by including the economic impacts of climate change, which means that defining $\mathbb{D}$ in the manner above is more likely to produce an *underestimation* of the risks to humanity and nature than vice versa.

1.3
Admissible Increases in CO₂-Concentration and Corresponding Global Emission Profiles

With the help of a simplified coupled climate-carbon cycle model devised by the *Max Planck Institute for Meteorology* in Hamburg (Hasselmann et al., 1995), the tolerable climate development domain $\mathbb{D}$ can be "reverse-translated" to obtain the size of the admissible global emission profile $E(t)$. The dynamics of the atmosphere-ocean-biosphere system and the radiative forcing on climate by CO₂ is taken account of by parameterizations. The increased concentration of CO₂ over the period under investigation is calculated by taking the atmospheric lifetime of the gas and the proportion of emissions that are not directly absorbed by the oceans and other ecosystems.

The emission profiles that are obtained are not wholly arbitrary functions, however. It would make sense to apply the following boundary conditions, for example:

$$E(0) = E_0 ,$$
$$\dot{E}(0) = E_1 ,$$
$$\dot{E}(\hat{t}) = 0 \text{ and } E(t) = E(\hat{t}) = \text{const. for } t > \hat{t} .$$

E_0 equals the current annual level of CO₂ emissions, $\dot{E}$ is the mean increase in emissions over the last few years and $\hat{t}$ is the planning horizon (e.g. 300 years) for the final stabilization or cessation of emissions. The boundary conditions could be defined differently, in principle; however, the final conditions would appear to have no major influence on the key results.

Let $\mathbb{E}$ be the set of all emission profiles $E(t)$ that are compatible with $\mathbb{D}$ (and hence with $\mathbb{A}$ as well) and that also fulfill the boundary conditions imposed. This set is obviously high-dimensional and cannot therefore be presented in the form of a simple graph.

Before carrying out a full analysis of $\mathbb{E}$ at a later point in time, one should attach greater priority in the meantime to simple issues that have direct relevance for environmental policymaking. Of special interest in this connection, for example, is the admissible boundary emission profile $E_*(t)$ that realizes the maximum level of global CO₂ emissions within a given (or indeed any) planning horizon.

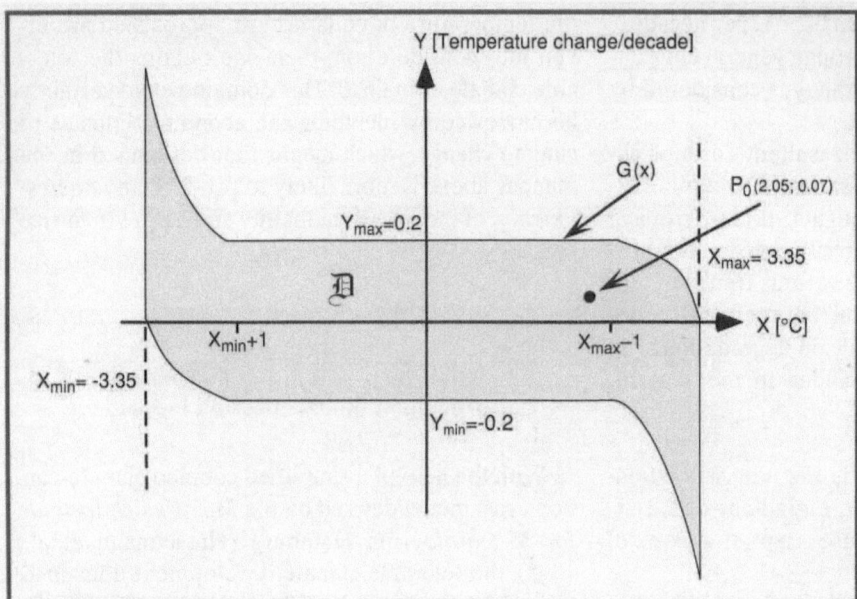

Figure 36
Tolerable climate domain
ⅅ. P_0 is the state of the
world's climate today.
Source: WBGU

E. can be selected from a total set 𝔈 with the help of optimization procedures. Optimal solutions to such a problem frequently display relatively bizarre forms, however, and do not usually match the available options for external control. Just as in the case of global emission profiles, one must make the realistic assumption that successful reduction agreements will be based on basic principles such as continuity and security of planning. One must therefore discuss the restrictions reflecting the technical and socioeconomic boundary conditions for the regulation of emissions that must be applied to 𝔈 as additional assumptions in this study.

In any case, however, E. defines an upper limit for the sum total of CO_2 emissions.

One can also calculate the benefit that a policy of rapid emission reductions would produce compared to a more reticent approach. To do this, one takes the set of emission profiles $\bar{E}(t)$ that result from initial trajectories as defined by the IPCC Business-as-Usual scenario, for example, or from a halt on all CO_2 emissions. With reference to the remaining trajectory of the function, one then searches for the optimum admissible profile, i.e. the curve $\bar{E}_*(t)$ that allows the maximum level of CO_2 emissions within the limited profile quantity. Comparing the different profiles enables conclusions to be drawn about the positive impacts produced by environmental policies geared to prevention rather than cure.

1.4
National Reduction Commitments

Let us assume an optimal global emission profile $\tilde{E}(t)$, i.e. the curve representing the first choice within the climatic and other boundary conditions. $\tilde{E}$ can, but need not, be equal to E_*. $\tilde{E}(t)$ has to be fractionated into dynamic national "pollution rights", $\in (L; t)$, according to politically defined allocation formulae. Here the index L numbers the various countries in the world, hence:

$$\sum_L \in (L;\ t) = \tilde{E}(t)$$

The dynamically adjusted national reduction commitments r (L;t) are then derived from the national pollution rights according to the formula

$$r\ (L; t) = 100 \left[1 - \frac{\in (L;\ t)}{\in_0 (L)}\right]\ ,$$

where $\in_0(L) = \ \in (L;\ 0)$ represents the current annual CO_2 emissions of country L, and the respective reduction commitment is expressed as a percentage of $\in_0(L)$. Note that r(L; 0) = 0 for all L, i.e. that national reduction commitments grow at a constant rate, commencing at zero.

What allocation principle should be used to determine the emission quotas $\in (L;t)$ of the individual states? Let P(t), B(t) be the global population and GGP expressed as a function of time, and p (L; t), b (L; t) the respective national values. The following factors in particular would play a role in determining the quotas of the individual states:

$f(L) = \dfrac{\epsilon_0 (L)}{E_0}$: current share of global emissions

$g(L; t) = \dfrac{p(L; t)}{P(t)}$: dynamic proportion of the world's population

$h(L; t) = \dfrac{b(L; t)}{B(t)}$: dynamic share in GGP.

This means that

$\in (L; t) = F (f(L1), f(L2) ...; g (L1; t), g (L2; t) ...; h (L1; t), h (L2; t) ...),$

where the allocation function F is the result of international negotiations.

Merely projecting the relative levels of pollution "owned" by the Parties would mean allocation according to the following formulae:

$\in (L; t) = f(L)\tilde{E}(t)$

and

$r(L; t) = 100 [1 - \dfrac{\tilde{E}(t)}{E_0}]$.

In this case, therefore, all states in the world would always have to reduce their CO_2 emissions synchronously.

Any form of allocation that releases the developing countries from any reduction commitment would necessitate greater cuts on the part of the industrialized countries: let

$S = \sum \epsilon_0 (L)$

be the current annual total emissions of the non-industrialized states. If this quota remains unaltered, then the Annex I-countries would have total emission rights amounting to

$\tilde{E}(t) - S$.

If equal obligations are imposed on all these states, their dynamically adjusted reduction obligation Ra(t) would be:

$R_a(t) = 100 [1 - (\dfrac{\tilde{E}(t) - S}{E_0 - S})]$.

Given the political preconditions discussed above, this reduction profile would also apply to Germany.

The industrialized countries would face a much greater challenge if the allocation formula were determined by the elementary humanitarian principle of "equal emission rights for all". In such a case,

$\in (L; t) = g(L; t) \tilde{E}(t)$

and

$r(L; t) = 100 [1 - g(L; t) \dfrac{\tilde{E}(t)}{\epsilon_0 (L)}]$.

However, one must note in this calculation that factors g(L; t) are not yet known precisely and can only be estimated at best with the help of demo-graphic projections. Inserting the current figures, however, provides an impression of how allocation by share of population leads to a shift in weightings.

If $g_0(D) = g(D; 0)$ is the current share of the world's population in Germany, the dynamically adjusted reduction commitment $R_b(t)$ for Germany under this static and egalitarian scenario would be as follows:

$R_b(t) = 100 [1 - g_0(D) \dfrac{\tilde{E}(t)}{\epsilon_0 (D)}]$.

The "worst case" for the OECD countries would be an allocation of emission rights that goes beyond demographic weighting and which also takes account of the need for sustainable development in Third World countries. In this case, the formula would have to be:

$\in (L; t) \sim g(L; t)/h(L; t)$

$= \left[\dfrac{p(L; t)}{b(L; t)}\right] / \left[\dfrac{P(t)}{b(t)}\right]$

and the scope for emissions on the part of the industrialized nations would tend towards zero! Since the chance of implementing such an allocation formula is similarly close to zero, this scenario is not given any further consideration here.

A much more realistic approach would be an allocation system according to which the developing countries are granted not only the right to maintain their absolute emission levels but even an increase, for a limited period, in their CO_2 emissions. For example, a linear increase in the emissions of the non-Annex I-states by 2% of the current value per annum over the next 50 years could form the basis for discussions. After a ten-year moratorium of this doubled allocation, the global reduction burden could then be distributed evenly among all the countries of the world

The dynamic reduction mode $R_c(t)$ for the industrialized countries would then have the following shape:

$R_c(t) = 100 [1 - \left(\dfrac{\tilde{E}(t) - S(t)}{E_0 - S_0}\right)]$,

where

$S(t) = \begin{cases} S_0 (1 + 0.02t) & , \quad 0 \le t \le 50 \\ 2 S_0 & , \quad 50 \le t \le 60 \\ \rho \tilde{E}(t) & , \quad t \le 60 \end{cases}$

and

$\rho = \dfrac{2 S_0}{\tilde{E}(60)}$.

Ainsworth, M (1994): The Socioeconomic Determinants of Fertility in Sub-Saharan Africa. Washington, DC: World Bank.

Altmann, J. (1992): Das Problem des Umweltschutzes im internationalen Handel. In: Sautter, H. (Ed.): Entwicklung und Umwelt. Berlin: Duncker & Humblot, 207-244.

Amelung, T. (1987): Zum Einfluß von Interessengruppen auf die Wirtschaftspolitik in Entwicklungsländern. Weltwirtschaft 38 (1), 158-171.

Anderson, K. (1992): The Standard Welfare Economics of Policies Affecting Trade and the Environment. In: Anderson, K. and Blackhurst, R. (Eds.): The Greening of World Trade Issues. New York: Harvester Wheatsheaf, 25-48.

Andreae, M.O., Anderson, B.E., Blake, D.R., Bradshaw, J.D., Collins, J.E., Gregory, G.L., Sachse, G.W. and Shipham, M.C. (1994): Influence of Plumes from Biomass Burning on Atmospheric Chemistry over the Equatorial and Tropical South Atlantic During CITE 3. Journal of Geophysical Research 99 (D6), 12793-12808.

Apel, H. and Reith, E. (1990): Umweltbildungsangebote an Volkshochschulen. Frankfurt a.M.: Pädagogische Arbeitsstelle des Deutschen Volkshochschul-Verbandes.

Archer, D. (1994): The Changing Roles of Non-governmental Organizations in the Field of Education (in the Context of Changing Relationships with the States). International Journal of Educational Development 14 (3), 223-232.

Arrow, K.J. (1962): The Economic Implications of Learning by Doing. Review of Economic Studies 29, 155-173.

Ashford, S. and Timms, N. (1992): What Europe Thinks. A Study of Western European Values. Aldershot: Dartmouth.

Athey, S. C. and Schmutzler, A. (1994): Flexibility, Coordination and the Organization of Work. Heidelberg: Universität Heidelberg.

Ausuble, J. H. (1991): Does Climate Still Matter? Nature 350, 649-652.

Ayres, R. U. and Walter, J. (1991): The Greenhouse Effect. Damages, Costs and Abatement. Environmental and Resource Economics 1 (3), 237-270.

Bach, W. (1995): Klimakonvention und Klimaschutzrichtwerte. Naturwissenschaften 82, 53-67.

Bago, A. L. and Velasquez, A. (1993): A Case of Environmental Education and Awareness Raising in the Philippines. In: Schneider, H. (Ed.): Environmental Education. An Approach to Sustainable Development. Paris: OECD, 177-184.

Ballwieser, W. (1993): Information und Umweltschutz aus der Sicht der betriebswirtschaftlichen Theorie. In: Wagner, G.R. (Ed.): Betriebswirtschaft und Umweltschutz. Stuttgart: Schäffer-Poeschel, 250-264.

Baumann, B. (1993): Marktprozeß und Staatsaufgaben: Möglichkeiten und Grenzen ökonomischer Theorien zur Erklärung der Funktionsweise offener Sozial-systeme und zur Legitimation staatlichen Handelns in offenen Gesellschaften. Baden-Baden: Nomos.

Bächler, G., Böge, V., Klötzli, S. and Libiszewski, S. (1993): Umweltzerstörung. Krieg oder Kooperation? Ökologische Konflikte im internationalen System und Möglichkeiten der friedlichen Bearbeitung. Münster: Agenda Verlag.

Bähr, J. (1992): Bevölkerungsgeographie. Verteilung und Dynamik der Bevölkerung in globaler, nationaler und regionaler Sicht. Stuttgart: Ulmer.

Becker-Soest, D. and Wink, R. (forthcoming): Reality Bites. Institutionen zum Schutz globaler Bodenfunktionen – Eine Analyse aus ökonomischer Sicht. Natural Resources Journal.

Benedick, R.E. (1991): Ozone Diplomacy. New Directions in Safeguarding the Planet. Cambridge, Ma.: Harvard University Press.

Benedick, R.E. (1992): Behind the Diplomatic Curtain. Inner Workings of the New Global Negotiations. Columbia Journal of World Business (Fall/Winter), 52-61.

Benz, A., Scharpf, F.W. and Zintl, R. (1992): Horizontale Politikverflechtung. Zur Theorie von Verhandlungssystemen. Frankfurt a.M.: Campus.

Bergmann, H. (1992): Die Umweltproblematik in der schulischen Grundausbildung – UNESCO-Schwerpunkt der neunziger Jahre. In: Becker, E. (Ed.): Umwelt und Entwicklung. Frankfurt a.M.: Verlag für Interkulturelle Kommunikation, 299-309.

Biermann, F. (1994a): Internationale Meeresumweltpolitik: Auf dem Weg zu einem Umweltregime für die Ozeane? Frankfurt a.M.: Peter Lang.

Biermann, F. (1994b): Schutz der Meere. Internationale Meeresumweltpolitik nach Inkrafttreten der Seerechtskonvention der Vereinten Nationen. Berlin: WZB.

Billig, A. (1994): Ermittlung des ökologischen Problembewußtseins der Bevölkerung. Berlin: UBA.

Birg, H. (1994): Auswirkungen möglicher Verzögerungen des Fertilitätsrückgangs auf das Weltbevölkerungswachstum – Alternative Berechnungen zu den UN-Bevölkerungsprojektionen. Bielefeld: Institut für Bevölkerungsforschung und Sozialpolitik.

BLK-Modellversuch "Energienutzung und Klima" (1991): Klimaschonendes Handeln. Materialien und Konzepte für den Unterricht in den Sekundarstufen, Teile A, B. Hamburg: Polycop.

BMBW – Bundesministerium für Bildung und Wissenschaft (1990): Schutz der Erdatmosphäre – eine Herausforderung an die Bildung. Zur Umsetzung der Empfehlungen der Bundestags-Enquete-Kommission "Vorsorge zum Schutz der Erdatmosphäre" in das Bildungssystem. Ergebnisbericht. Bonn: Economica.

BMI – Bundesministerium des Innern (1994): Bericht der Regierung der Bundesrepublik Deutschland für die Internationale Konferenz für Bevölkerung und Entwicklung. Bonn: BMI.

BML – Bundesministerium für Ernährung, Landwirtschaft und Forsten (1993): Waldzustandsbericht der Bundesregierung. Ergebnisse der Waldschadenserhebung. Bonn: BML.

BMU – Bundesministerium für Umwelt, Naturschutz und Reaktorsicherheit (1992a): Konferenz der Vereinten Nationen für Umwelt und Entwicklung im Juni 1992 in Rio de Janeiro – Dokumente. Bonn: BMU.

BMU – Bundesministerium für Umwelt, Naturschutz und Reaktorsicherheit (1992b): Bericht der Bundesregierung über die Konferenz der Vereinten Nationen für Umwelt und Entwicklung im Juni 1992 in Rio de Janeiro. Bonn: BMU.

BMZ – Bundesministerium für wirtschaftliche Zusammenarbeit und Entwicklung (1992): Förderung von Grundbildung in Entwicklungsländern. Bonn: BMZ.

BMZ – Bundesministerium für wirtschaftliche Zusammenarbeit und Entwicklung (1993a): Förderung von Bildung und Wissenschaft in der Entwicklungszusammenarbeit. Bonn: BMZ.

BMZ – Bundesministerium für wirtschaftliche Zusammenarbeit und Entwicklung (1993b): Internationale Agrarforschung. Bonn: BMZ.

Bodansky, D. (1993): The United Nations Framework Convention on Climate Change. A Commentary. Yale Journal of International Law 18 (2), 451-558.

Boisson, P. (1994): New Dispensation for Safety at Sea: Reasons and Conditions. Journal of Maritime Law and Commerce 25 (2), 309-317.

Bolscho, D. (1986): Umwelterziehung in der Schule. Ergebnisse aus der empirischen Forschung. Kiel: Institut für die Pädagogik der Naturwissenschaften (IPN).

Bolscho, D. (1989): Umwelterziehung in der Schule. Ergebnisse einer empirischen Studie. Die Deutsche Schule 81 (1), 61-72.

Bolscho, D. (1991): Empirische Forschung zur Umwelterziehung: Neue Trends? In: Eulefeld, G., Bolscho, D. and Seybold, H. (Eds.): Umweltbewußtsein und Umwelterziehung. Ansätze und Ergebnisse empirischer Forschung. Kiel: Institut für die Pädagogik der Naturwissenschaften (IPN), 7-21.

Bolscho, D. (1993): Praxis der Umwelterziehung. Ergebnisse empirischer Studien. In: Ermert, K. (Ed.): Umweltkrise, Umweltbildung und die Zukunft der Schule. Loccum: Loccumer Protokolle, 79-99.

Bolscho, D., Eulefeld, G. and Seybold, H. (1994): Bildung und Ausbildung im Umweltschutz. Bonn: Economica.

Bonus, H. (1991): Umweltpolitik in der Sozialen Marktwirtschaft. Aus Politik und Zeitgeschichte (März), 37-46.

Borrmann, A. and Weber, H. (1983): Meeresforschung und Meeresfreiheit. Perspektiven nach der dritten UN-Seerechtskonferenz. Hamburg: Verlag Weltarchiv.

Borrmann, A. and Koopmann, G. (1994): Regionalisierung und Regionalismus im Welthandel. Wirtschaftsdienst 74 (7), 365-372.

Brennan, G. and Buchanan, J. M. (1993): Die Begründung von Regeln. Konstitutionelle Politische Ökonomie. Tübingen: J.C.B. Mohr.

Brown-Weiss, E. (1992): Environmental Change and International Law. New Challenges and Dimensions. Tokyo: United Nations University Press.

Brown-Weiss, E. (1993): Plädoyer für einen ökologischen Generationenvertrag. In: Altner, G., Mettler-Meibom, B., Simonis, U.E. and von Weizsäcker, E.U. (Eds.): Jahrbuch Ökologie 1994. München: C.H. Beck, 31-36.

Brühl, C. and Crutzen, P.J. (1989): On the Disproportionate Role of Tropospheric Ozone as Filter Against Solar UV-B Radiation. Geophysical Research Letters 16, 703-706.

Caldwell, L.K. (1972): In Defense of Earth. International Protection of the Biosphere. London: Indiana University Press.

Caldwell, L.K. (1990): International Environmental Policy. Emergence and Dimensions. Durham, NC: Duke University Press.

Cameron, J., Makuch, Z. and Ward, H. (1994): Sustainable Development and Integrated Dispute Settlement in GATT 1994. Gland: WWF.

Cansier, D. (1991): Bekämpfung des Treibhauseffektes aus ökonomischer Sicht. Berlin: Springer.

Carter, T.R. (1994): IPCC Technical Guidelines for Assessing Climate Change Impacts and Adaptations. London: University College London.

Chameides, W.L., Kasibhatla, P.S., Yienger, J. and Levy, I.H. (1994): Growth of Continental-scale Metro-agro-plexes, Regional Ozone Pollution, and World Food Production. Science (264), 74-77.

Charlson, R J., Schwartz, S.E., Hales, J.M., Cess, R.D., Coakley, J.A., Hansen, J.E. and Hofmann, D.J. (1992): Climate Forcing by Anthropogenic Aerosols. Science 255, 423-430.

Chayes, A., Skolnikoff, E.B. and Victor, D.G. (1992): A Prompt Start. Implementing the Framework Convention on Climate Change (A Report from the Bellagio Conference on Institutional Aspects of International Cooperation on Climate Change, 28-30 January 1992). Cambridge, Ma.: MIT Press.

Chayes, A. and Chayes, A.C. (1993): On Compliance. International Organization 47 (2), 175-205.

Claude, H. and Köhler, U. (1994): Ozonbulletin des Deutschen Wetterdienstes 7.

Claude, H. (1995): Ozonbulletin des Deutschen Wetterdienstes 2+3.

Cline, W.R. (1992): The Economics of Global Warming. Washington, DC: Institute for International Economics.

Cohen, S.J. (1994): Mackenzie Basin Impact Study. Interim Report 2. Toronto: Environment Canada.

Cousteau-Society (1991): Protecting the Rights of Future Generations. Calypso Log, August 1991.

CSD – Commission on Sustainable Development (1994): CSD Holds Second Session. Environmental Policy and Law 24 (5), 206-218.

Dahl, T.E. (1990): Wetlands Losses in the United States 1780s to 1980s. Washington, DC: US Department of the Interior, Fish and Wildlife Service.

Dahl, A. L. (1994): Global Sustainability and its Implications for Trade. GATT Symposium on Trade, Environment and Sustainable Development. Geneva: GATT, 87-90.

Daly, H.E. and Goodland, R. (1994): An Ecological-economic Assessment of Deregulation of International Commerce under GATT. Ecological Economics 9 (1), 73-92.

Datta, A. (1992): Grundbildung – ja, aber welche und wie? Anmerkungen zu dem Sektorkonzept des BMZ. epd-Entwicklungspolitik 14, m-q.

Deutsche Stiftung Weltbevölkerung (1994): Vatikan von eigener Akademie der Wissenschaften angegriffen. ICPD-Newsletter (10), 1.

Deutscher Bundestag (1994): 3. Bericht der Bundesregierung an den Deutschen Bundestag über Maßnahmen zum Schutz der Ozonschicht. Bonn: Deutscher Bundestag.

Dexel, B. (1995): Internationaler Artenschutz: Neuere Entwicklungen. Berlin: WZB.

DGVN – Deutsche Gesellschaft für die Vereinten Nationen (1992): Megastädte. Zeitbombe mit globalen Folgen? Bonn: DGVN.

DGVN – Deutsche Gesellschaft für die Vereinten Nationen (1993): Weltbevölkerungsbericht 1993. Das Individuum und die Welt: Bevölkerung, Migration und Entwicklung in den neunziger Jahren. Bonn: DGVN.

DGVN – Deutsche Gesellschaft für die Vereinten Nationen (1994): Weltbevölkerungskonferenz 1994. Die besondere Rolle der NRO. UNFPA Informationsdienst (20), 1.

DIESA – Department of International Economic and Social Affairs (1991): World Urbanization Prospects 1990. New York: UN.

DIW – Deutsches Institut für Wirtschaftsforschung (1994): Uruguay-Runde: Weitere Marktöffnung für Entwicklungsländer. DIW-Wochenberichte 61 (23), 385-393.

Dolzer, R. (1992): Umweltschutz im Völkerrecht und Kollisionsrecht. Heidelberg: C.F. Müller.

Dörner, D. (1987): Denken und Wollen: ein systemtheoretischer Ansatz. In: Heckhausen, H., Gollwitzer, P.M. and Weinert, F.E. (Eds.): Jenseits des Rubikon: Der Wille in den Humanwissenschaften. Berlin: Springer, 238-250.

Dörner, D. (1989): Die Logik des Mißlingens. Strategisches Denken in komplexen Situationen. Reinbek: Rowohlt.

Draguhn, W. (1991): Asiens Schwellenländer. Dritte Weltwirtschaftsregion? Hamburg: Institut für Überseewirtschaft.

DSE – Deutsche Stiftung für Internationale Entwicklung (1993): Population Policies and Programmes: The Impact of HIV/AIDS. Recommendations to the Secretariat of the International Conference on Population and Development Cairo 1994. Berlin: DSE

Dudley, N. (1992): Forests in Trouble: A Review of the Status of Temperate Forests Worldwide. Gland: WWF.

Dugan, P.J. (1990): Wetland Conservation: A Review of Current Issues and Required Action. Gland: IUCN.

Dunlap, R.E., Gallup jr., G.H. and Gallup, A.M. (1993): Health of the Planet. A George H. Gallup Memorial Survey. Results of a 1992 International Environmental Opinion Survey of Citizens in 24 Nations. Princeton, NJ: The George H. Gallup International Institute.

Durning, A.T. and Ayres, E. (1994): Bittere Bohnen. World Watch 3 (5), 44-46.

DW – Deutsche Welthungerhilfe (1994a): Weltbevölkerung und Welternährung. Positionspapier. Bonn: DW.

DW – Deutsche Welthungerhilfe (1994b): Weltbevölkerung und Welternährung. Thesenpapier zum Positionspapier. Bonn: DW.

Edmonds, J., Barns, D.W. and Ton, M. (1993): The Regional Costs and Benefits of Participation in Alternative Hypothetical Fossil Fuel Carbon Emissions Reduction Protocols. In: Kaya, Y., Nakicenovic, N., Nordhaus, W.D. and Toth, F.L. (Eds.): Costs, Impacts, and Benefits of CO_2 Mitigation. Laxenburg: IIASA, 291.

Edom, E., Rapsch, H.-J. and Veh, G.M. (1986): Reinhaltung des Meeres. Nationale Rechtsvorschriften und internationale Übereinkommen. Köln: Carl Heymanns.

Eglin, R. (1994): International Economics, International Trade, International Environmental Protection. Inter-American Development Bank, 89-101.

Ehrlich, P. R. (1992): Der Verlust der Vielfalt: Ursachen und Konsequenzen. In: Wilson, E.O. (Ed.): Ende der biologischen Vielfalt? Heidelberg: Spektrum, 39-45.

Eisner, T. (1989): Prospecting for Nature's Chemical Riches. Issues in Science and Technology 6, 31-34.

Elger, U., Hönigsberger, H. and Schluchter, W. (1992a): Evaluierung von Maßnahmen der Umwelterziehung. Band 1: Fortbildung an der Evangelischen Akademie. Berlin: UNESCO-Verbindungsstelle für Umwelterziehung im Umweltbundesamt.

Elger, U., Hönigsberger, H. and Schluchter, W. (1992b): Evaluierung von Maßnahmen der Umwelterziehung. Band 2: Kurzzeitbildung in der Ökostation. Berlin: UNESCO-Verbindungsstelle für Umwelterziehung im Umweltbundesamt.

Elger, U., Hönigsberger, H. and Schluchter, W. (1992c): Evaluierung von Maßnahmen der Umwelterziehung. Band 3: Umwelterziehung an der Hauptschule. Berlin: UNESCO-Verbindungsstelle für Umwelterziehung im Umweltbundesamt.

Elger, U., Hönigsberger, H. and Schluchter, W. (1992d): Evaluierung von Maßnahmen der Umwelterziehung. Band 4: Wirkungen der Umwelterziehung. Berlin: UNESCO-Verbindungsstelle für Umwelterziehung im Umweltbundesamt.

Enquete Commission "Protecting the Earth's Atmosphere" of the 11th German Bundestag (1990): Protecting the Earth. A Status Report with Recommendations for a New Energy Policy. Bonn: Economica.

Enquete Commission "Preventive Measures to Protect the Earth's Atmosphere" of the 11th German Bundestag (1990a): Schutz der Erdatmosphäre. Eine internationale Herausforderung. Zwischenbericht der Enquete-Kommission "Vorsorge zum Schutz der Erdatmosphäre" Bonn: Economica.

Enquete Commission "Preventive Measures to Protect the Earth's Atmosphere" of the 11th German Bundestag (1990b): Schutz der Tropenwälder. Eine internationale Schwerpunktaufgabe. Bonn: Economica.

Enquete Commission "Preventive Measures to Protect the Earth's Atmosphere" of the 11th German Bundestag (1990c): Protecting the Earth. A Status Report with Recommendations for a New Energy Policy. Bonn: German Bundestag.

Enquete Commission "Protecting the Earth's Atmosphere" of the 12th German Bundestag (1992): Climate Change – A Threat to Global Development. Acting now to Safeguard the Future. Bonn: Economica.

Enquete Commission "Protecting the Earth's Atmosphere" of the 12th German Bundestag (1995): Protecting Our Green Earth. How to Manage Global Warming through Environmentally Sound Farming and Preservation of the World's Forests. Bonn: Economica.

Enquete Commission "Protecting the Earth's Atmosphere" of the 12th German Bundestag (1995a): Mobility and Climate. Bonn: Economica.

Enquete Commission "Protecting the Earth's Atmosphere" of the 12th German Bundestag (1995b): Securing Our Earth's Future. Long-term Global Warming Management through Sustainable Energy Policies. Bonn: Economica.

Erdmann, G. (1993): Elemente einer evolutorischen Innovationstheorie. Tübingen: J.C.B. Mohr.

Erdmann, K.-H. and Nauber, J. (1995): Der deutsche Beitrag zum UNESCO-Programm "Der Mensch und die Biosphäre" (MAB) im Zeitraum Juli 1992 bis Juni 1994. Bonn: BMU.

Erichsen, S. (1993): Der ökologische Schaden im internationalen Umwelthaftungsrecht. Völkerrecht und Rechtsvergleichung. Frankfurt a.M.: Peter Lang.

Ester, P., Halman, L. and de Moor, R. (1993): The Individualizing Society. Value Change in Europe and North America. Tilburg: Tilburg University Press.

Estrada-Oyuela, R. (1995): Rede anläßlich der Eröffnung der 1. Vertragsstaatenkonferenz der Klimarahmenkonvention am 28.3.1995 in Berlin.

Esty, D.C. (1994a): Greening the GATT: Trade, Environment and the Future. Harlow Essex: Longman.

Esty, D.C. (1994b): GATTing the Greens – Not Just Greening the GATT. Foreign Affairs 72 (5), 32-36.

Eulefeld, G., Bolscho, D., Rost, J. and Seybold, H. (1988): Praxis der Umwelterziehung in der Bundesrepublik Deutschland. Kiel: Institut für die Pädagogik der Naturwissenschaften (IPN).

Eulefeld, G., Bolscho, D., Rode, H., Rost, J. and Seybold, H. (1993): Entwicklung der Praxis schulischer Umwelterziehung in Deutschland. Ergebnisse empirischer Studien. Kiel: Institut für die Pädagogik der Naturwissenschaften (IPN).

European Chemical Industry Council (1994): Jahresbericht. European Chemical Industry Council.

Eurostep (1994): International Conference on Population and Development. Eurostep Position Paper on Population and Development. Brussels: Eurostep.

Ewringmann, D. and Schafhausen, F. (1985): Abgaben als ökonomischer Hebel in der Umweltpolitik. Ein Vergleich von 75 praktizierten oder erwogenen Abgabenlösungen im In- und Ausland. Berlin: Erich Schmidt.

Fankhauser, S. (1993a): The Economic Costs of Global Warming: Some Monetary Estimates. In: Kaya, Y., Nakicenovic, N., Nordhaus, W.D. and Toth, F.L. (Eds.): Costs, Impacts, and Benefits of CO_2 Mitigation. Laxenburg: IIASA, 85.

Fankhauser, S. (1993b): The Social Costs of Greenhouse Gas Emissions: An Expected Value Approach. London: CSERGE, University College London.

Fietkau, H.-J. and Kessel, H. (1981): Umweltlernen: Veränderungsmöglichkeiten des Umweltbewußtseins. Modelle – Erfahrungen. Königstein/Ts.: Hain.

Fikentscher, W. and Heinemann, A. (1994): Der "Draft International Antitrust Code" – Initiative für ein Weltkartellrecht im Rahmen des GATT. Wirtschaft und Wettbewerb 44 (2), 97-107.

Finlayson, M. and Moser, M. (1991): Wetlands. Oxford: Facts on File Limited.

Fishman, J., Fakhruzzaman, K., Cros, B. and Nganga, D. (1991): Identification of Widespread Pollution in the Southern Hemisphere Deduced from Satellite Analyses. Science 252, 1693-1696.

Forschungsgruppe Wahlen (1992-1995): Politbarometer. Monatliche repräsentative Umfrage. Mannheim: Forschungsgruppe Wahlen e.V.

Fraenkel, E. (1964): Deutschland und die westlichen Demokratien. Frankfurt a.M.: Suhrkamp.

Freestone, D. and IJlstra, T. (1990): The North Sea: Perspectives on Regional Environmental Co-operation. International Journal of Estuarine and Coastal Law (Special Issue) 5 (1-3).

Frey, B.S. (1985): Internationale Politische Ökonomie. München: Vahlen.

Fromm, O. and Hansjürgens, B. (1994): Erfolgsbedingungen von Zertifikatelösungen in der Umweltpolitik – am Beispiel der Novelle des US Clean Air Act von 1990. Zeitschrift für Umweltpolitik und Umweltrecht. 17 (4), 473-505.

Galenson, W. (1985): Foreign Trade and Investment: Economic Growth in the Newly Industrialized Countries. Madison, Wi.: University of Wisconsin.

Gámez, R., Piva, A., Sittenfeld, A., Leon, E., Jimenez, J. and Mirabelli, G. (1993): Costa Rica's Conservation Program and National Biodiversity Institute (INBio). In: Reid, W.V., Laird, S.A., Meyer, C.A. and Gámez, R. (Eds.): Biodiversity Prospecting. Washington, DC: WRI, 53–67.

Ganapin jr., D.J. (1991): Effective Environmental Regulation: The Case of the Philippines. In: Eröcal, D. (Ed.): Environmental Management in Developing Countries. Paris: OECD, 255-273.

GATT (1992): International Trade 1990-91, Including Trade and the Environment. Geneva: GATT.

GATT Secretariat (1993): Background Paper to the Meeting of the Contracting Parties. Geneva: GATT Secretariat.

Gehring, T. (1991): International Environmental Regimes. Dynamic Sectoral Legal Systems. Yearbook of International Environmental Law 1, 35-56.

Gehring, T. and Oberthür, S. (1993): The Copenhagen Meeting. Environmental Policy and Law 23 (1), 6-12.

Gerybadze, A. (1982): Innovation, Wettbewerb und Evolution: eine mikro- und makrotheoretische Untersuchung der Anpassungsprozesse von Herstellern und Anwendern neuer Produzentengüter. Tübingen: J.C.B. Mohr.

GESAMP – Joint Group of Experts on the Scientific Aspects of Marine Pollution (1990): The State of the Marine Environment. Oxford: Basil Blackwell.

GESAMP – Joint Group of Experts on the Scientific Aspects of Marine Pollution (1993): Impact of Oil and Related Chemicals and Wastes on the Marine Environment. London: IMO.

Glowka, L., Burhenne-Guilmin, F. and Synge, H. (1994): A Guide to the Convention on Biological Diversity. Environmental Policy and Law Paper 30. Gland: IUCN.

Glöge, M. (1993): Kampagnen und Bildungsarbeit am Beispiel der "Aktion Dritte Welt Handel". Eine kritische Bestandsaufnahme. In: BUNTSTIFT e.V., Verein Niedersächsischer Bildungsinitiativen (VNB) und Fachbereich internationale und interkulturelle Arbeit (Eds.): Bildung + Aktion = Veränderung? Barnstorf: VNB, 59-61.

Göhler, G. (1987): Grundfragen der Theorie politischer Institutionen. Forschungsstand, Probleme, Perspektiven. Opladen: Westdeutscher Verlag.

Göhler, G. (1994): Eigenart der Institutionen. Baden-Baden: Nomos.

Göpfert, J. (1995): Fingerzeig auf Klimawandel. VDI Nachrichten (12), 26.

Goldin, I., Knudsen, O. and van der Mensbrugghe, D. (1993): Trade Liberalisation: Global Economic Implications. Paris, Washington, DC: OECD and World Bank.

Goldschmidt, D. (1993): Education Through the Nonformal Sector – 12 Theses. In: Deutsche Stiftung für Internationale Entwicklung (DSE) (Ed.): Out-of-school Education, Work and Sustainability in the South: Experiences and Strategies. International Conference in Berlin, 30.3.-4.4.1993. Bonn: Zentralstelle für Erziehung, Wissenschaft und Dokumentation (ZED), 355-358.

Gonzalez, A. (1992): Die Umweltbewegung in Venezuela – Eine Bestandsaufnahme. Universität Karlsruhe: Institut für Regionalwissenschaft.

Greenpeace (1994a): Adressing the Loss of Forest Biodiversity. Amsterdam: Greenpeace International.

Greenpeace (1994b): Kinder- und Jugendarbeit bei Greenpeace. Hamburg: Polycop.

Grossmann, G.M. and Helpman, E. (1991): Innovation and Growth in the Global Economy. Cambridge, Ma.: MIT Press.

Großmann, H., Koopmann, G. and Michaelowa, A. (1994): Die neue Welthandelsorganisation: Schrittmacher für den Welthandel? Wirtschaftsdienst 74 (5), 256-264.

Grubb, M. (1990): The Greenhouse Effect. Negotiating Targets. International Affairs 66 (1), 67-89.

GTZ – Deutsche Gesellschaft für Technische Zusammenarbeit (1993): Environmental Education in Kenya, Tanzania and Uganda. Eschborn: GTZ.

Güth, W. and Pethig, R. (1992): Illegal Pollution and Monitoring of Unknown Quality – a Signalling Game Approach. In: Pethig, R. (Ed.): Conflicts and Cooperation in Managing Environmental Resources. Berlin: Springer, 276-330.

Haas, P.M. (1990): Saving the Mediterranean. The Politics of International Environmental Cooperation. New York: Columbia University Press.

Haas, P.M. (1992): Banning Chlorofluorocarbons. Epistemic Community Efforts to Protect Stratospheric Ozone. International Organization 46 (1), 187-224.

Haas, P.M., Levy, M.A. and Parson, E.A. (1992): Appraising the Earth Summit. How Should We Judge UNCED's Success? Environment 34 (8), 6-15, 26-36.

Haas, P.M., Keohane, R.O. and Levy, M.A. (1993): Institutions for the Earth. Sources of Effective International Environmental Protection. Cambridge, Ma.: MIT Press.

Hadley Center for Climate Prediction and Research (1995): Modelling Climate Change 1860-2050. Bracknell: U.K. Meteorological Office.

Häder, D. P. (1992): UV-Strahlung. Ein weiteres globales Umweltproblem. In: Altner, G., Mettler-Meibom, B., Simonis, U.E. and von Weizsäcker, E.U.(Eds.): Jahrbuch Ökologie 1993. München: C.H. Beck, 127-135.

Häusser, E. (1995): Patentwesen und Forschung. Forschung & Lehre (3), 136-138.

Hampicke, U., Tampe, K., Kiemstedt, H., Horlitz, T., Walters, M. and Timp, D. (1991): Kosten und Wertschätzung des Arten- und Biotopschutzes. Berlin: UBA.

Hampicke, U. (1991): Naturschutz-Ökonomie. Stuttgart: Ulmer.

Handelsblatt (8.8.1994): Bei Bonns Hilfe reiben sich die Exporteure die Hände. Handelsblatt, 7.

Hartje, V.J. (1983): Theorie und Politik der Meeresnutzung. Eine ökonomisch-institutionelle Analyse. Frankfurt a.M.: Campus.

Hauff, V. (1987): Unsere gemeinsame Zukunft. Der Brundt-land-Bericht der Weltkommission für Umwelt und Ent-wicklung. Greven: Eggenkamp.

Hausman, J.A. (1993): Contingent Valuation: A Critical As-sessment. Amsterdam: North Holland.

Hayek, F.A. v. (1969): Die Anschauungen der Menschheit und die zeitgenössische Demokratie. Tübingen: J.C.B. Mohr.

Hayek, F.A. v. (1979): Die drei Quellen menschlicher Werte. Tübingen: J.C.B. Mohr.

Hayek, F.A. v. (1991): Die Verfassung der Freiheit. Tübin-gen: J.C.B. Mohr.

Heck, W.W., Heagle, A.S. and Shriner, D.S. (1988): Assess-ment of Crop Loss from Air Pollutants. London: Else-vier.

Hegerl, G.C., Storch, H. v., Hasselmann, K., Santer, B.D., Cubasch, U. and Jones, P.D. (1994): Detecting Anthropo-genic Climate Change with an Optimal Fingerprint Me-thod. Hamburg: MPI für Meteorologie.

Heinze, M. and Kaiser, G. (1994): Ökologie-Dialog. Düssel-dorf: Econ.

Helland-Hansen, E. (1991): The Global Environment Facil-ity. International Environmental Affairs 3 (1), 137-144.

Helm, C. (1995): Sind Freihandel und Umweltschutz ver-einbar? Ökologischer Reformbedarf des GATT/WTO-Regimes. Berlin: edition sigma.

Hempel, G. (1994): Deutsche Meeresforschung unter dem neuen Seerecht. Berichte des Bundesamtes für Seeschif-fahrt und Hydrographie (4), 65-70.

Hendrickx, F., Koester, V. and Pripp, C. (1994): Access to Genetic Resources: A Legal Analysis. In: Sánchez, V. and Juma, C. (Eds.): Biodiplomacy: Genetic Resources and International Relations. Nairobi: ACTS-Press, 139-153.

Henke, J. (1990): Moderne Hochtechnologien und natio-nale Selbständigkeit. Berlin: Duncker & Humblot.

Heuvels, K. (1993): Die EG-Öko-Verordnung im Praxis-test. Erfahrungen aus einem Pilot-Audit-Programm der Europäischen Gemeinschaften. Umwelt-Wirtschafts-Forum 1 (3), 41-48.

Hickling-Hudson, A. (1994): The Environment as Radical Politics: Can "Third World" Education Rise to the Chal-lenge? International Review of Education 40 (1), 19-36.

Höll, O. (1994): Environmental Cooperation in Europe. The Political Dimension. Boulder: Westview Press.

Hoering, U. (01.09.1992): Umweltschutz und Demokratie sollen zwei Seiten einer Medaille sein. Frankfurter Rundschau.

Hoering, U. (1994): Dorfrepublik für Gemeinschaftsgüter. Umweltschutz und Demokratie. Der Überblick 30 (1), 44-46.

Hofheinz, R. J. and Calder, K. E. (1982): The Eastasia Edge. New York: Praeger.

Hong, Y.S. (1994): Technology Transfer: The Korean Expe-rience. Seoul: Korean Institute for International Eco-nomic Policy.

Horlacher, B. and Urban, A. (1992): Ozonentstehung und Ozonabbau in einem einfachen Demonstrationsver-such. Praxis der Naturwissenschaften – Chemie 41 (3), 18-20.

House, E. (1994): Environment and School Initiatives: En-vironmental Education Policies. Konferenzunterlage (94) 9 zur International Conference on Environmental Education Policy and Practice (Braunschweig, 06.-11.03.1994). Centre for Educational Research and Inno-vation (CERI).

Howe, C.W. (1994): Taxes Versus Tradable Discharge Per-mits. A Review in the Light of the U.S. and European Experience. Environmental and Resource Economics 4 (2), 151-169.

Hurrel, A. and Kingsbury, B. (1991): The International Pol-itics of the Environment. Actors, Interests and Institu-tions. Oxford: Clarendon Press.

IAE – International Agency of the Environment (1995): Policy Dialogue on Industrialized Country Commit-ments under the Framework Convention on Climate Change. Geneva: IAE.

ICCBD – Intergovernmental Committee on the Conven-tion on Biological Diversity (1994): Farmers' Rights and Rights of Similar Groups. Note by the Interim Secretar-iat of the CBD. Document UNEP/ CBD/IC/2/14. Nai-robi: UNEP.

IEA – International Energy Agency (1993): Energy Bal-ances of OECD Countries 1990-1991. Paris: IEA.

IEA – International Energy Agency (1994a): World Energy Outlook. Paris: IEA.

IEA – International Energy Agency (1994b): Climate Change Policy Initiatives. 1994 Update. OECD Coun-tries. Paris: OECD.

Illing, R., Merten, D., Zeitler, H., Sander, U. and Traub, U. (1994): Fruchtbarer Energieunterricht. AnSchUB. Das Forum für Schulische Umweltbildung in Berlin (4), 3-6.

IMO – International Maritime Organization (1993): IMO News – Magazine of the International Maritime Organ-ization (3).

IMO – International Maritime Organization (1994): IMO News – Magazine of the International Maritime Organ-ization (1).

INBio – Instituto Nacional de Biodiversidad (1994): Summary of Terms – Collaboration Agreement INBio-Merck & Co., Inc. Santo Domingo (Costa Rica): INBio.

INF – Institut für Naturschutzforschung (1994): Ökologische Chancen und Risiken des Einsatzes biotechnischer Verfahren zur nachhaltigen Nutzung biologischer Ressourcen in "Entwicklungsländern". Studie im Auftrag des Büros für Technikfolgenabschätzung beim Deutschen Bundestag. Regensburg: INF.

Inglehart, R. (1977): The Silent Revolution. Changing Values and Political Styles Among Western Publics. Princeton, NJ: Princeton University Press.

Inglehart, R. and Abramson, P.R. (1994): Economic Security and Value Change. American Political Science Review 88 (2), 336-354.

International Court for the Environment (1994): Environmental Policy and Law 24 (4), 173-187.

IPCC – Intergovernmental Panel on Climate Change (1990): Climate Change. The IPCC Scientific Assessment. Cambridge: Cambridge University Press.

IPCC – Intergovernmental Panel on Climate Change (1992): Climate Change 1992. The Supplementary Report to the IPCC Scientific Assessment. Cambridge: Cambridge University Press.

IPCC – Intergovernmental Panel on Climate Change (1994a): Radiative Forcing of Climate Change. The 1994 Report of the Scientific Assessment Working Group of IPCC. Summary for Policymakers. Geneva: WMO and UNEP.

IPCC – Intergovernmental Panel on Climate Change (1994b): Preparing to Meet the Coastal Challenges of the 21st Century. Den Haag: IPCC.

IPCC – Intergovernmental Panel on Climate Change (1995): Climate Change 1994. Radiative Forcing of Climate Change and an Evaluation of the IPCC IS92 Emission Scenarios. Cambridge: Cambridge University Press.

ipos – Institut für Praxisorientierte Sozialforschung (1993): Einstellungen zu aktuellen Fragen der Innenpolitik 1993 in Deutschland. Ergebnisse jeweils einer repräsentativen Bevölkerungsumfrage in den alten und neuen Bundesländern. Mannheim: ipos.

IUCN – The World Conservation Union, UNEP – United Nations Environment Programme and WWF – World Wide Fund for Nature (1980): World Conservation Strategy: Living Resource Conservation for Sustainable Development. Gland: IUCN, UNEP and WWF.

Jacobson, J.L. (1992): Die Vertreibung aus den Wäldern. World Watch 6 (1), 30-35.

Jäger, J. and Loske, R. (1994): Handlungsmöglichkeiten zur Fortschreibung und Weiterentwicklung der Verpflichtungen innerhalb der Klimarahmenkonvention. Wuppertal: Wuppertal Institut für Klima, Umwelt, Energie.

Jänicke, M. (1993): Ökologische und politische Modernisierung in entwickelten Industriegesellschaften. In: Prittwitz, V. v. (Ed.): Umweltpolitik als Modernisierungsprozeß. Politikwissenschaftliche Umweltforschung und -lehre in der Bundesrepublik. Opladen: Westdeutscher Verlag, 31-50.

Johnson, N. and Cabarle, B. (1993): Surviving the Cut: Natural Forest Management in the Humid Tropics. Washington, DC: WRI.

Jonas, H. (1984): Das Prinzip Verantwortung. Versuch einer Ethik für die technologische Zivilisation. Frankfurt a.M.: Insel.

Jones, A., Roberts, D.L. and Slingo, A. (1994): A Climate Model Study of the Indirect Radiative Forcing by Anthropogenic Sulphate Aerosols. Nature 370, 450-453.

Joos, F. and Sarmiento, J.L. (1995): Der Anstieg des atmosphärischen Kohlendioxids. Physikalische Blätter 51 (5), 405-411.

Jorgenson, D.W. and Wilcoxen, P.J. (1993): Reducing US Carbon Dioxide Emissions: An Assessment of Different Instruments. In: Kaya, Y., Nakicenovic, N., Nordhaus, W.D. and Toth, F.L. (Eds.): Costs, Impacts, and Benefits of CO_2 Mitigation. Laxenburg: IIASA, 387.

Jost, P.-J. (1993): Economic Analysis of Procedural Aspects in the German Environment Liability Law. Journal of Institutional and Theoretical Economics 149 (4), 609-633.

Kaas, K.P. (1993): Informationsprobleme auf Märkten für umweltfreundliche Produkte. In: Wagner, G.R. (Ed.): Betriebswirtschaft und Umweltschutz. Stuttgart: Metzler-Poeschel, 29-43.

Kahn, H. (1979): World Economic Development: 1979 and Beyond. London: Routledge.

Kahneman, D., Ritov, I., Jacowitz, K.E. and Grant, P. (1993): Stated Willingness to Pay for Public Goods: A Psychological Perspective. Psychological Science 4 (5), 310-315.

Kammer der EKD – Evangelischen Kirche in Deutschland für Kirchlichen Entwicklungsdienst (1993): Wie viele Menschen trägt die Erde? Ethische Überlegungen zum Wachstum der Welt. Hannover: EKD.

Karcher, W. (1994): Lernen im "Informellen Sektor in der Dritten Welt" – Entwicklungspolitische Relevanz. Zeitschrift für Entwicklungspädagogik 17 (1), 32-34.

Karl, H. (1992): Umweltschutz mit Hilfe zivilrechtlicher und kollektiver Haftung. RWI-Mitteilungen 43 (3), 183-199.

Karl, H. (1994): Marktsystem und risikoreiche Produktionstechnik – ein Beitrag zum ordnungspolitischen Umgang mit Umwelt- und Gesundheitsrisiken. Berlin: Duncker & Humblot.

Kaufman, Y.J. and Chou, M.D. (1993): Model Simulations of the Competing Climatic Effects of SO_2 and CO_2. Journal of Climate Change 6, 1241-1252.

Kay, D.A. and Jacobson, H.K. (1983): Environmental Protection. The International Dimension. Totowa, NJ: Allanhold, Osmun.

Kaya, Y., Nakicenovic, N., Nordhaus, W.D. and Toth, F.L. (Eds.) (1993): Costs, Impacts, and Benefits of CO_2 Mitigation. Laxenburg: IIASA.

Kerber, W. (1991): Zur Entstehung von Wissen: Grundsätzliche Bemerkungen zu Möglichkeiten und Grenzen staatlicher Förderung der Wissensproduktion aus der Sicht der Theorie evolutionärer Marktprozesse. In: Oberender, P. and Streit, M.E. (Eds.): Marktwirtschaft und Innovation. Baden-Baden: Nomos, 9-52.

Kiehl, J.T. and Briegleb, B.P. (1993): The Relative Role of Sulphate Aerosols and Greenhouse Gases in Climate Forcing. Science 260, 311-314.

Kilian, M. (1987): Umweltschutz durch internationale Organisationen. Die Antwort des Völkerrechts auf die Krise der Umwelt. Berlin: Duncker & Humblot.

Kimball, L.A. (1992): Forging International Agreements. Strengthening Inter-Governmental Institutions for Environment and Development. Washington, DC: WRI.

King, S.R. (1994): Establishing Reciprocity: Biodiversity, Conservation and New Models for Cooperation Between Forest-dwelling Peoples and the Pharmaceutical Industry. In: Greaves, T. (Ed.): Intellectual Property Rights for Indigenous Peoples: A Sourcebook. Oklahoma City: The Society for Applied Anthropology, 69-82.

Klages, H. (1992): Die gegenwärtige Situation der Wert- und Wertwandelforschung – Probleme und Perspektiven. In: Klages, H., Hippler, H.-J. and Herbert, W. (Eds.): Werte und Wandel. Ergebnisse und Methoden einer Forschungstradition. Frankfurt a.M.: Campus, 5-39.

Klemmer, P., Werbeck, N. and Wink, R. (1993): Institutionenökonomische Aspekte globaler Umweltveränderungen. Berlin: Analytica.

Klemmer, P. (1994): Nachhaltige Entwicklung – aus ökonomischer Sicht. Zeitschrift für angewandte Umweltforschung 7 (1), 14-19.

Klima-Bündnis/Alianza del Clima (1993): Klima – lokal geschützt. Aktivitäten europäischer Kommunen. München: Raben.

Klinger, J. (1994): Debt-for-Nature Swaps and the Limits to International Cooperation on Behalf of the Environment. Environmental Politics 3 (2), 229-246.

Kloppenburg Jr., J. and Kleinman, D.L. (1987): The Plant Germplasm Controversy: Analysing Empirically the Distribution of World's Plant Genetic Resources. BioScience 37 (3).

Koch, L. (1995): Vandana Shiva. Natur (1), 108-113.

Kochanek, H.-M. (1991): Umweltzentren in Deutschland. Eine Dokumentation der Tagung der Arbeitsgemeinschaft Natur und Umweltschutz (ANU) vom 21.-23.09.1990 im Naturschutzzentrum Ökowerk Berlin. Mülheim/Ruhr: Verlag an der Ruhr.

Kommission der Europäischen Gemeinschaften (1975): Empfehlung des Rates vom 3. März 1975 über die Kostenzurechnung und die Intervention der öffentlichen Hand bei Umweltschutzmaßnahmen. ABl. Nr. L 194. Brüssel: EC.

Kommission der Europäischen Gemeinschaften (1992): Die Umweltpolitik in der Europäischen Gemeinschaft. Brüssel: EC.

Kommission der Europäischen Gemeinschaften (1994): Die Europäische Gemeinschaft vor dem Problem des Bevölkerungswachstums: Vorschlag für die Position der Gemeinschaft auf der Weltkonferenz für Bevölkerung und Entwicklung in Kairo. Köln: Bundesverlagsgesellschaft.

Kommission Weltkirche der Deutschen Bischofskonferenz (1993): Bevölkerungswachstum und Entwicklungsförderung. Ein kirchlicher Beitrag zur Diskussion. Bonn: Sekretariat der Deutschen Bischofskonferenz.

Krasner, S.D. (1983): International Regimes. Ithaca, NY: Cornell University Press.

Krause, G.H.M. (1989): Luftverunreinigungen und neuartige Waldschäden, Ursachen und Wirkungen. In: Stiftung "Wald in Not" (Ed.): Fakten, Forschung, Hypothesen; Ursachen des Waldsterbens.

Krugmann, P.R. (1993): The Current Case for Industrial Policy. In: Salvatore, D. (Ed.): Protectionism and Economic Welfare. Cambridge, Ma.: MIT Press, 160-179.

Kruize, R.R. (1991): North Sea Pollution – Technical Strategies for Improvement. Water Science and Technology 24 (10).

Kulessa, M. (1990): The Newly Industrializing Economies of Asia. Berlin: Duncker & Humblot.

Künzel, W. and Künzel, G. (1992): FCKW und Ozonloch. Ein fächerübergreifendes Thema aus der Sicht der Chemie. Praxis der Naturwissenschaften – Chemie 41 (3), 11-15.

Kwiatkowska, B. (1994): Ocean-related Impact of Agenda 21 on International Organizations of the United Nations System in Follow-up to the Rio Summit. In: International Organizations and the Law of the Sea. Documentary Yearbook 1992. London, Dordrecht, Boston: Graham & Trottman and Martinus Nijhoff, xiii-lii.

Lahaye, N. and Llerena, D. (1994): Technology and Sustainability: An Organizational and Institutional Change. In: AFCET (Ed.): Modèles de developpement soutenable. Des approches exclusives ou complementaires de la soutenabilité? Paris: author and publisher, 1115-1131.

Laird, S.A. (1993): Contracts for Biodiversity Prospecting. In: Reid, W.V., Laird, S.A., Meyer, C.A. and Gámez, R. (Eds.): Biodiversity Prospecting. Washington, DC: WRI, 99-130.

LANA – Länderarbeitsgemeinschaft für Naturschutz, Landschaftspflege und Erholung (1993): Lübecker Grundsätze des Naturschutzes. Schriftenreihe des Ministeriums für Natur, Umwelt und Landesentwicklung Schleswig-Holstein. Lübeck: LANA.

Landsberg-Becher, J.W. (1991): Bericht über einen Modell-versuch in Berlin. Berlin: Senatsverwaltung für Schule, Berufsbildung und Sport.

Lantermann, E.-D. and Döring-Seipel, E. (1990): Umwelt und Werte. In: Kruse, L., Graumann, C.-F. and Lantermann, E.-D. (Eds.): Ökologische Psychologie. Ein Handbuch in Schlüsselbegriffen. München: Psychologie Verlags Union, 632-639.

Lantermann, E.-D., Döring-Seipel, E. and Schima, P. (1992): Ravenhorst. Gefühle, Werte und Unbestimmtheit im Umgang mit einem ökologischen Scenario. München: Quintessenz.

Lass, W. and Schuldt, N. (1994): Die Bedeutung der um-weltökonomischen Prinzipien im Bereich der globalen Umweltveränderungen. In: Zimmermann, H. and Hansjürgens, B. (Eds.): Prinzipien der Umweltpolitik aus ökonomischer Sicht. Bonn: Economica, 108-151.

Lecher, T. and Hoff, E.-H. (1993): Ökologisches Bewußt-sein. Theoretische Grundlagen für ein Teilkonzept im Projekt "Industriearbeit und ökologisches Verantwor-tungsbewußtsein". Berlin: Psychologisches Institut der FU Berlin.

Lehrer- und Schülergruppe (1990): Das Askanische Gymnasium. Berlin: Lehrer- und Schülergruppe.

Lersner, H. v. (1994): Interview zum Thema "Generationen-kammer". Natur (1), 5.

Lesser, W. and Krattiger, A. F. (1994): Marketing "Genetic Technologies" in South-North and South-South Exchanges: The Proposed Role of a Facilitating Mechanism. In: Krattiger, A.F. et al. (Eds.): Widening Perspectives on Biodiversity. Gland, Geneva: IUCN and International Academy of the Environment, 291-304.

Lewin, K. and Bajah, S.T. (1991): Teaching and Learning in Environmental and Agricultural Science: An Evaluation. In: Deutsche Stiftung für Internationale Entwicklung (DSE) (Ed.): Teaching and Learning in Environmental and Agricultural Science: Meeting Basic Educational Needs in Zimbabwe. An Evaluation. Bonn: Zentralstelle für Erziehung, Wissenschaft und Dokumentation (ZED), 1-89.

Lieschke, L.H. (1985): Technischer Fortschritt und Außenhandel. Analyse der Förderung "internationaler Wettbewerbsfähigkeit" durch die Forschungspolitik. Frankfurt a.M.: Peter Lang.

Loske, R. and Oberthür, S. (1994): Joint Implementation under the Climate Change Convention. International Environmental Affairs 6 (1), 45-58.

Lucas, R.E. (1988): On the Mechanics of Economic Development. Journal of Monetary Economics 22, 3-42.

Lucas, R. E. (1990): Why Doesn't Capital Flow from Rich to Poor Countries. American Economic Review 80 (3), 92-96.

Lyke, J. and Fletcher, S. R. (1992): Deforestation: An Overview of Global Programs and Agreements. Washington, DC: Congressional Research Service.

Mace, G. M. and Stuart, S. N. (1994): Draft IUCN Red List Categories. Species 21/22, 13-24.

Maddison, D. (1993): The Shadow Price of Greenhouse Gases and Aerosols. In: Nakicenovic, N., Nordhaus, W. D., Richels, R. and Toth, F. L. (Eds.): Integrative Assessments of Mitigation, Impacts, and Adaptation to Climate Change. Laxenburg: IIASA.

Madronich, S. (1992): Implications of Recent Total Atmospheric Ozone Measurements for Biologically Active Ultraviolet Radiation Reaching the Earth's Surface. Geophysical Research Letters 19, 37-40.

Maier-Reimer, E. and Hasselmann, K. (1987): Transport and Storage of Carbon Dioxide in the Ocean – An Inorganic Ocean Circulation Carbon Cycle Model. Climate Dynamics 2, 63-90.

Manalo, J. A. (1994): The Philippine Ecovolunteers. Tao-Kalikasan (Newsletter of Lingkod Tao-Kalikasan) 9 (3), 1-4.

Manne, A.S. and Richels, R.G. (1992): Buying Greenhouse Insurance – The Economic Costs of CO_2 Emission Limits. Cambridge, Ma.: MIT Press.

Manne, A S. and Rutherford, T.F. (1993): International Trade in Oil, Gas and Carbon Emission Rights: An Intertemporal General Equilibrium Model. In: Kaya, Y., Nakicenovic, N., Nordhaus, W.D. and Toth, F.L. (Eds.): Costs, Impacts, and Benefits of CO_2 Mitigation. Laxenburg: IIASA, 315.

Manne, A.S., Mendelsohn, R. and Richels, R. (1993): MERGE – A Model For Evaluating Regional and Global Effects of GHG Reduction Policies. In: Kaya, Y., Nakicenovic, N., Nordhaus, W.D. and Toth, F.L. (Eds.): Costs, Impacts, and Benefits of CO_2 Mitigation. Laxenburg: IIASA, 143.

March, J.G. and Olsen, J.P. (1984): The New Institutionalism. Organizational Factors in Political Life. American Political Science Review 78 (3), 734-749.

Markham, A. (1995): Climate Change and Biodiversity Conservation. Gland: WWF.

Martens, J. (1993): Dabeisein ist noch nicht alles. Die NGOs in den Vereinten Nationen. Vereinte Nationen 41 (5), 168-171.

Mavroidis, P.C. (1993): Handelspolitische Abwehrmechanismen der EWG und der USA und ihre Vereinbarkeit mit den GATT-Regeln: eine rechtsvergleichende Analyse der Verordnung 2641/84 (EWG) und "Section 301" des "Omnibus Trade and Competitiveness Act" von 1988 (USA). Stuttgart: Verlagsgesellschaft Internationales Recht.

Mayr, T. (1991): Der "ethnische Konflikt" in Ruanda. Pogrom (157), 36-39.

McNeely, J.A., Miller, K.R., Reid, W.V., Mittermeier, R.A. and Werner, T.B. (1990): Conserving the World's Biological Diversity. Gland, Washington, DC: IUCN, WRI, CI, WWF and World Bank.

Mehta, P.S. (1994): Existing Inequities in Trade: A Challenge to GATT. GATT-Symposium on Trade, Environment and Sustainable Development. Geneva: GATT.

Merkel, A. (1995): Rede der Bundesministerin für Umwelt, Naturschutz und Reaktorsicherheit Dr. Angela Merkel zur Eröffnung der ersten Vertragsstaatenkonferenz am 28.3.1995 in Berlin. Bonn: BMU.

Merten, P. (1987): Know-how-Transfer durch multinationale Unternehmen in Entwicklungsländer. Ein System Dynamics Modell zur Erklärung und Gestaltung von Internationalisierungsprozessen der Montageindustrien. Berlin: Erich Schmidt.

Metzen, H. (1994): Schlankheitskur für den Staat. Lean Management in der öffentlichen Verwaltung. Frankfurt a.M.: Campus.

Michaelowa, A. (1995): Internationale Kompensationsmöglichkeiten zur CO_2-Reduktion unter Berücksichtigung steuerlicher Anreize und ordnungsrechtlicher Maßnahmen. Bonn: BMWi.

Mims, F.M. (1994): Cumulus Clouds and UV-B. Nature 371, 291.

Mitchell, R.B. (1994): Regime Design Matters. Intentional Oil Pollution and Treaty Compliance. International Organization 48 (3), 425-458.

Montgomery, M. and Kouamé, A. (1994): "Fertility and Child Schooling in Côte d'Ivoire: Is There a Tradeoff?" LSMS Working Paper 112. Washington, DC: World Bank.

MORI – Market and Opinion Research International (1994): 12 European Union Countries' Attitudes Toward European Union. World Opinion Update 18 (8), 86-88.

Moyo, S. (1991): Zimbabwe's Environmental Dilemma. Balancing Resource Inequities. Harare: ZERO.

Muyanda-Mutebi, P. and Yiga-Matovu, M. (1993): Environmental Education for Sustainable Development for Primary Teachers and Teacher Educators in Africa. Nairobi: African Social and Environmental Studies Programme (ASEP).

Myers, N. (1992): Tropische Wälder und ihre Arten: dem Ende entgegen? In: Wilson, E.O. (Ed.): Ende der biologischen Vielfalt? Heidelberg: Spektrum, 46-53.

Naess, A. (1986): Intrinsic Value: Will the Defenders of Nature Please Rise? In: Soulé, M.E. (Ed.): Conservation Biology: The Science of Scarcity and Diversity. Sunderland: Sinauer Publishers, 504-515.

Nakicenovic, N., Nordhaus, W.D., Richels, R. and Toth, F. (1994): Integrative Assessment of Mitigation, Impacts, and Adaptation to Climate Change. Laxenburg:IIASA.

Naschold, F. (1993): Modernisierung des Staates. Zur Ordnungs- und Innovationspolitik des öffentlichen Sektors. Berlin: edition sigma.

Naya, S. and Takayama, A. (1990): Economic Development in East and Southeast Asia – Essays in Honor of Shinichi Ichimura. Singapore: author and publisher.

Nelson, R.R. and Winter, S.G. (1982): An Evolutionary Theory of Economic Change. Cambridge, Ma.: Harvard University Press.

Newport, F. (1993): Health Care, Crime Escalate as "Most Important Problem". The Gallup Poll Monthly (9), 4-5.

Nguyen, T., Perroni, C. and Wigle, R. (1993): An Evaluation of the Draft Final Act of the Uruguay Round. The Economic Journal 103, 1540-1549.

Niedersächsisches Kultusministerium (1993): Empfehlungen zur Umweltbildung in allgemeinbildenden Schulen. Teil I: Rahmenkonzept und Informationsmaterialien. Teil II: Ideenbörse. Beispielhafte Unterrichtsvorhaben und -ideen für eine handlungsorientierte Umweltbildung. Hannover: Niedersächsisches Kultusministerium.

Nishioka, S. (1993): The Potential Effects of Climate Change in Japan. Onegawa: Center for Global Environmental Research.

Nordhaus, W.D. (1991a): The Cost of Slowing Climate Change: A Survey. The Energy Journal 12 (1), 37-65.

Nordhaus, W.D. (1991b): To Slow or Not to Slow: The Economics of the Greenhouse Effect. The Economic Journal 101 (6), 920-937.

Nordhaus, W.D. (1993a): Optimal Greenhouse Gas Reductions and Tax Policy in the DICE Model. American Economic Review, Papers and Proceedings 83 (2), 313-317.

Nordhaus, W.D. (1993b): Rolling the DICE: An Optimal Transition Path for Controlling Greenhouse Gases. Resource and Energy Economics 15, 27-50.

NSTF – North Sea Task Force (1993): North Sea Quality Status Report 1993. Fredensborg: Olsen & Olsen.

O'Connor, D. and Turnham, D. (1992): Managing the Environment in Developing Countries. Paris: OECD.

Oberthür, S. (1993): Politik im Treibhaus. Die Entstehung des internationalen Klimaschutzregimes. Berlin: edition sigma.

Oberthür, S. (1994): Climate Change Convention. Preparations for the First Conference of the Parties. Environmental Policy and Law 24 (6), 299-303.

OECD – Organisation for Economic Co-operation and Development (1974): Problems in Transfrontier Pollution. Paris: OECD.

OECD – Organization for Economic Co-operation and Development (1975): The Polluter Pays Principle. Paris: OECD.

OECD – Organisation for Economic Co-operation and Development (1976): Economics of Transfrontier Pollution. Paris: OECD.

OECD – Organisation for Economic Co-operation and Development (1977): Legal Aspects of Transfrontier Pollution. Paris: OECD.

OECD – Organisation for Economic Co-operation and Development (1979): Protection of the Environment in Frontier Regions. Paris: OECD.

OECD – Organisation for Economic Co-operation and Development (1991): State of the Environment 1991. Paris: OECD.

OECD – Organisation for Economic Co-operation and Development (1993a): Environmental Performance Reviews: Norway, Germany etc. Paris: OECD.

OECD – Organization for Economic Co-operation and Development (1993b): Umwelt, Schule und handelndes Lernen. Frankfurt a.M.: Peter Lang.

Ojwang, J.B. (1994): National Domestication of the Convention on Biological Diversity. In: Sánchez, V. and Juma, C. (Eds.): Biodiplomacy. Nairobi: ACTS-Press, 289-309.

Oldfield, M. (1984): The Value of Conserving Genetic Resources. Washington, DC: US Department of the Interior, National Park Service.

Oliver, R. (1994): Contraceptive Use in Ghana: The Role of Service Availability, Quality, and Price. LSMS Working Paper 111. Washington, DC: World Bank.

Oppermann, T. and Baumann, J. (1993): Handelsbezogener Schutz geistigen Eigentums ("TRIPS") im GATT. Ordo 44, 121-137.

Ostrom, E. (1990): Governing the Commons. The Evolution of Institutions for Collective Action. Cambridge: Cambridge University Press.

Ostrom, E., Schröder, L. and Wynne, S. (1993): Institutional Incentives and Sustainable Development. Infrastructure Policies in Perspective. Boulder, Co.: Westview Press.

OTA – Office of Technology Assessment (1992): Trade and the Environment – Conflicts and Opportunities. Washington, DC: OTA.

Ott, H. (1991): The New Montreal Protocol. A Small Step for the Protection of the Ozone Layer, a Big Step for International Law and Relations. Verfassung und Recht in Übersee (24), 188-208.

Ott, H. (1994): Tenth Session of the INC/FCCC. Results and Options for the First Conference of the Parties. ELNI Newsletter (Environmental Law Network International) (2), 3-7.

Oye, K.A. (1986): Cooperation Under Anarchy. Princeton, NJ: Princeton University Press.

PAI – Population Action International (1994): Global Migration. People in the Move. Washington, DC: PAI.

Parr, T. and Eatherall, A. (1994): Demonstrating Climate Change Impacts in the UK: The DoE Core Model Programme. London: UK Department of the Environment.

Parry, M.L., Carter, T.R. and Konijn, N.T. (1988): The Impact of Climatic Variations on Agriculture. Dordrecht: Kluwer.

Parson, E.A. (1993): Protecting the Ozone Layer. In: Haas, P.M., Keohane, R.O. and Levy, M.A. (Eds.): Institutions for the Earth. Sources of Effective International Environmental Protection. Cambridge, Ma.: MIT Press, 27-73.

Pascha, W. (1990): Dritte Welt im Aufbruch: Ostasiatische Schwellenländer als neue weltwirtschaftliche Entwicklungspole. In: Cassel, D. (Ed.): Wirtschaftssysteme im Umbruch. München: Vahlen, 95-111.

Pearce, D. and Moran, D. (1994): The Economic Value of Biodiversity. London: IUCN and Earthscan.

Permpongsacharoen, W. (1993): Environmental Education Alternatives from the Thai Environmental Movement. In: Schneider, H. (Ed.): Environmental Education. An Approach to Sustainable Development. Paris: OECD, 185-198.

Pinzler, P. (02.12.1994): Moral statt Markt. Hamburg, Die Zeit, 37.

Plaza, F. (1994): Port State Control: Towards Global Standardization. IMO News – Magazine of the International Maritime Organization (1), 13-20.

Popper, K.R. (1984a): Die Logik der Forschung. Tübingen: J.C.B. Mohr.

Popper, K.R. (1984b): Objektive Erkenntnis. Ein evolutionärer Entwurf. Hamburg: Hoffmann und Campe.

Posey, D. (1994): International Agreements for Protecting Indigenous Knowledge. In: Sánchez, V. and Juma, C. (Eds.): Biodiplomacy. Nairobi: ACTS-Press, 119-137.

Presse- und Informationsamt der Bundesregierung (1995): Anreize für klimaschonende Investitionen schaffen. Rede des Bundeskanzlers in Berlin am 5.4.1995. Bulletin des Presse- und Informationsamtes der Bundesregierung (30), 249.

Prittwitz, V. v. (1984): Umweltaußenpolitik. Grenzüberschreitende Luftverschmutzung in Europa. Frankfurt a.M.: Campus.

Prittwitz, V. v. (1994a): Politikanalyse. Opladen: Westdeutscher Verlag.

Prittwitz, V. v. (1994b): Affluence and Scarceness. The Effect of Economic and Sociocultural Capacities on Environmental Cooperation. In: Höll, O. (Ed.): Environmental Cooperation in Europe. The Political Dimension. Boulder: Westview Press, 71-83.

Projektstelle UNCED '92 (1992): Anforderungen an internationale Verhandlungen zum Schutz des Klimas, der Wälder, der biologischen Vielfalt und der Meere. Ergebnisse des UNCED-Workshops vom 18. und 19. November 1991 in Bonn. Bonn: Projektstelle UNCED.

Prose, F. (1995): Nordlicht – Die Klimaschutzaktion zum Mitmachen (Handzettel). Kiel: Institut für Psychologie der Christian-Albrechts-Universität.

Raaflaub, P. (1994): Subventionsregeln der EU und des GATT – Theorie und Politik für die Hochtechnologie. Chur: Rüegger.

Reed, D. (1993): The Global Environment Facility. Sharing Responsibility for the Biosphere. Washington, DC: WWF.

Reichert, P. (1994): Evolution und Innovation. Berlin: Duncker & Humblot.

Reid, W. V. (1992): How Many Species Will be There? In: Whitmore, T.C. and Sayer, J.A. (Eds.): Tropical Deforestation and Species Extinction. London: Chapman and Hall, 55-73.

Reid, W.V., Laird, S.A., Gámez, R., Sittenfeld, A., Janzen, D.H., Gollin, M.A. and Juma, C. (1993): A new Lease of Life. In: Reid, W.V., Laird, S.A., Meyer, C.A. and Gámez, R. (Eds.): Biodiversity Prospecting. Washington, DC: WRI, 1-52.

Rest, A. (1994): Neue Mechanismen der Zusammenarbeit und Sanktionierung im internationalen Umweltrecht. Natur und Recht (6), 271-279.

Richels, R. and Edmonds, J. (1994): The Economics of Stabilizing Atmospheric CO_2 Concentrations. In: Nakicenovic, N., Nordhaus, W.D., Richels, R. and Toth, F.L. (Eds.): Integrative Assessment of Mitigation, Impacts, and Adaptation to Climate Change. Laxenburg: IIASA.

Rittberger, V. (1993): Regime Theory and International Relations. Oxford: Clarendon Press.

Rode, H. (1995): Schuleffekte in der Umwelterziehung. Universität Hannover, Fachbereich Erziehungswissenschaften I.

Romer, P.M. (1986): Increasing Returns and Long-Run Growth. Journal of Political Economy 94, 1002-1037.

Romer, P.M. (1987): Growth Based on Increasing Returns Due to Specialization. Journal of Political Economy 77 (1), 55-62.

Romer, P.M. (1990a): Endogenous Technological Change. Journal of Political Economy 98 (5), 71-102.

Romer, P.M. (1990b): Are Nonconvexities Important for Understanding Growth? American Economic Review 80 (2), 97-101.

Rosenau, J.M. and Czempiel, E.O. (1992): Governance without Government: Order and Change in World Politics. Cambridge: Cambridge University Press.

Rosenberg, M.J. and Hovland, C.I. (1960): Cognitive, Affective, and Behavioral Components of Attitudes. In: Rosenberg, M.J., Hovland, C.I., McGuire, W.J., Abelson, R.P. and Brehm, J.W. (Eds.): Attitude Organization and Change. New Haven, Ct.: Yale University Press, 1-14.

Rosenberg, N.J. (1993): Towards an Integrated Impact Assessment of Climate Change: The Mink Study. Climatic Change 24 (1), 10.

Round, R. (1992): At the Crossroads – The Multilateral Fund of the Montreal Protocol. London: Friends of the Earth.

Rowlands, I.H. (1993): The Fourth Meeting of the Parties to the Montreal Protocol. Report and Reflection. Environment 35 (6), 25-34.

Röpke, J. (1977): Die Strategie der Innovation. Eine systemtheoretische Untersuchung der Interaktion von Individuum, Organisation und Markt im Neuerungsprozeß. Tübingen: J.C.B. Mohr.

RWI – Rheinisch-Westfälisches Institut für Wirtschaftsforschung (1994a): Grundlagen eines mittelfristigen umweltpolitischen Aktionsplans. Essen: RWI.

RWI – Rheinisch-Westfälisches Institut für Wirtschaftsforschung (1994b): Der Umweltsektor in der Bundesrepublik Deutschland. Essen: RWI.

RWI – Rheinisch-Westfälisches Institut für Wirtschaftsforschung (1995): Gesamtwirtschaftliche Beurteilung von CO_2-Minderungsstrategien. Gutachten im Auftrag des BMWi. 2. Zwischenbericht. Essen: RWI.

Saenger, P., Hegerl, E.J. and Davie, J.D.S. (1983): Global Status of Mangrove Ecosystems. Gland: IUCN.

Sample Institut (1994): Pressemitteilung zur Untersuchung: Umwelt und Verbraucher '94 – Das grüne GeWissen der Verbraucher. Mölln: Sample Institut GmbH.

Sand, P.H. (1990): Lessons Learned in Global Environmental Governance. Washington, DC: WRI.

Sand, P.H. (1994): Trusts for the Earth. New Financial Mechanisms for International Environmental Protection. Hull: University of Hull.

Schaart, F.M., Garbe, C. and Orfanos, C.E. (1993): Ozonabnahme und Hautkrebs: Versuch der Risikoabschätzung. Hautarzt 44, 63-68.

Schahn, J. and Holzer, E. (1990): Konstruktion, Validierung und Anwendung von Skalen zur Erfassung des individuellen Umweltbewußtseins. Zeitschrift für Differentielle und Diagnostische Psychologie 11 (3), 185-204.

Scharpf, F.W. (1991): Political Institutions, Decision Styles, and Policy Choices. In: Czada, R.M. and Windhoff-Heritier, A. (Eds.): Political Choice, Institutions, Rules and the Limits of Rationality. Boulder, Co.: Westview Press, 53-86.

Scharpf, F.W. (1992): Zur Theorie von Verhandlungssystemen. In: Benz, A., Scharpf, F.W. and Zintl, R. (Eds.): Horizontale Politikverflechtung. Zur Theorie von Verhandlungssystemen. Frankfurt a.M.: Campus, 11-27.

Schelling, T.C. (1992): Some Economics of Global Warming. The American Economic Review 82 (1), 1-14.

Schellnhuber, H.J. and Sterr, H. (1993): Klimaänderung und Küste. Berlin: Springer.

Schellnhuber, H.J., Enke, W. and Flechsig, M. (1994): Nordsommer 92. PIK Report 2. Potsdam: PIK.

Scheraga, J.D., Leary, N.A., Goettle, R.J., Jorgenson, D.W. and Wilcoxen, P.J. (1993): Macroeconomic Modeling and the Assessment of Climate Change Impacts. In: Kaya, Y., Nakicenovic, N., Nordhaus, W.D. and Toth, F.L. (Eds.): Costs, Impacts, and Benefits of CO_2 Mitigation. Laxenburg: IIASA, 107-133.

Scherhorn, G. (1994): Pro- und post-materielle Werthaltungen in der Industriegesellschaft. In: Altner, G., Mettler-Meibom, B., Simonis, U.E. and von Weizsäcker, E.U. (Eds.): Jahrbuch Ökologie 1995. München: C.H. Beck, 186-198.

Scheunpflug-Peetz, A., Seitz, K. and Treml, A.K. (1992): Die ökologische Dimension des Lernbereichs "Dritte Welt" – Zwischenergebnisse aus einem Forschungsprojekt zur Geschichte der entwicklungspolitischen Bildung. In: Becker, E. (Ed.): Umwelt und Entwicklung. Frankfurt a.M.: Verlag für Interkulturelle Kommunikation, 311-329.

Schluchter, W., Elger, U. and Hönigsberger, H. (1991): Die psychosozialen Kosten der Umweltverschmutzung. Berlin: UBA.

Schmidheiny, S. (1992): Kurswechsel. Globale unternehmerische Perspektiven für Entwicklung und Umwelt. München: Artemis & Winkler.

Schmidt, M. (1993): Results of Ozone Measurements in Northern Germany – A Case Study. Ozone in the Troposphere and Stratosphere, Part 1 NASA Conference Publ. 3266, 170-173.

Schmidt, M. (1994): Evidence of a 50-year Increase in Tropospheric Ozone in Upper Bavaria. Annales Geophysicae 12, 1197-1206.

Schneider, H. (1993): Environmental Education. An Approach to Sustainable Development. Paris: OECD.

Schönwiese C.-D. (1987): Climate Variations. In: Etling, D., Hantel, M., Kraus, H. and Schönwiese, C.-D. (Eds.): Landolt-Boernstein – Zahlenwerte und Funktionen aus Naturwissenschaften und Technik. Neue Serie Gruppe V, Band 4, Teilband c, Teil 1. Heidelberg: Springer, 93-150.

Schulz, W. (1985): Bessere Luft, was ist sie uns wert? Eine gesellschaftliche Bedarfsanalyse auf der Basis individueller Zahlungsbereitschaften. Berlin: UBA.

Schumacher, E.F. (1973): Small is Beautiful. London: Blond and Briggs.

Schumann, U. (1994): On the Effect of Emissions from Aircraft Engines on the State of the Atmosphere. Annales Geophysicae 12, 12365-12384.

Schumpeter, J.A. (1912): Theorie der wirtschaftlichen Entwicklung. Leipzig: Duncker & Humblot.

Sebenius, J.K. (1991): Designing Negotiations Toward a New Regime. The Case of Global Warming. International Security 15 (4), 110-148.

Sebenius, J.K. (1992): Challenging Conventional Explanations of International Cooperation. Negotiation Analysis and the Case of Epistemic Communities. International Organization 46 (1), 323-365.

SEI – Stockholm Environment Institute (1994): A Clearing-House Mechanism to Promote and Facilitate Technical and Scientific Cooperation Under the Convention of Biological Diversity. Stockholm: SEI.

Senti, R. (1994): Die neue Welthandelsordnung – Ergebnisse der Uruguay-Runde. Chancen und Risiken. Ordo 45, 301-314.

Sessions, K.G. (1992): Institutionalizing the Earth Summit. United Nations Association of the USA.

Setlow, R.B., Grist, E., Thompson, K. and Woodhead, A.D. (1993): Wavelengths Effective in Unduction of Malignant Melanoma. Proceedings of the National Academy of Sciences USA 90, 6666-6670.

Sharma, N.P. (1992): Managing the World's Forests. Dubuque: Kendall and Hunt.

Shepsle, K.A. (1989): Studying Institutions. Some Lessons from the Rational Choice Approach. Journal of Theoretical Politics 1 (2), 131-147.

Shiva, V. (1994a): Farmers' Rights and the Convention on Biological Diversity. In: Sánchez, V. and Juma, C. (Eds.): Biodiplomacy: Genetic Resources and International Relations. Nairobi: ACTS-Press, 107-118.

Shiva, V. (1994b): Einige sind immer globaler als andere. In: Sachs, W. (Ed.): Der Planet als Patient. Über die Widersprüche globaler Umweltpolitik. Basel: Birkhäuser, 173-183.

Sietz, M. and Saldern, A. v. (1993): Umweltschutz-Management und Öko-Auditing. Berlin: Springer.

Simmons, I.G. (1993): Ressourcen- und Umweltmanagement. Heidelberg: Spektrum.

Simonis, U. E. (1993): Towards a Houston Protocol. International Journal of Social Economics 20, 32-48.

Simonis, U.E. (1994): Towards a "Houston Protocol". How to Allocate CO_2 Emissions Reductions between North and South. In: Ferré, F. and Hartel, P. (Eds.): Ethics and Environmental Policy. Theory Meets Practice. Athens: The University of Georgia Press, 106-124.

Sittenfeld, A. and Gámez, R. (1993): Biodiversity Prospecting by INBio. In: Reid, W.V., Laird, S.A., Meyer, C.A. and Gámez, R. (Eds.): Biodiversity Prospecting. Washington, DC: WRI, 69-97.

Sittenfeld, A. and Lovejoy, A. (1994): Biodiversity Prospecting. Our Planet 6 (4), 20-21.

Solomon, S., Garcia, R.R. and Ravishankara, A.R. (1994a): On the Role of Iodine in Ozone Depletion. Journal of Geophysical Research 99 (D10), 20491-20499.

Solomon, S., Burkholder, J.B., Ravishankara, A.R. and Garcia, R.R. (1994b): On the Ozone Depletion and Global Warming Potentials of CF3I. Journal of Geophysical Research 99 (D10), 20929-20935.

Soroos, M.S. (1986): Beyond Sovereignity. The Challenge of Global Policy. Columbia, SC: South Carolina University Press.

Spada, H. (1990): Umweltbewußtsein: Einstellung und Verhalten. In: Kruse, L., Graumann, C.-F. and Lantermann, E.-D. (Eds.): Ökologische Psychologie. Ein Handbuch in Schlüsselbegriffen. München: Psychologie Verlags Union, 623-631.

SRU – Rat von Sachverständigen für Umweltfragen (1994): Umweltgutachten 1994. Für eine dauerhaft-umweltgerechte Entwicklung. Stuttgart: Metzler-Poeschel.

Staehelin, J., Thudium, J., Bühler, R., Volz-Thomas, A. and Graber, W. (1994): Trends in Surface Ozone Concentrations at Arosa (Switzerland). Atmospheric Environment 28, 75-87.

Staehelin-Witt, E. and Spillmann, A. (1994): Emissionshandel. Erfahrungen in der Region Basel und neue Ansätze. Zeitschrift für Umweltpolitik und Umweltrecht 17 (2), 207-223.

Stern, P.C. and Oskamp, S. (1987): Managing Scarce Environmental Resources. In: Stokols, D. and Altman, I. (Eds.): Handbook of Environmental Psychology. Volume 2. New York: Wiley, 1043-1088.

Stevenson, J.R. and Oxman, B.H. (1994): The Future of the United Nations Convention on the Law of the Sea. American Journal of International Law 88 (3), 488-499.

Stolpe, M. (1993): Industriepolitik aus Sicht der neuen Wachstumstheorie. Weltwirtschaft 44 (3), 361-377.

Strübel, M. (1991): Internationale Umweltpolitik. Entwicklungen, Defizite, Aufgaben. Opladen: Leske & Budrich.

Strzepek, K. M. and Smith, J. B. (1995): As Climate Changes: International Impacts and Implications. Cambridge: Cambridge University Press.

Susskind, L.E. (1994): Environmental Diplomacy. Negotiating More Effective Environmental Agreements. Oxford: Oxford University Press.

Swiss Reinsurance (1994): Global Warming: Element of Risk. Zurich: Swiss Reinsurance Company.

Tahuanen, H., Storch, H. v. and Storch, J. v. (1994):Economic Efficiency of CO_2 Reduction Programs. Climate Research 4, 127.

Taylor, D.E. (1988): Environmental Education in Jamaica: The Gap Between Policymakers and Teachers. Journal of Environmental Education 20 (1), 22-28.

Theierl, H. (1989): Technologien für Entwicklungsländer. Die Konkurrenz zwischen Gegenwart und Zukunft. Bonn: BMWi.

Thiel, H. and Schriever, G. (1994): Environmental Consequences of Using the Deep Sea. Exemplified by Mining of Polymetallic Nodules. Nord-Süd aktuell VIII (3), 404-408.

Toulmin, C. (1994): Empowering the People. Our Planet 6 (5), 21-22.

Treitz, W. (1990): Kriterien für eine Weiterentwicklung der internationalen Agrarforschung. Entwicklung und ländlicher Raum (5), 11-14.

Treml, A.K. (1992): Desorientierung überall oder Entwicklungspolitik und Entwicklungspädagogik in neuer Sicht. Zeitschrift für Entwicklungspädagogik 15 (1), 6-17.

Tucker, M. (1994): A Proposed Debt-for-Nature Swap in Madagascar and the Larger Problem of LDC Debt. International Environmental Affairs 6 (1), 59-68.

UBA – Umweltbundesamt and StBA – Statistisches Bundesamt (1995): Umweltdaten Deutschland 1995. Berlin: UBA.

Umweltstiftung WWF-Deutschland (1995): WWF-Ozon-Kampagne 1993/1994. Auswertung, Perspektiven. Frankfurt a.M.: WWF.

UN – United Nations (1987): Umweltperspektiven der Vereinten Nationen – bis zum Jahr 2000 und danach (A/RES/42/186 vom 11. Dezember 1987). Geneva: UN.

UN – United Nations (1991): Preparations for the United Nations Conference on Environment and Development on the Basis of General Assembly Resolution 44/228 and Taking into Account Other Relevant General Assembly Resolutions. Cross-Sectoral Issues. Geneva: UN.

UN – United Nations (1992): Institutional Arrangements to Follow up the United Nations Conference on Environment and Development. Geneva: UN.

UN – United Nations (1992a): Law of the Sea. Progress Made in the Implementation of the Comprehensive Legal Regime Embodied in the UN Convention on the Law of the Sea. (Report of UN Secretary General to 47th UN General Assembly, New York, Nov. 5, 1992): Geneva: UN.

UN – United Nations (1992b): Law of the Sea. (Report of UN Secretary General to 47th UN General Assembly, New York, Nov. 5, 1992): Geneva: UN.

UN – United Nations (1993): Progress in the Incorporation of Recommendations UNCED in the Activities of International Organizations, and Measures Undertaken by the Administrative Committee on Coordination to Ensure that Sustainable Development Principles are Incorporated in Programmes and Processes Within the United Nations System. Geneva: UN.

UN – United Nations (1994a): Financial Resources and Mechanisms for Sustainable Development. Overview of Current Issues and Developments. Geneva: UN.

UN – United Nations (1994b): General Discussion on Progress in the Implementation of AGENDA 21. Focusing on the Cross-Sectoral Components of AGENDA 21 and the Critical Elements of Sustainability. Geneva: UN.

UN – United Nations (1994c): Report of the High-level Advisory Board on Sustainable Development on its Second Session, New York 17-22 March 1994. Geneva: UN.

UN – United Nations (1994d): Report of the Commission on Sustainable Development on its Second Session (New York 16-27 May 1994). (UN-Doc (E/CN.17/1994/20). Geneva: UN.

UN – United Nations (1994e): Economic and Environmental Questions: Reports of Subsidiary Bodies, Conferences and Related Questions. Sustainable Development. Geneva: UN.

UNCED – United Nations Conference on Environment and Development (1992): Konferenzdokumente. Rio de Janeiro: UNCED.

UNCED – United Nations Conference on Environment and Development (1992): AGENDA 21. Agreements on Environment and Development. Rio de Janeiro: UNCED.

UNCLOS Commentary (1990): United Nations Convention on the Law of the Sea. A Commentary. Dordrecht: Martinus Nijhoff.

UNCTAD (1994): World Investment Report. Geneva: UNO-Verlag.

UNDP – United Nations Development Programme (1991): Human Development Report 1991. Oxford: Oxford University Press.

UNDP – United Nations Development Programme (1992): Human Development Report 1992. Oxford: Oxford University Press.

UNDP – United Nations Development Programme (1993): Human Development Report 1993. Oxford: Oxford University Press.

UNDP – United Nations Development Programme (1994): Human Development Report 1994. Oxford: Oxford University Press.

UN-ECE – United Nations Economic Commission for Europe and CEE – Commission of the European Communities (1992): Forest Conditions in Europe. 1992 Report. Geneva: UN-ECE and CEE.

UNEP – United Nations Environment Programme (1991): Register of International Treaties and Other Agreements in the Field of the Environment. Nairobi: UNEP.

UNEP – United Nations Environment Programme (1992a): The World Environment 1972-1992. Two Decades of Challenge. London: UNEP and Chapman & Hall.

UNEP – United Nations Environmental Programme (1992b): Convention on Biological Diversity. Nairobi: UNEP.

UNEP – United Nations Environment Programme (1993): Report of the Tenth Meeting of the Executive Committee of the Multilateral Fund for the Implementation of the Montreal Protocol. Nairobi: UNEP.

UNEP – United Nations Environment Programme (1994): Preparation of the Participation of the Convention on Biological Diversity in the Third Session of the Commission on Sustainable Development. Dokument UNEP/CBD/COP/1/L.10. Nassau: UNEP.

UNEP – United Nations Environment Programme (1994a): Scientific Assessment of Ozone Depletion: 1994; Executive Summary. Nairobi: UNEP.

UNEP – United Nations Environment Programme (1994b): Environmental Effects of Ozone Depletion: 1994 Assessment. Nairobi: UNEP.

UNEP – United Nations Environmental Programme and IUCN – The World Conservation Union (1988): Coral Reefs of the World. UNEP Regional Seas Directories and Bibliographies. Nairobi, Gland: UNEP and IUCN.

UNESCO-Verbindungsstelle für Umwelterziehung im Umweltbundesamt (1988): Internationaler Aktionsplan für Umwelterziehung in den neunziger Jahren. Ergebnisse des Internationalen UNESCO/ UNEP-Kongresses über Umwelterziehung (Moskau 1987). Berlin: UNESCO-Verbindungsstelle für Umwelterziehung im Umweltbundesamt.

Urban, D. (1986): Was ist Umweltbewußtsein? Exploration eines mehrdimensionalen Einstellungskonstruktes. Zeitschrift für Soziologie 15 (5), 363-377.

Urban, D. (1991): Die kognitive Struktur von Umweltbewußtsein. Ein kausalanalytischer Modelltest. Zeitschrift für Sozialpsychologie 22 (3), 166-180.

Vandersee, W. (1994): Ozonbulletin des Deutschen Wetterdienstes, 6.

Vinke, J. (1993): Actors and Approaches in Environmental Education in Developing Countries. In: Schneider, H. (Ed.): Environmental Education. An Approach to Sustainable Development. Paris: OECD, 39-77.

Volz, A. and Klug, D. (1988): Evaluation of the Montsouris Series of Ozone Measurements Made in the Nineteenth Century. Nature 332, 240-242.

Voppel, G. (1970): Stadt als geographischer Begriff. In: Akademie für Raumforschung und Landesplanung (Ed.): Handwörterbuch der Raumordnung und Raumforschung. Hannover: Gebrüder Jänecke, 3079-3089.

WBGU – German Advisory Council on Global Change (1994): World in Transition: Basic Structure of Global Human-Environment Interactions. 1993 Annual Report. Bonn: Economica.

WBGU – German Advisory Council on Global Change (1995): World in Transition: The Threat to Soils. 1994 Annual Report. Bonn: Economica.

WBGU – German Advisory Council on Global Change (1995): Scenario for the derivation of global CO_2 reduction targets and implementation strategies. Statement on the occasion of the First Conference of the Parties to the Framework Convention on Climate Change in Berlin. Bremerhaven: WBGU.

WCMC – World Conservation Monitoring Centre (1992): Global Biodiversity: Status of the Earth's Living Resources. London: Chapman & Hall.

WCMC – World Conservation Monitoring Centre (1994a): Biodiversity Data Sourcebook. Cambridge: World Conservation Press.

WCMC – World Conservation Monitoring Centre (1994b): Priorities for Conserving Global Species Richness and Endemism. Cambridge: World Conservation Press.

Weijers, E.P. and Vellinga, P. (1995): Climate Change and River Flooding. Amsterdam: Free University Amsterdam.

Weiss, E. (1989): In Fairness to Future Generations. International Law, Common Patrimony, and Intergenerational Equity. Tokyo: United Nations University Press.

Weissmahr, J.A. (1992): The Factors of Production of Evolutionary Economics. In: Witt, U. (Ed.): Process and Change – Approaches to Evolutionary Economics. Ann Arbor, Mi.: University of Michigan, 67-79.

Wießner, E. (1991): Umwelt und Außenhandel. Der Einbau von Umweltgütern in die komparativ-statische und dynamische Außenwirtschaftstheorie. Baden-Baden: Nomos.

Wilkinson, C.R. and Buddemeier, R.W. (1994): Global Climate Change and Coral Reefs: Implications for People and Reefs. Report of the UNEP-IOC-ASPEI-IUCN Global Task Team on the Implications of Climate Change on Coral Reefs. Gland: IUCN.

Wilson, E.O. (1992): Der gegenwärtige Stand der biologischen Vielfalt. In: Wilson, E.O. (Ed.): Ende der biologischen Vielfalt? Heidelberg: Spektrum, 19-36.

Wissenschaftsrat (1994): Stellungnahme zur Umweltforschung in Deutschland. Köln: Wissenschaftsrat.

Witt, U. (1994): Wirtschaft und Evolution. WiSt-Wirtschaftswissenschaftliches Studium 25 (10), 503-512.

WMO – World Meteorological Organization (1993): Press release dated 21.03.1993. Geneva: WMO.

WMO – World Meteorological Organization (1995): Scientific Assessment of Ozone Depletion: 1994. Geneva: WMO. Global Ozone Research and Monitoring Project.

Wöhlke, M. (1992): Umweltflüchtlinge. Ursachen und Folgen. München: C.H. Beck.

Wolf, H.C. (1994): Wachstumstheorien im Widerstreit. Konvergenz oder Divergenz? WiSt-Wirtschaftswissenschaftliches Studium (4), 187-193.

Wolf, P., Donawho, C. and Kripke, M. (1994): Effect of Sunscreen UV-radiation Induced Enhancement of Melanoman Growth in Mice. Journal of the National Cancer Institute (86), 99-105.

Wolfrum, R. (1991): Law of the Sea at the Crossroads: The Continuing Search for a Universally Accepted Regime. Berlin: Duncker & Humblot.

Wood, A. (1993): The Interim Multilateral Fund for the Implementation of the Montreal Protocol. In: Reed, D. (Ed.): The Global Environmental Facility. Sharing Responsibility for the Biosphere. Washington, DC: WWF, 79-92.

World Bank (1991): Aide-memoire: Identification Mission for Primary and Secondary Education Project. Polycop.

World Bank (1991): World Development Report 1991. The Challenge of Development. Oxford, New York: Oxford University Press.

World Bank (1992): World Development Report 1992. Development and the Environment. Oxford, New York: Oxford University Press.

World Bank (1993): World Development Report 1993. Investing in Health. Oxford, New York: Oxford University Press.

World Bank (1993a): The East Asian Miracle: Economic Growth and Public Policy. Oxford: Oxford University Press.

World Bank (1994): World Development Report 1994. Infrastructure for Development. Oxford, New York: Oxford University Press.

World Values Study Group (1994): World Values Survey, 1981-1984 and 1990-1993. Ann Arbor, Mi.: Interuniversity Consortium for Political and Social Research (ICPSR).

WRI – World Resources Institute (1992): World Resources 1992/93. Oxford: Oxford University Press.

WRI – World Resources Institute (1994): World Resources 1994/95. Oxford: Oxford University Press.

WRI – World Resources Institute, IUCN – The World Conservation Union and UNEP – United Nations Environmental Programme (1992): Global Biodiversity Strategy. Washington, DC: WRI.

Yomiuri Shimbun (1991): International Relations (Japan, the United States, Great Britain, Germany, France, Russia). In: Hastings, E.H. and Hastings, P.K. (Eds.): Index to International Public Opinion. New York: Greenwood Press, 643-647.

Young, O.R. (1979): Compliance and Public Authority. A Theory with International Applications. Baltimore, Ml.: Johns Hopkins University Press.

Young, O.R. (1989): International Cooperation. Building Regimes for Natural Resources and the Environment. Ithaca, NY: Cornell University Press.

Young, O.R. (1991): Political Leadership and Regime Foundation. On the Development of Institutions in the International Society. International Organization 45 (3), 281-308.

Young, O.R., Demko, G.J. and Ramakrishna, K. (1991): Global Environmental Change and International Governance. Hanover, NH: Dartmouth College.

Young, O.R. (1992): The Effectiveness of International Institutions. Hard Cases and Critical Variables. In: Rosenau, J. N. and Czempiel, E. O. (Eds.): Governance without Government: Order and Change in World Politics. Cambridge: Cambridge University Press, 160-194.

Young, M.D. (1994): Ecologically-accelerated Trade Liberalisation: A Set of Disciplines for Environment and Trade Agreements. Ecological Economics 9 (1), 43-52.

Zachow, E. (1993): Ozonkampagne. Umweltlernen (65), 18-25.

Zahn, B. (1993): Vom Mauerblümchen zur Lotusblüte: Vorschläge zu einer Gesamtkonzeption. In: Zentrum für Entwicklungsbezogene Bildung (ZEB) (Ed.): Lernen für die "Eine Welt" in der Grundschule. Bad Honnef: Horlemann, 41-47.

Zerner, C. and Kennedy, K.J. (1994): What is Equity in Biodiversity Exploration? Institutional Approaches to the Return of Benefits to Developing Nations, Communities and Persons. Manuskript eines Vortrags auf dem Internationalen Symposium "Patents, Genes and Butterflies", 20.-21.10.1994. Bern: Swissaid and WWF.

Zulehner, P.M. and Denz, H. (1992): Wie Europa lebt und glaubt. Europäische Wertestudie. Düsseldorf: Patmos.

THE COUNCIL
Prof. Horst Zimmermann, Marburg
(Chairperson)
Prof. Hans-Joachim Schellnhuber, Potsdam
(Vice Chairperson)
Prof. Friedrich O. Beese, Göttingen
Prof. Gotthilf Hempel, Bremen
Prof. Lenelis Kruse-Graumann, Hagen
Prof. Paul Klemmer, Essen
Prof. Karin Labitzke, Berlin
Prof. Heidrun Mühle, Leipzig
Prof. Udo Ernst Simonis, Berlin
Prof. Hans-Willi Thoenes, Wuppertal
Prof. Paul Velsinger, Dortmund

STAFF TO THE COUNCIL MEMBERS
Dr. Arthur Block, Potsdam
Dipl.-Ing. Sebastian Büttner, Berlin
Dr. Svenne Eichler, Leipzig
Dipl.-Volksw. Oliver Fromm, Marburg
Dipl. Psych. Gerhard Hartmuth, Hagen
Dipl.-Met. Birgit Köbbert, Berlin
Dipl.-Geol. Udo Kubitz, Essen
Dr. Gerhard Lammel, Hamburg
Dipl.-Volksw. Wiebke Lass, Marburg
Dipl.-Ing. Roger Lienenkamp, Dortmund
Dr. Heike Schmidt, Bremen
Dr. Rüdiger Wink, Bochum
Dr. Ingo Wöhler, Göttingen

THE SECRETARIAT OF THE COUNCIL,
BREMERHAVEN*
Prof. Meinhard Schulz-Baldes
(Director)
Dr. Marina Müller
(Deputy Director)
Dipl. Geoök. Holger Hoff
Vesna Karic
Ursula Liebert
Dr. Carsten Loose
Dipl.-Volksw. Barbara Schäfer
Martina Schneider-Kremer, M.A.

---

* Secretariat WBGU
  c/o Alfred-Wegener-Institute for Polar- and
  Marine Research
  P.O. Box 12 01 61
  D-27615 Bremerhaven, Germany
  Tel. ++49/471-4831-349   Fax: ++49/471-4831-218
  Email: wbgu@awi-bremerhaven.de

Joint Decree on the Establishment of the German Advisory Council on Global Change

Article 1

In order to periodically assess global environmental change and its consequences and to help all institutions responsible for environmental policy as well as the public to form an opinion on these issues, an Advisory Council on "Global Environmental Change" shall be established with the Federal Government.

Article 2

(1)
The Council shall annually submit a report to the Federal Government by the first of June giving an updated description of the state of global environmental change and its consequences, specifying quality, size and range of possible changes and giving an analysis of the latest research findings. In addition, the report should contain indications on how to avoid or correct maldevelopments. The report will be published by the Council.

(2)
While preparing the reports, the Council shall allow for the Federal Government to give an opinion on central questions arising from this task.

(3)
The Federal Government may ask the Council to prepare special reports and opinions on specified topics.

Article 3

(1)
The Council shall consist of up to twelve members with special knowledge and experience regarding the tasks assigned to the Council.

(2)
The members of the Council shall be jointly appointed by the two ministries in charge, the Federal Ministry for Research and Technology and the Federal Ministry for the Environment, Nature Conservation and Reactor Safety, in agreement with the departments concerned, for a period of four years. Reappointment is possible.

(3)
Members may declare their resignation from the Council in writing at any time.

(4)
If a member resigns before the end of his or her term of office, a new member shall be appointed for the retired member's term of office.

Article 4

(1)

The Council is only subject to the fulfilment bound only to the brief defined by this Decree and is otherwise independent to determine its own activities.

(2)

Members of the Council must neither belong to the Government or a legislative body of the Federal Republic or of a Land nor to the public service of the Federal Republic, of a Land or of any other juristic person of the Public Law other than as a university professor or as staff member of a scientific institute. Furthermore, they may not be representatives of an economic association or an employer's or employee's organisation, or be attached to these by the permanent execution of services and business in their favour. They must not have held any such position during the last year prior to their appointment as member of the Council.

Article 5

(1)

The Council shall elect a Chairperson and a Vice-Chairperson from its midst for a term of four years by secret ballot.

(2)

The Council shall set up its own rules of procedure. These must be approved by the two ministries in charge.

(3)

If there is a differing minority with regard to individual topics of the report then this minority opinion can be expressed in the report.

Article 6

In the execution of its work the Council shall be supported by a Secretariat which shall initially be located at the Alfred Wegener Institut (AWI) in Bremerhaven.

Article 7

Members of the Council as well as the staff of the Secretariat are bound to secrecy with regard to meeting and conference papers considered confidential by the Council. This obligation to secrecy is also valid with regard to information given to the Council and considered confidential.

Article 8

(1)

Members of the Council shall receive an all-inclusive compensation as well as a reimbursement of their travel expenses. The amount of the compensation shall be fixed by the two ministries in charge in agreement with the Federal Ministry of Finance.

(2)

The costs of the Council and its Secretariat shall be shared equally by the two ministries in charge.

Dr. Heinz Riesenhuber
Federal Minister for Research and Technology
Prof. Klaus Töpfer
Federal Minister for Environment, Nature Conservation and Reactor Safety
May 1992

— Appendix to the Council Mandate —

TASKS TO BE PERFORMED BY THE Advisory Council pursuant to Article 2, para. 1

The tasks of the Council include:

(1)

Summarising and continuous reporting on current and acute problems in the field of global environmental change and its consequences, e.g. with regard to climate change, ozone depletion, tropical forests and fragile terrestrial ecosystems, aquatic ecosystems and the cryosphere, biological diversity and the socioeconomic consequences of global environmental change; Natural and anthropogenic causes (industrialisation, agriculture, overpopulation, urbanisation, etc.) should be considered, and special attention should be given to possible feedback effects (in order to avoid undesired reactions to measures taken).

(2)

Observation and evaluation of national and international research activities in the field of global environmental change (with special reference to monitoring programmes, the use and management of data, etc.).

(3)

Identification of deficiencies in research and coordination.

(4)

Suggestions on how to avoid and correct maldevelopments.

In its reporting the Council should also consider ethical aspects of global environmental change.

Index

Publications of the German Advisory Council on Global Change

Welt im Wandel: Grundstruktur globaler Mensch-Umwelt-Beziehungen: Jahresgutachten 1993. Bonn: Economica Verlag, 1993.

World in Transition: Basic Structure of Global Human-Environment Interactions. 1993 Annual Report. Bonn: Economica Verlag, 1994.

Welt im Wandel: Die Gefährdung der Böden. Jahresgutachten 1994. Bonn: Economica Verlag, 1994.

World in Transition: The Threat to Soils. 1994 Annual Report. Bonn: Economica Verlag, 1995.

Szenario zur Ableitung globaler CO_2-Reduktionsziele und Umsetzungsstrategien. Stellungnahme zur ersten Vertragsstaatenkonferenz der Klimarahmenkonvention in Berlin. Bremerhaven: WBGU, 1995.

Scenario for the derivation of global CO_2 reduction targets and implementation strategies. Statement on the occasion of the First Conference of the Parties to the Framework Convention on Climate Change in Berlin. Bremerhaven: WBGU, 1995.

Welt im Wandel: Wege zur Lösung globaler Umweltprobleme. Jahresgutachten 1995. Berlin–Heidelberg–New York: Springer Verlag, 1995.

Springer-Verlag
and the Environment

We at Springer-Verlag firmly believe that an international science publisher has a special obligation to the environment, and our corporate policies consistently reflect this conviction.

We also expect our business partners – paper mills, printers, packaging manufacturers, etc. – to commit themselves to using environmentally friendly materials and production processes.

The paper in this book is made from low- or no-chlorine pulp and is acid free, in conformance with international standards for paper permanency.